Advanced Concepts in Biophysics

Advanced Concepts in Biophysics

Edited by **Betty Karasek**

R CALLISTO REFERENCE

New York

Published by Callisto Reference,
106 Park Avenue, Suite 200,
New York, NY 10016, USA
www.callistoreference.com

Advanced Concepts in Biophysics
Edited by Betty Karasek

International Standard Book Number: 978-1-63239-015-8 (Hardback)

Printed in the United States of America.

Contents

Preface

Biophysics is a field of scientific study which stands on the pillars of biology, computational biology, and physics circumscribing the research and study of all living organisms. The ever innovative and endlessly emerging field of biophysics also relates to the study of agriculture and medicine. One of the major topics that fall under biophysics is photobiology which deals with the study of electromagnetic radiation and its impact and resourcefulness in life of different organisms, bio-elementological point of view; Laser correlation spectroscopy approach to bio-elementology problems: nutritional, ecological and toxic aspects; Chlorophyll fluorescence in plant biology, Thermo-luminescence in chloroplast thylakoid. Topics on biotechnology and medical biophysics deals with molecular motors, micotools and application of hemodynamic theory in mechanical cardiac assist devices. Also in this field of biophysics, fluorescence techniques are frequently used as a tool to characterize the structure and dynamics of bio-membranes. However, a novel computer simulation method for the study of molecular dynamics (MD) of membranes for atomic-scale information using fluorescent membrane probes is also a part of biophysics. The studies on chlorophyll fluorescence in plant biology and thermoluminescence in chloroplast thylakoid are considered to deal with the aspects of techniques and applications of bio-membranes.

Each chapter in this book includes an extensive review of the literature as well as current techniques and applications relevant in the field of biophysics. I have been fortunate to have an outstanding group of biophysics specialists from all over the world contributing to this publication. Some current topics have also been discussed by several authors. However, as the problems are illustrated from different points of view, this only brings an added value to the book. I would like to thank all the contributors for their time and dedication. I also wish to thank my publisher for considering me worthy of this opportunity and supporting me at every step.

Editor

Impacts of Temperature on the Stability of Tropical Plant Pigments as Sensitizers for Dye Sensitized Solar Cells

Aiman Yusoff,[1] **N. T. R. N. Kumara,**[2] **Andery Lim,**[1]
Piyasiri Ekanayake,[2] **and Kushan U. Tennakoon**[1]

[1] *Faculty of Science & Institute for Biodiversity & Environmental Research, Universiti Brunei Darussalam,*
Tungku Link, Gadong, BE1410, Brunei Darussalam
[2] *Applied Physics Program, Faculty of Science, Universiti Brunei Darussalam, Jalan Tungku Link, Gadong BE1410, Brunei Darussalam*

Correspondence should be addressed to Kushan U. Tennakoon; kushan.tennakoon@ubd.edu.bn

Academic Editor: Andrei B. Rubin

Natural dyes have become a viable alternative to expensive organic sensitizers because of their low cost of production, abundance in supply, and eco-friendliness. We evaluated 35 native plants containing anthocyanin pigments as potential sensitizers for DSSCs. *Melastoma malabathricum* (fruit pulp), *Hibiscus rosa-sinensis* (flower), and *Codiaeum variegatum* (leaves) showed the highest absorption peaks. Hence, these were used to determine anthocyanin content and stability based on the impacts of storage temperature. *Melastoma malabathricum* fruit pulp exhibited the highest anthocyanin content (8.43 mg/L) followed by *H. rosa-sinensis* and *C. variegatum*. Significantly greater stability of extracted anthocyanin pigment was shown when all three were stored at 4°C. The highest half-life periods for anthocyanin in *M. malabathricum*, *H. rosa-sinensis*, and *C. variegatum* were 541, 571, and 353 days at 4°C. These were rapidly decreased to 111, 220, and 254 days when stored at 25°C. The photovoltaic efficiency of *M. malabathricum* was 1.16%, while the values for *H. rosa-sinensis* and *C. variegatum* were 0.16% and 1.08%, respectively. Hence, *M. malabathricum* fruit pulp extracts can be further evaluated as an alternative natural sensitizer for DSSCs.

1. Introduction

Dye sensitized solar cell (DSSC) is a new derivative of a solar cell, developed by Grätzel [1]. It is based on semiconductor electrode-adsorbed dye, a counter electrode, and an electrolyte containing iodide and triiodide ions [2]. This device is capable of generating energy by converting the light absorbed into electrical energy.

Numerous metal complexes and organic dyes have been used and utilized as sensitizers [3]. Previously, it has been reported that the highest efficiency from a metal as sensitizer has been achieved from a compound containing Ruthenium, with a total of 11-12% efficiency [4]. Recent findings have found that perovskite sensitized solar cells have achieved a power conversion efficiency of approximately 15% [5]. Although such results provide better efficiency and high durability, the advantages are often offset by their high cost of production, complicated synthetic routes, environmental

impact, and the tendency to undergo degradation in presence of water [6].

In contrast, the natural organic dyes are widely available and involve simple preparation, nontoxic, and complete biodegradation [7]. The use of nontoxic natural pigments as sensitizer would definitely enhance the environmental and economic benefits of this alternative form of solar energy conversion [8]. Due to these reasons, natural dyes are becoming attractive inexpensive candidates for renewable energy resources. The natural dye sensitizer may still produce very low efficiency, but with continuous advanced studies and research, improvisation of the efficiency of DSSCs has become a reality and hopeful.

Anthocyanins are the most abundant, naturally occurring flavonoid pigments which often give a bright red, blue, or violet color to plant petals, fruits, and stems [9]. Sometimes, they are present in a range of tissues including roots, tubers, and stems [4]. Since anthocyanin shows the red to blue color

TABLE 1: List of plants studied to determine the anthocyanin content.

Number	Family	Species	Plant part analyzed for pigments
1	Anacardiaceae	*Mangiferaindica* L.	Leaves
2	Myrtaceae	*Syzygium campanulatum*	Leaves
3	Lamiaceae	*Coleus blumei*	Leaves
4	Amaranthaceae	*Alternantheradentata* var 1	Leaves
5	Amaranthaceae	*Alternantheradentata* var 2	Leaves
6	Euphorbiaceae	*Acalyphawilkesiana*	Leaves
7	Euphorbiaceae	*Codiaeumvariegatum*	Leaves
8	Agavaceae	*Cordylineterminalis*	Leaves
9	Heliconiaceae	*Heliconiarostrata*	Flowers
10	Malvaceae	*Hibiscus rosa-sinensis*	Flowers
11	Convolvulaceae	*Ipomoea* sp.	Flowers
12	Nyctaginaceae	*Bougainvillea* spp.	Flowers
13	Leguminosae	*Caesalpinia pulcherrima*	Flowers
14	Bignoniaceae	*Jacaranda obtusifolia*	Flowers
15	Papilionaceae	*Andirainermis*	Flowers
16	Lythraceae	*Lagerstroemia* sp.	Flowers
17	Verbenaceae	*Durantaerecta/repens*	Flowers
18	Melastomataceae	*Melastomamalabathricum*	Fruit pulp
19	Dilleniaceae	*Dilleniasuffruticosa*	Fruits
20	Palmaceae	*Licuala orbicularis*	Fruits
21	Solanaceae	*Solanumtuberosum*	Tubers
22	Amaranthaceae	*Spinacia oleracea*	Stem
23	Dioscoreaceae	*Dioscorea villosa*	Tubers
24	Costaceae	*Costuswoodsonii*	Flowers
25	Heliconiaceae	*Heliconiarostrata*	Flowers
26	Verbenaceae	*Durantaerecta*	Flowers
27	Clusiaceae	*Garciniamangostana*	Fruits
28	Fabaceae	*Delonixregia*	Flowers
29	Nepenthaceae	*Nepenthes rafflesiana*	Modified leaves
30	Nepenthaceae	*Nepenthes ampullaria*	Modified leaves
31	Amaranthaceae	*Gomphrenaglobosa*	Flowers
32	Myrtaceae	*Rhodomyrtustomentosa*	Flowers
33	Musaceae	*Musa paradisiacal*	Flowers
34	Leguminosae	*Mimosa pudica*	Flowers
35	Bignoniaceae	*Tabebuiapentaphylla*	Flowers

of the visible spectrum, it is considered as one of the best sensitizers for wide bandgap semiconductors [3].

The performance of the cell mainly depends on the dye used as sensitizer [10]. Optimizing the structure of a natural dye is necessary to improve DSSC efficiency [4]. Although anthocyanin pigments are abundant in plants, isolated anthocyanin pigments are highly instable and degradable [11]. Their stability is affected by several factors including pH, storage temperature, and sunlight exposure levels [12]. Hence, it is important to evaluate the optimum conditions required to maintain the anthocyanin stability over a long period of time.

Storage temperature plays a critical role for anthocyanin stability [13]. Investigating the effects of storage temperature on anthocyanin degradation will be highly beneficial because

one of the vital steps in the procedure of manufacturing DSSCs involves storage of the extracted pigments.

In this study, a range of plants grown in Brunei Darussalam were tested for anthocyanin pigments. Special emphasis was paid to study the stability of promising pigments stored under different storage temperature regimes. Potential dye extracts were further tested as natural sensitizers in DSSCs.

2. Materials and Methods

2.1. Plant Materials. Brightly red/purple colored plant parts (flowers, fruits, tubers, and leaves) were harvested to determine the presence of anthocyanin (Table 1).

2.2. Anthocyanin Extraction. The anthocyanin extractions of the above plant parts were made following the procedure described by Rodriguez-Soana and Wrolstad [14]. 5 g of each freshly collected plant samples was used to extract the anthocyanin pigments. The pigments were initially extracted using 150 mL of 70% ethanol (w/v%) and stored overnight at 4°C. On the following day, the extraction was mixed thoroughly by using a magnetic stirrer for two hours under air-conditioned room temperature (25°C). The extraction was filtered using Whatman's ashless 110 mm filter paper to remove any solid residues. Subsequently, the extracts were centrifuged at 4500 rpm using a Denley BS400 (UK) centrifuge machine for five minutes to separate all residues. Lastly the supernatant of the ethanolic extracts was gently mixed with equal volumes of petroleum ether to separate polar and nonpolar pigments. The final ethanolic extract was assumed to carry only the polar anthocyanin pigments. This component was carefully poured to a 10 mL glass bottle, tightly stoppered and wrapped in aluminum foil to avoid exposure to light and treatments for different temperature regimes.

2.3. Plant Screening for Anthocyanin Pigments. Screening of separated anthocyanin pigments was done by measuring their absorbance spectra using UV-vis spectrophotometer (Shimadzu UV-1800, Japan). Before the commencement of absorbance measurements, each of the samples was treated with 45 μL of concentrated HCl [15]. This acidification process converts anthocyanin derivatives to anthocyanidin that gives absorption spectra in the region of 490–550 nm [11, 15, 16]. Plant extracts that showed higher absorption spectra were selected for further investigations to evaluate the impacts of varying temperature regimes. All measurements were done in three replicates per sample.

2.4. Determination of Anthocyanin Content. To finalize the sample selection for DSSCs, those extracts that showed the highest UV-vis absorbance reading were chosen, and their anthocyanin contents were determined following the pH differential method described by Giusti and Wrolstad [11]. The results were expressed as micrograms per gram fresh weight.

Anthocyanin content was calculated according to the following equation:

$$\text{Anthocyanin pigment content} = \frac{A \times \text{MW} \times \text{DF} \times 10^3}{\varepsilon \times L},$$

(1)

where $A = (A_{520\,\text{nm}} - A_{700\,\text{nm}})$ pH 1.0 $- (A_{520\,\text{nm}} - A_{700\,\text{nm}})$ pH 4.5, MW (Molecular Weight) = 449.2 g/mol for cyanidin-3-glucoside, DF = Dilution factor, $\varepsilon = 26900\,\text{L mol}^{-1}\,\text{cm}^{-1}$, 10^3 is the factor for converting g to mg, and L is the assumed path length in cm.

Aliquots of plant extracts were brought to pH 1 and 4.5 and allowed to equilibrate for one hour. The absorbance of each equilibrated solution was then measured at 520 nm (λ_{max}) and 700 nm for haze correction. Spectroscopic absorbance readings were repeated against 70% ethanol as the reference. All measurements were done in three replicates per sample.

The MW used in this formula corresponds to the predominant anthocyanin in the sample. In some cases, predominant anthocyanin in a material may be known and could be different from cyanidin-3-glucoside. However, throughout the years, there has been a lack of uniformity in the values of absorptivity of purified anthocyanin, mainly due to difficulties of obtaining pure crystalline anthocyanin in adequate quantities [11, 17]. Since there is a huge variety of anthocyanin spread in nature, it has been suggested that if the major anthocyanin is unknown, it can be expressed as cyanidin-3-glucoside because that is the most abundant anthocyanin in nature [11, 12, 17–20].

2.5. Impacts of Storage Temperature on Anthocyanin Stability. The anthocyanin extracts of *M. malabathricum*, *H. rosa-sinensis*, and *C. variegatum* were stored in a tightly stoppered glass bottle fully covered with aluminum foil to avoid exposure to light. Extracts were stored at three different storage temperatures, namely, 4°C, −20°C, and 25°C, to evaluate the stability during storage. In order to determine the anthocyanin contents, the spectroscopic absorbance of the extracts were initially determined for three consecutive days followed by weekly measurements over a period of four months from September 2012 to January 2013.

2.6. Degradation Rate of Anthocyanin during Storage. The first-order reaction constant rate (k) and half-life ($t_{1/2}$) were calculated using the following equation [21]:

$$\ln\left(\frac{C_t}{C_o}\right) = -k \times t,$$

$$t_{1/2} = \ln(0.5) \times k^{-1},$$

(2)

where C_o is the initial monomeric anthocyanin content and C_t is the monomeric anthocyanin content after t minute storage at a given temperature.

2.7. Photovoltaic Test of DSSC. The preparations of TiO$_2$ anode are described elsewhere [22]. The anodes were dipped in the dye extract for overnight at room temperature (25°C) and dried out [15]. The cell was assembled using Dyesol's Test Cell Assembly Machine with the Surlyn (50 μm, Dyesol). The electrolyte solution containing tetrabutylammonium iodide (TBAI; 0.5 M)/I$_2$ (0.05 M), acetonitrile, and ethylene carbonate (6 : 4, v/v) [16] was introduced through a predrilled hole in platinum counter electrode. The cell was kept under irradiation of about 3-4 h for light soaking.

Finally *I-V* characteristic of the DSSC was measured under 1 sun level (DYESOL Solar Simulator LP-156B). The effective irradiated area of solar cell was 0.25 cm^2. The performance of DSSC sensitized with anthocyanin pigments extracted from *M. malabathricum*, *H. rosa-sinensis*, and *C. variegatum* was evaluated by short circuit current (J_{sc}), open circuit voltage (V_{oc}), fill factor (ff), and energy conversion efficiency (η).

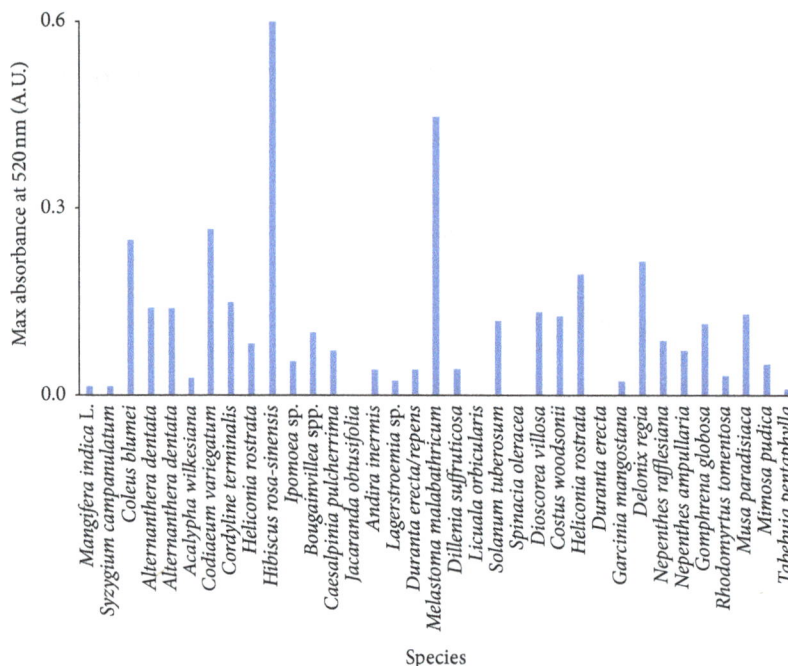

FIGURE 1: The absorbance spectra of anthocyanin pigments extracted from study species ($n = 35$) observed at 520 nm during the initial screening for the presence of anthocyanin pigments.

The absorbance spectra of the dye adsorbed on TiO_2 electrodes were also measured. Before the commencement of absorbance measurements, each of the TiO_2 electrodes were dipped in the dye extract overnight at room temperature ($25°C$) and air dried.

3. Results and Discussion

3.1. Plant Selection for DSSCs. As shown in Figure 1, the maximum absorbance of anthocyanin varied significantly in different species. *Jacaranda obtusifolia, Licuala orbicularis, Spinacia oleracea,* and *Durantaerecta* flower extracts showed no absorbance at 520 nm; hence it can be concluded that they do not possess anthocyanin. Among the rest, 17 other plant extracts showed maximum absorbance of 0.1 or lower and therefore they were not selected to further investigations. On the other hand, the remaining sample extracts showed absorbance maxima greater than 0.1. However, only three species, each representing fruit, flower, and leaves (*Melastoma malabathricum, Hibiscus rosa-sinensis,* and *Codiaeum variegatum*), which showed that highest absorbance maxima were selected for further investigations.

3.2. Determination of Anthocyanin Content of Selected Plant Extracts for the Evaluation of DSSCs. Table 2 showed that among the samples investigated after preliminary screening, the highest anthocyanin concentration was found to be in the fruit pulp of *M. malabathricum* ($8.43 \, \text{mg L}^{-1}$), followed by *H. rosa-sinensis* ($4.63 \, \text{mg L}^{-1}$) then *C. variegatum* ($2.22 \, \text{mg L}^{-1}$).

TABLE 2: Anthocyanin content of promising species that showed higher absorbance reading at 520 nm during the preliminary screening process.

Study species	Plant part used for pigment extraction	Anthocyanin content (mg/L fresh weight)[*]
Hibiscus rosa-sinensis	Flower	4.63
Melastomamalabathricum	Fruit pulp	8.43
Codiaeumvariegatum	Leaf	2.22

[*] $n = 3$.

3.3. The Absorbance Spectrum. All three extracts showed prominent peaks at 490–550 nm after the extracts were acidified with HCl (Figure 2(a)). This result indicated and proved once again that more anthocyanidin presents in the extracts [11, 15, 16].

On the other hand, Figure 2(b) showed that *M. malabathricum* extract exhibited the best absorbance after being adsorbed into the TiO_2 electrode. This extract also gave the best efficiencies in DSSCs, while *C. variegatum* in TiO_2 gave the second best absorbance, followed by *H. rosa-sinensis*. The absorbance results of the dye adsorbed TiO_2 electrodes were consistent with *I-V* characteristics data.

3.4. The Effect of Storage Temperature on Anthocyanin Stability. The storage temperature had a strong influence on the degradation of anthocyanins extracted from all three extracts (see Figure 3 and Table 3).

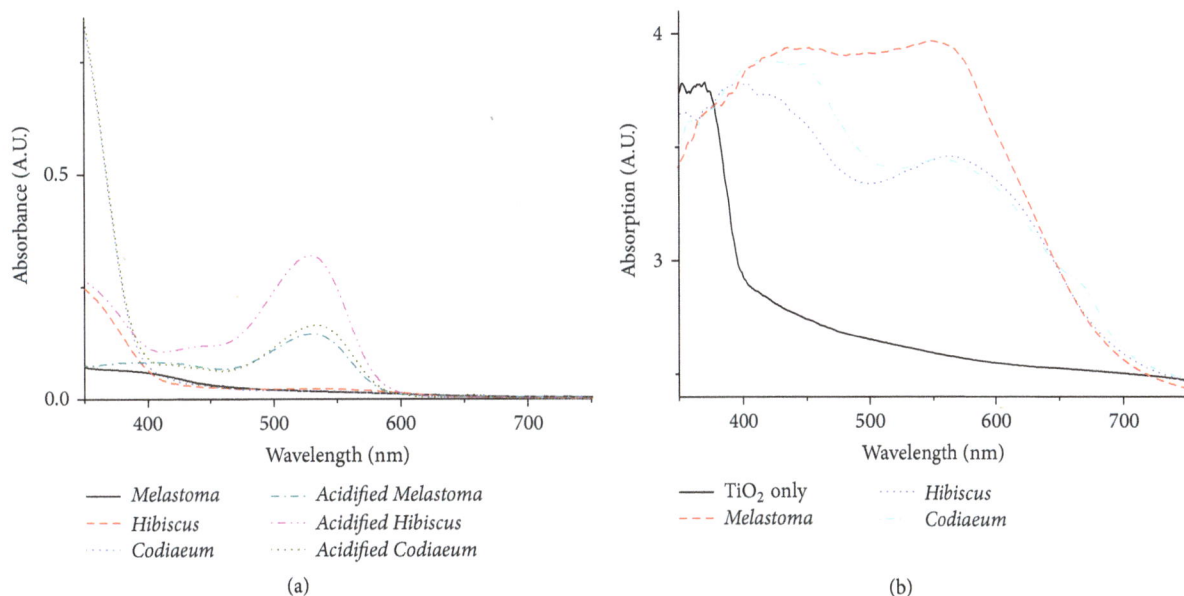

FIGURE 2: (a) The absorption spectra of the extracts of *Melastoma malabathricum*, *Hibiscus rosa-sinensis*, and *Codiaeum variegatum* in original and acidified extract and (b) absorption spectra of *Melastoma malabathricum*, *Hibiscus rosa-sinensis*, and *Codiaeum variegatum* dye onto TiO_2 film.

The most distinctive pattern that was found in all three species was that anthocyanin pigments decreased progressively when stored at 25°C over a three-month period. However the stability of all three pigments was relatively high when the temperature was maintained at 4°C.

The degradation rates are represented by the half-life values; the higher the number, the more stable the anthocyanin extract. Result showed significantly greater stability of anthocyanin in all three species when they were stored at 4°C, and storage at 25°C resulted in much faster degradation. The highest half-life periods for anthocyanin in *M. malabathricum*, *H. rosa-sinensis*, and *C. variegatum* were 540.77, 571.19, and 352.86 days at 4°C, respectively, and it decreased rapidly to 110.71, 219.74, and 254.25 days at 25°C over a period of three months.

Similar results were reported by Janna et al. [23], who also studied the stability of *Melastoma malabathricum* and found that the suitable storage condition for anthocyanin pigment is acidic solution in dark and low temperature (4°C). The result of this investigation was also consistent with other similar studies where they found that anthocyanin pigments degrade faster as the temperature increases to 25°C and the stability is maintained at low temperatures (i.e., 4°C) [12, 21, 23].

A previous study on the anthocyanin degradation in black carrot showed that the $t_{1/2}$ value in shalgam drinks maintained at 4 and 25°C were 34 and 11 weeks, respectively [24]. A similar study also found that the $t_{1/2}$ value of monomeric anthocyanin of black carrot showed a distinct difference of 71.8 and 18.7 weeks, respectively, when maintained at 4 and 20°C, respectively [21]. Our investigation showed that frozen anthocyanin extracts maintained at −20°C also ensure a good stability over a period of three months; however, the best storage temperature was still 4°C.

3.5. *The Efficiency of Natural Dye.* The current-voltage characteristics of the DSSCs sensitized with the anthocyanin pigment extracted from *M. malabathricum* fruit pulp, *H. rosa-sinensis* flowers, and *C. variegatum* leaves are shown in Figure 4. The conversion efficiencies (η) of DSSCs were 1.16, 0.16, and 1.08%, respectively (Table 4). The highest effciency was obtained from DSSC sensitized with *M. malabathricum* fruit pulp extract with the open curcuit voltage (V_{oc} = 0.383 V), short curcuit current density (I_{sc} = 6.17 mA/cm^2), and fill factor (ff = 0.44).

Natural pigments extracted from fruits and vegetables such as chlorophyll and anthocyanins have been extensively investigated as sensitizers for DSSCs. By far, the best performance reported was obtained from beet roots with an efficiency of 2.71% [25, 26].

Other studies include *Punicagranatum*, *Hibiscus sabdariffa*, pomegranate juice, wild Silicon prickly pear (*Opuntia vulgaris*), *Rhoeospathacea*, Mangosteen pericarp, red turnip, *Ficus reusa*, and *Hibiscus surattensis* with conversion efficiencies of 1.86, 1.6, 1.5, 2.06, 1.49, 1.17, 1.70, 1.18, and 1.14%, respectively [6, 7, 27–31].

Our study has shown that extract from *M. malabathricum* yielded the highest efficiency, 1.16%. The result is encouraging and the methods employed to maintain its stability is extremely promising. High efficiency obtained in the fruit pulps of *M. malabathricum* can be attributed to the carbonyl and hydroxyl groups of anthocyanin molecules present [3, 6, 7, 25]. This ability favours photoelectric conversion as it allows effective binding with the surface of TiO_2 porous film. Further improvements in refinement of extraction and application methods will no doubt increase the efficiency of this dye in DSSCs.

(a)

(b)

(c)

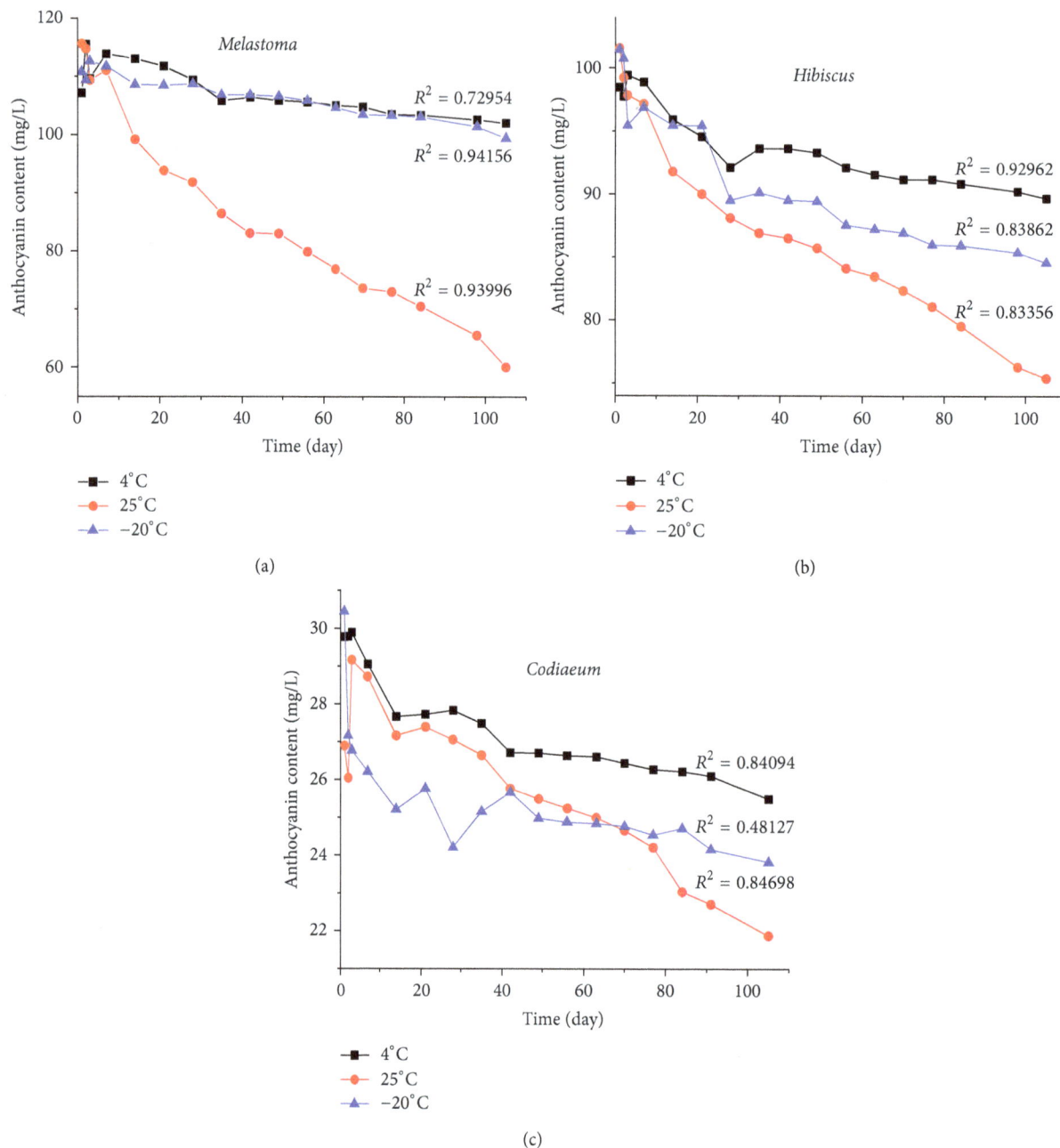

FIGURE 3: Degradation of anthocyanin pigments extracted from *M. malabathricum* fruit pulp (a), *H. rosa-sinensis* flowers, (b) and *C. variegatum* leaves (c) at three different storage temperatures (−20°C, 4°C and 25°C) over a three-month period.

4. Conclusion

Out of the 35 different species that were tested for the presence of anthocyanin pigments, *Melastoma malabathricum, Hibiscus rosa-sinensis*, and *Codiaeum variegatum* were selected as potential candidates in DSSCs. Among the three species, *M. malabathricum* extract exhibited the highest anthocyanin content. Based on the studies of anthocyanin stability on storage temperature, 4°C was the best to ensure pigment stability during storage. Among the three different species investigated, dye obtained from *M. malabathricum* fruit pulp also gave the highest efficiency. The photovoltaic

performance of this dye was encouraging (1.16%). With further refinement of extraction and application methods, the efficiency of this dye can be further improved. Furthermore, due to the simple and cost-effective preparation techniques involved in the dye extraction of this species, it makes a promising alternative sensitizer for DSSCs.

Conflict of Interests

The authors declare that there is no conflict of interests regarding the publication of this paper.

TABLE 3: Kinetic parameters of anthocyanin degradation in *M. malabathricum* fruit pulp, *H. rosa-sinensis* flowers, and *C. variegatum* leaves at three different storage temperatures.

Species	Original pH	Temp./°C	$k/10^{-3}$ (day^{-1})	$t_{1/2}$ (day)
Melastoma malabathricum	pH 5.23	25	6.261	110.71
		4	1.282	540.77
		−20	1.286	539.13
Hibiscus rosa-sinensis	pH 5.73	25	3.154	219.74
		4	1.34	571.19
		−20	2.061	336.37
Codiaeumvariegatum	pH 5.93	25	2.726	254.25
		4	1.964	352.86
		−20	1.708	405.72

TABLE 4: The photoelectric parameters of DSSCs sensitized with natural dye extracted from the fruit pulp of *M. malabathricum*, *H. rosa-sinensis* flowers, and *C. variegatum*.

Sensitizer	I_{sc} (mA cm^{-2})	V_{oc} (V)	ff	η (%)
Melastoma malabathricum	6.17	0.383	0.44	1.16
Hibiscus rosa-sinensis	3.31	0.145	0.30	0.16
Codiaeumvariegatum	4.03	0.435	0.55	1.08

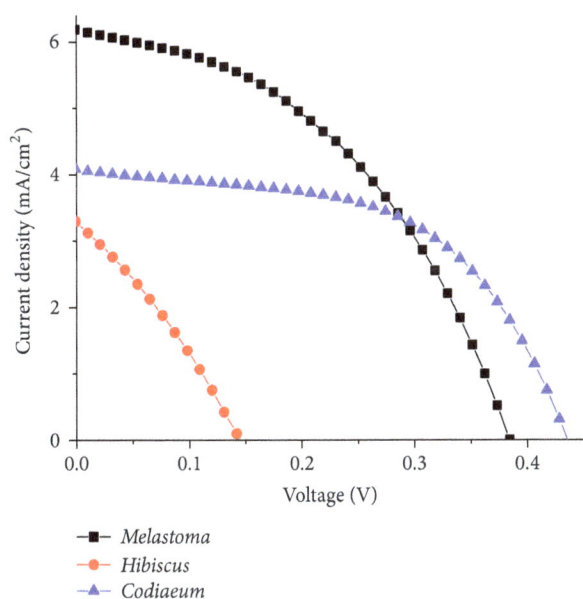

FIGURE 4: Current-voltage characteristics of the DSSCs sensitized with anthocyanins extracted from *Melastoma malabathricum*, *Hibiscus rosa-sinensis*, and *Codiaeum variegatum*.

Acknowledgments

Universiti Brunei Darussalam (UBD) Research Grant UBD/ PNC2/2/RG/1(176) and Brunei Research Council Science and Technology Research Grant (S & T 17) are acknowledged for financial support.

References

[1] M. Grätzel, "Dye-sensitized solar cells," *Journal of Photochemistry and Photobiology C*, vol. 4, no. 2, pp. 145–153, 2003.

[2] H. Zhou, L. Wu, Y. Gao, and T. Ma, "Dye-sensitized solar cells using 20 natural dyes as sensitizers," *Journal of Photochemistry and Photobiology A*, vol. 219, no. 2-3, pp. 188–194, 2011.

[3] S. Hao, J. Wu, Y. Huang, and J. Lin, "Natural dyes as photosensitizers for dye-sensitized solar cell," *Solar Energy*, vol. 80, no. 2, pp. 209–214, 2006.

[4] M. R. Narayan, "Review: dye sensitized solar cells based on natural photosensitizers," *Renewable and Sustainable Energy Reviews*, vol. 16, no. 1, pp. 208–215, 2012.

[5] J. Burschka, N. Pellet, S. J. Moon et al., "Sequential deposition as a route to high-performance perovskite-sensitized solar cells," *Nature*, vol. 499, no. 7458, pp. 316–319, 2013.

[6] A. R. Hernandez-Martinez, M. Estevez, S. Vargas, F. Quintanilla, and R. Rodriguez, "Natural pigment-based dye-sensitized solar cells," *Journal of Applied Research and Technology*, vol. 10, no. 1, pp. 38–47, 2012.

[7] H. Zhou, L. Wu, Y. Gao, and T. Ma, "Dye-sensitized solar cells using 20 natural dyes as sensitizers," *Journal of Photochemistry and Photobiology A*, vol. 219, no. 2-3, pp. 188–194, 2011.

[8] D. Zhang, S. M. Lanier, J. A. Downing, J. L. Avent, J. Lum, and J. L. McHale, "Betalain pigments for dye-sensitized solar cells," *Journal of Photochemistry and Photobiology A*, vol. 195, no. 1, pp. 72–80, 2008.

[9] E. Młodzińska, "Survey of plant pigments: molecular and environmental determinants of plant colors," *Acta Biologica Cracoviensia Series Botanica*, vol. 51, no. 1, pp. 7–16, 2009.

[10] K. Wongcharee, V. Meeyoo, and S. Chavadej, "Dye-sensitized solar cell using natural dyes extracted from rosella and blue pea

flowers," *Solar Energy Materials and Solar Cells*, vol. 91, no. 7, pp. 566–571, 2007.

[11] M. M. Giusti and R. E. Wrolstad, "Characterization and measurement of anthocyanin by UV-visible spectroscopy," in *Current Protocols in Food Analytical Chemistry*, pp. F1.2.1–F1.2.13, John Wiley & Sons, New York, NY, USA, 2001.

[12] A. Castañeda-Ovando, M. D. L. Pacheco-Hernández, M. E. Páez-Hernández, J. A. Rodríguez, and C. A. Galán-Vidal, "Chemical studies of anthocyanins: a review," *Food Chemistry*, vol. 113, no. 4, pp. 859–871, 2009.

[13] A. Patras, N. P. Brunton, B. K. Tiwari, and F. Butler, "Stability and degradation kinetics of bioactive compounds and colour in strawberry jam during storage," *Food and Bioprocess Technology*, vol. 4, no. 7, pp. 1245–1252, 2011.

[14] L. E. Rodriguez-Saona and R. E. Wrolstad, "Extraction, isolation and purifications of anthoyanins," in *Current Protocols in Food Analytical Chemistry*, pp. F1.1.1–F1.1.11, John Wiley & Sons, New York, NY, USA, 2001.

[15] N. T. R. N. Kumara, P. Ekanayake, A. Lim, M. Iskandar, and L. C. Ming, "Study of the enhancement of cell performance of dye sensitized solar cells sensitized with *Nephelium lappaceum* (F: Sapindaceae)," *Journal of Solar Energy Engineering*, vol. 135, no. 3, Article ID 031014, 5 pages, 2013.

[16] N. T. R. N. Kumara, P. Ekanayake, A. Lim et al., "Layered co-sensitization for enhancement of conversion efficiency of natural dye sensitized solar cells," *Journal of Alloys and Compounds*, vol. 581, pp. 186–191, 2013.

[17] J. Lee, K. W. Barnes, T. Eisele et al., "Determination of total monomeric anthocyanin pigment content of fruit juices, beverages, natural colorants, and wines by the pH differential method: collaborative study," *Journal of AOAC International*, vol. 88, no. 5, pp. 1269–1278, 2005.

[18] P. M. Dey and J. B. Harborne, *Plant Phenolics Methods in Plant Biochemistry*, Academic Press, London, UK, 2nd edition, 1993.

[19] F. J. Francis, "Food colorants: anthocyanins," *Critical Reviews in Food Science and Nutrition*, vol. 28, no. 4, pp. 273–314, 1989.

[20] J.-M. Kong, L.-S. Chia, N.-K. Goh, T.-F. Chia, and R. Brouillard, "Analysis and biological activities of anthocyanins," *Phytochemistry*, vol. 64, no. 5, pp. 923–933, 2003.

[21] A. Kirca, M. Özkan, and B. Cemeroğlu, "Effects of temperature, solid content and pH on the stability of black carrot anthocyanins," *Food Chemistry*, vol. 101, no. 1, pp. 212–218, 2006.

[22] P. Ekanayake, M. R. R. Kooh, N. T. R. N. Kumara et al., "Combined experimental and DFT–TDDFT study of photo-active constituents of *Canarium odontophyllum* for DSSC application," *Chemical Physics Letters*, vol. 585, pp. 121–127, 2013.

[23] O. A. Janna, A. Khairul, M. Maziah, and Y. Mohd, "Flower pigment analysis of *Melastoma malabathricum*," *African Journal of Biotechnology*, vol. 5, no. 2, pp. 170–174, 2006.

[24] N. Turker, S. Aksay, and H. I. Ekiz, "Effect of storage temperature on the stability of anthocyanins of a fermented black carrot (*Daucus carota var. L.*) beverage: shalgam," *Journal of Agricultural and Food Chemistry*, vol. 52, no. 12, pp. 3807–3813, 2004.

[25] M. Shahid, Shahid-ul-Islam, and F. Mohammad, "Recent advancements in natural dye applications: a review," *Journal of Cleaner Production*, vol. 53, pp. 310–331, 2013.

[26] C. Sandquist and J. L. McHale, "Improved efficiency of betanin-based dye-sensitized solar cells," *Journal of Photochemistry and Photobiology A*, vol. 221, no. 1, pp. 90–97, 2011.

[27] G. Calogero, J.-H. Yum, A. Sinopoli, G. Di Marco, M. Grätzel, and M. K. Nazeeruddin, "Anthocyanins and betalains as light-harvesting pigments for dye-sensitized solar cells," *Solar Energy*, vol. 86, no. 5, pp. 1563–1575, 2012.

[28] H. Hug, M. Bader, P. Mair, and T. Glatzel, "Biophotovoltaics: natural pigments in dye-sensitized solar cells," *Applied Energy*, vol. 115, pp. 216–225, 2014.

[29] G. Calogero, G. Di Marco, S. Cazzanti et al., "Efficient dye-sensitized solar cells using red turnip and purple wild sicilian prickly pear fruits," *International Journal of Molecular Sciences*, vol. 11, no. 1, pp. 254–267, 2010.

[30] M. H. Bazargan, "Performance of nano structured dye-sensitized solar cell utilizing natural sensitizer operated with platinum and carbon coated counter electrodes," *Digest Journal of Nanomaterials & Biostructures*, vol. 4, no. 4, pp. 723–727, 2009.

[31] W. H. Lai, Y. H. Su, L. G. Teoh, and M. H. Hon, "Commercial and natural dyes as photosensitizers for a water-based dye-sensitized solar cell loaded with gold nanoparticles," *Journal of Photochemistry and Photobiology A*, vol. 195, no. 2-3, pp. 307–313, 2008.

2

Analysis of Flow Characteristics of the Blood Flowing through an Inclined Tapered Porous Artery with Mild Stenosis under the Influence of an Inclined Magnetic Field

Neetu Srivastava

Department of Mathematics, Amrita Vishwa Vidyapeetham (University), Karnataka 560 035, India

Correspondence should be addressed to Neetu Srivastava; s_neetu@blr.amrita.edu

Academic Editor: Jianwei Shuai

Analytical investigation of MHD blood flow in a porous inclined stenotic artery under the influence of the inclined magnetic field has been done. Blood is considered as an electrically conducting Newtonian fluid. The physics of the problem is described by the usual MHD equations along with appropriate boundary conditions. The flow governing equations are finally transformed to nonhomogeneous second-order ordinary differential equations. This model is consistent with the principles of magnetohydrodynamics. Analytical expressions for the velocity profile, volumetric flow rate, wall shear stress, and pressure gradient have been derived. Blood flow characteristics are computed for a specific set of values of the different parameters involved in the model analysis and are presented graphically. Some of the obtained results show that the flow patterns in converging region ($\xi < 0$), diverging region ($\xi > 0$), and nontapered region ($\xi = 0$) are effectively influenced by the presence of magnetic field and change in inclination of artery as well as magnetic field. There is also a significant effect of permeability on the wall shear stress as well as volumetric flow rate.

1. Introduction

Blood is a thick red liquid circulating in the blood vessels. It has a strong nourishing effect on the human body and serves as one of the basic substances constituting the human body. Blood is a suspension of cellular elements—red blood cells (erythrocytes), white blood cells (leukocytes), and platelets in an aqueous electrolyte solution called plasma. Plasma contains 90% of water and 7% of principal proteins (albumin, globulin, lipoprotein, and fibrinogen). The volumetric fraction of the erythrocytes is 45% of the total volume of blood in normal blood, which defines an important variable called hematocrit. Of the remaining blood cells, the leukocytes are less than (1/600)th of the red blood cells and the platelets concentration is (1/20)th of the red blood cells.

Hemodynamics describes the mechanism that affects the dynamics of blood circulation in the human body. Under this, atherosclerosis, a leading cause of death in many countries, is one of the phenomena in which the blood circulation will get affected by the intimal thickening of stenos artery. A blockage

by atherosclerosis, which is a progressive vascular disease that causes accumulation of fatty substances, calcium, fibrin, cellular waste, and cholesterol, also known as plaque, inside the walls of the arteries, leads to the narrowing of the walls of the arteries and causes the condition known as carotid artery disease.

Severe stenosis may cause critical flow conditions of blood by reducing the blood supply and resulting in serious consequences called carotid artery blockage which is one of the major contributing factors to strokes. This is because when the plaque hardens and narrows down the arteries completely, the blood supply and oxygen to the brain are reduced. Without proper blood or oxygen supply, the cells in the brain start to die. This leads to the loss of brain function and permanent brain damage or the death of an individual. However, the chances of developing carotid artery disease may increase as a result of carotid aneurysm disease, fibromuscular dysplasia, or diabetes. In certain cases, plaque built up in the arteries can break off, travel through the bloodstream, and get lodged in some blood vessels in the

(a)

(b)

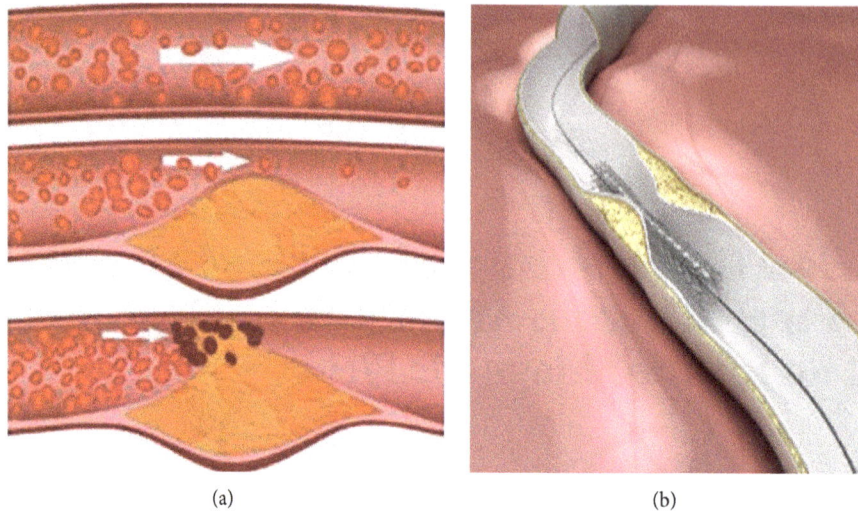

FIGURE 1: Atherosclerosis.

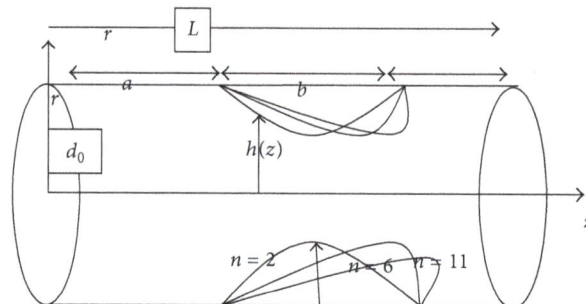

FIGURE 2: Schematic diagram for the porous artery with the different stenosis.

brain. This can trigger off a transient ischemic attack. It is thus vital to watch out for the symptoms of a carotid artery blockage, so that necessary steps can be taken before the individual's condition worsens.

The problem of fluid mechanics has attracted the attention of many investigators. If fluid is considered as blood, then the equation of motion suggested by the [1, 2] can be modified and can be used to gather an accurate knowledge of the mechanical properties of the vascular wall together with the flow characteristics of blood, in order to have a full understanding of the development of these diseases, which will assist bioengineers who are engaged in the design and construction of improved artificial organs and also helps in the treatment of vascular diseases. Perhaps the actual cause of abnormal growth in arteries is not completely clear to the theoretical investigators but its effect on the cardiovascular system has been determined by studying the flow characteristics of blood in the stenosed area. Although the applicability of a purely mechanical model for such a physiological problem has obvious limitations, vascular rheology together with hemodynamic factors is predominant in the development and progression of arterial stenosis.

The idea of electromagnetic fields in medical research was firstly given by Kolin [3], and later Korchevskii and Marochnik [4] discussed the possibility of regulating the movement of blood in human system by applying magnetic field. Vardanyan [5] studied the effect of magnetic field on blood flow theoretically and his work was later corroborated by Sud et al. [6] and Suri and Suri Pushpa [7] by considering different models. It was observed by these authors that the effect of magnetic field is to slow down the speed of blood. However, the published literature lacks the analysis of magnetic effect on blood flow through an inclined branched artery. If a magnetic field is applied to a moving electrically conducting liquid, it induces electric and magnetic fields. The interaction of these fields produces a body force known as Lorentz force which has a tendency to oppose the movement of the liquid. For the flow of blood in arteries with arterial disease like arterial stenosis or arteriosclerosis, the influence of magnetic field may be utilized as a blood pump in carrying out cardiac operations, and in addition to this, the effects of vessels tapering together with the shape of stenosis on the flow characteristics seem to be equally important and deserve special attention.

Chakraborty et al. [8] discussed the suspension model blood flow through an inclined tube with an axially non-symmetric stenosis. Eldesoky [9] studied the slip effect of unsteady magnetic field on pulsatile blood flow through porous medium in an artery under the effect of body acceleration. Bali and Awasthi [10] studied the Newtonian blood flow

Analysis of Flow Characteristics of the Blood Flowing through an Inclined Tapered Porous Artery with Mild Stenosis
under the Influence of an Inclined Magnetic Field

11

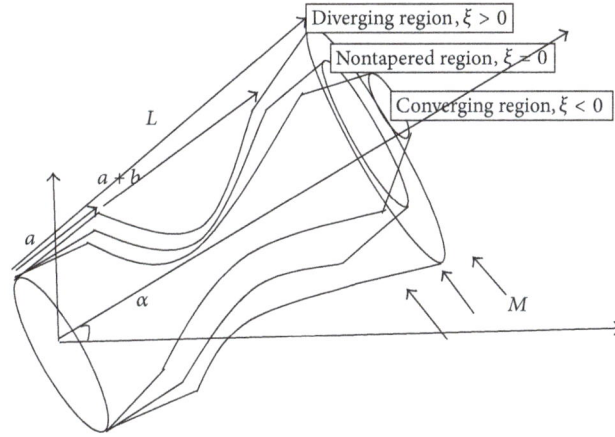

FIGURE 3: Schematic diagram for the tapered porous artery inclined at an angle α and with the applied magnetic field inclined at θ.

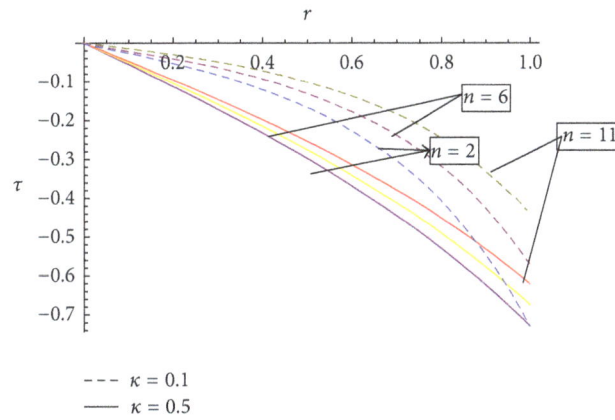

FIGURE 4: Effect of permeability on the shearing stress for different values of shape parameter n and permeability.

through the tapered arteries. The important contributions of recent years to the topic are referenced in the literature [11–19]. However, the published work still lacks the analysis of the blood flow characteristics through an inclined artery with the applied magnetic field. This problem has taken a care for this.

In this paper, we have considered the effects of arterial wall parameters on the flow of Newtonian fluid (as a blood model) through an inclined tapered porous artery with axially nonsymmetric mild stenosis under effect of an external inclined magnetic field (Figure 2).

2. Formulation of the Problem

Consider an axisymmetric steady flow of blood with a constant viscosity μ and density ρ in a cylindrical porous artery of radius d_0 and length L inclined at an angle α represented in Figure 1. Let θ be an angle of inclined magnetic field externally applied to an inclined porous artery (Figure 3). Let (r, ϕ, z) be the cylindrical polar coordinates with $r = 0$ as the axis of symmetry of the tube. Let us define V_r, V_ϕ, V_z as components of velocity. Since the flow is axisymmetric, so

variation in the blood flow characteristic is independent of an azimuthal angle. The geometry of stenosis is defined by

$$h(z) = \begin{cases} d(z)\left[1 - \eta\left(b^{n-1}(z-a) \\ \qquad\qquad - (z-a)^2\right)\right], & a \le z \le a+b, \\ d(z), & \text{"otherwise"}, \end{cases} \quad (1)$$

with

$$d(z) = d_0 + \xi z, \quad (2)$$

where $d(z)$ is the radius of tapered arterial segment in the stenotic region, d_0 is the radius of the nontapered artery in the nonstenotic region, ξ is the tapering parameter, b is the length of the stenosis, and $n(\ge 2)$ is a parameter determining the shape constriction profile and referred to as a shape parameter. The parameter η is defined by

$$\eta = \frac{\delta n^{(1/(n-1))}}{d_0 b^n (n-1)}, \quad (3)$$

where δ denotes the maximum height of the stenosis located at

$$z = a + \frac{b}{n^{(1/(n-1))}}. \quad (4)$$

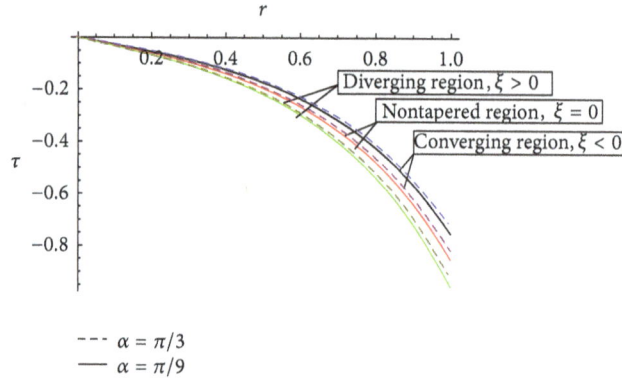

FIGURE 5: Effect of artery inclination on the shearing stress for different values of tapering angle ξ.

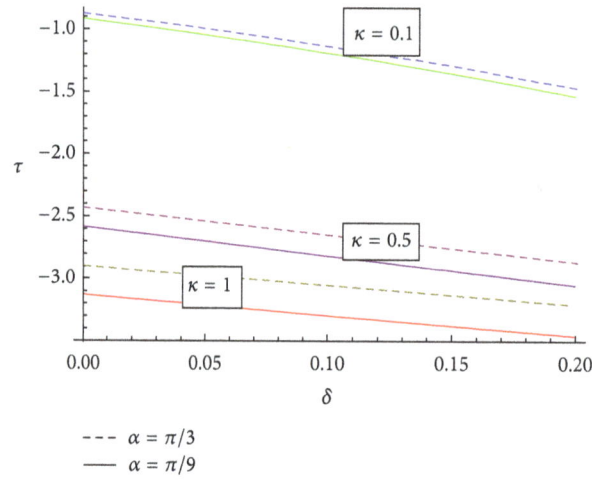

FIGURE 6: Variation of shearing stress at the stenosis throat δ with different inclination for "α" different values of permeability parameter "κ".

The equation describing the steady flow of blood through the cylindrical artery inclined at an angle α can be written as follows:

$$\nabla \cdot \overline{V} = 0, \tag{5}$$

$$\rho \overline{F} = -\nabla p + \mu \nabla^2 \overline{V} - \frac{\mu \overline{V}}{\kappa} + \overline{J} X \overline{B}. \tag{6}$$

Boundary conditions for the given problem can be written as follows:

$$u = 0 \quad \text{at } r = h(z),$$

$$\frac{\partial u}{\partial r} = 0 \quad \text{at } r = 0, \tag{7}$$

$$u \text{ is finite at } r = 0.$$

Based on the assumption made for the flow, (5) can be reduced to

$$\rho g \sin \alpha = -\frac{\partial p}{\partial z} + \mu \left(\frac{\partial^2 u}{\partial r^2} + \frac{1}{r}\frac{\partial u}{\partial r} \right) - \frac{\mu u}{\kappa} - \sigma B_0^2 u, \tag{8}$$

where α is the angle of inclination of an artery, ρ is the fluid density, B_0 is an applied magnetic field with an inclination θ, κ is the permeability of the porous medium, g is the acceleration due to gravity, and μ is the viscosity of the blood.

3. Equation of Motion

Introducing the nondimensional variables as follows:

$$z' = \frac{z}{d}, \qquad r' = \frac{r}{d_0}, \qquad h' = \frac{h}{d_0},$$

$$v' = \frac{bv}{\delta u_0}, \qquad u' = \frac{u}{u_0}, \qquad h' = \frac{h}{d_0},$$

$$p' = \frac{pd_0^2}{b\mu u_0}, \tag{9}$$

$$\text{Re} = \frac{\rho u_0 d_0}{\mu}, \qquad \kappa' = \frac{\kappa}{d_0^2},$$

$$\text{Fr} = \frac{u_0^2}{gd_0}, \qquad M = \frac{\sigma d_0^2 B_0^2}{\mu},$$

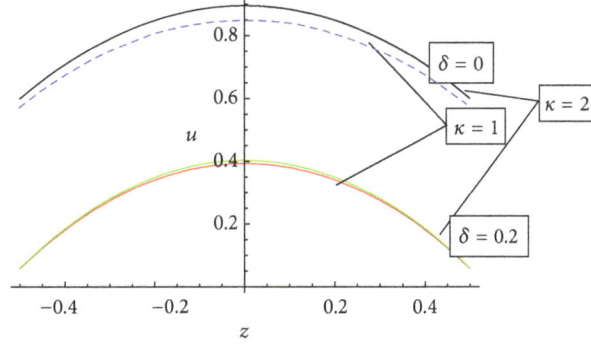

FIGURE 7: Variation of axial velocity u with z and height of stenosis δ for different values of permeability.

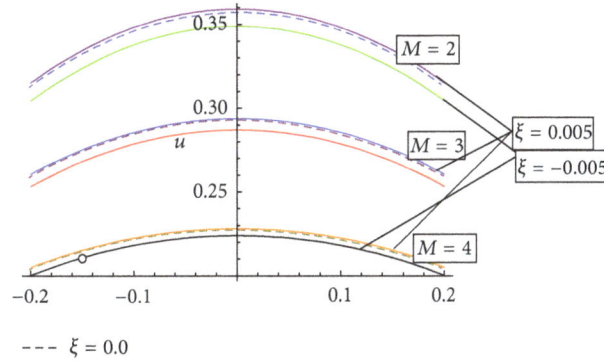

$--- \xi = 0.0$

FIGURE 8: Variation of axial velocity u with magnetic field for different values of tapering angle.

where Re, Fr, and M define the Reynolds number, Froude number, and Hartmann number, respectively. Substituting (9) in (8), we can have a dimensionless form for (8) as follows:

$$\frac{\text{Re}}{\text{Fr}} \sin\alpha = -\frac{\partial p}{\partial z} + \mu\left(\frac{\partial^2 u}{\partial r^2} + \frac{1}{r}\frac{\partial u}{\partial r}\right) - \left(\frac{1}{\kappa} + M^2\cos^2\theta\right)u. \tag{10}$$

As the flow is steady and axisymmetric, let the solution for $u(r,t)$ and p be set in the forms:

$$u(r,t) = \bar{u}(r), \qquad -\frac{\partial p}{\partial z} = P, \tag{11}$$

where P is a constant. Substituting (11) in (10), we can have an ordinary differential equation as follows:

$$\frac{d^2 u}{dr^2} + \frac{1}{r}\frac{du}{dr} - \beta^2 u = \frac{\text{Re}}{\text{Fr}} \sin\alpha - P, \tag{12}$$

where $\beta^2 = (1/\kappa + M^2\cos^2\theta)$.

The solution for the equation of motion (12) can be written as follows:

$$u(r) = \left(\frac{\kappa}{\kappa + M^2\cos^2\theta}\right)\left(\frac{\text{Re}}{\text{Fr}}\sin\alpha - P\right)\left[\frac{I_0(\beta r)}{I_0(\beta h)} - 1\right], \tag{13}$$

where I_0 is the modified Bessel function of the zero order. The volumetric flow rate can be given by

$$Q = 2\pi \int_0^h ru\,dr. \tag{14}$$

Substituting (13) with (14) and integrating it, we get

$$Q = \frac{1}{\beta^2}\left(\frac{\text{Re}}{\text{Fr}}\sin\alpha - P\right)\left(-\frac{h^2}{2} + \frac{h}{\beta}\frac{I_1(\beta h)}{I_0(\beta h)}\right), \tag{15}$$

where I_1 is the Bessel function of the order one. The expression for the wall shear stress at $r = h$ can be written as follows:

$$\tau_r = \frac{\mu}{\beta}\left(\frac{\text{Re}}{\text{Fr}}\sin\alpha - P\right)\left(\frac{I_1(\beta h)}{I_0(\beta h)}\right). \tag{16}$$

4. Results

In a converging and diverging region, for the analysis of the salient features of the blood flow, the effect of vital parameters defining flow geometries and fluid behaviour such as permeability, Hartmann number, and inclination angle of artery (α) as well as the inclination of magnetic field (θ) on the flow characteristic such as wall shear stress, shear stress at stenosis throat, axial velocity, and the volumetric flow rate are discussed numerically with computational illustrations. All graphs are plotted for the values of $\sigma = b/a = 0$, $b = 1$,

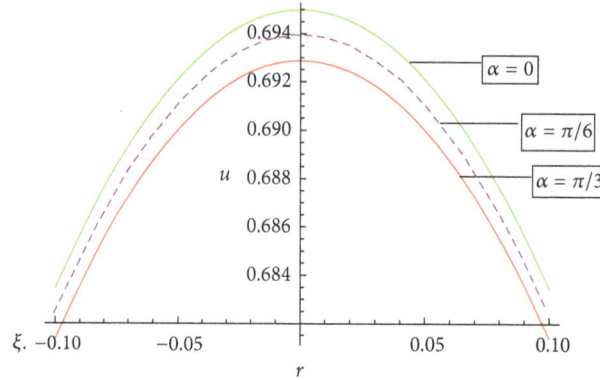

FIGURE 9: Variation of axial velocity u with z for different values of artery angle α.

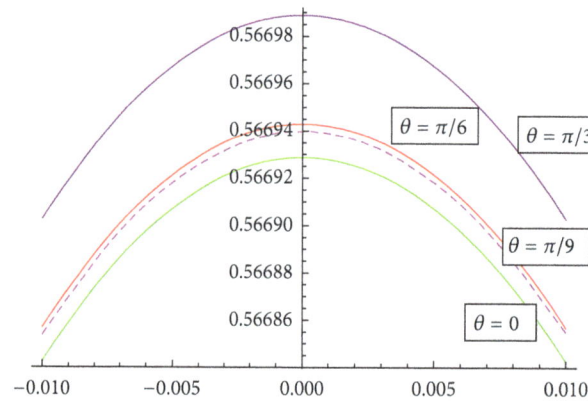

FIGURE 10: Axial velocity with the inclination of magnetic field θ for Re = 0.1, Fr = 0.1, κ = 2, M = 2, and $\alpha = \pi/3$.

$\xi = \tan\phi = 0.002, 0, -0.002$, Re = 0.1, Fr = 0.1, M = 2, 3, and 4, shape parameter n = 2, 6, and 11, and stenosis height = 0 to .20 using Mathematica software.

Figure 4 indicates that the behaviour of wall shear stress is inversely proportional to the permeability, and with the increase of shape parameter, the wall shear stress increases. For the symmetric stenosis (n = 2), the wall shear stress is less as compared to the asymmetric stenosis (n = 6, 11). Also, with the decrease of permeability parameter, the graph tends towards the parabolic nature, and also shear stress decreases with an increase of permeability parameter.

Figure 5 represents the behaviour of shearing stress with the inclination of artery for the different values of tapering angle. It has been observed that for the converging region ($\xi < 0$), the stress will be more as compared to the diverging region ($\xi > 0$) and the nontapered region ($\xi = 0$) that is. Wall shear stress increases with an increase in tapering angle (ϕ) which is in agreement with Chaturani and Prahlad's model [20].

Figure 6 represents the behaviour of shearing stress with the stenosis throat for the different values of inclination of artery. It has been observed that with the increase of "α", the shearing stress "τ" increases and also the variation of permeability parameter affects the stress inversely. This graph also reveals that with the increase of stenosis height the shearing stress increases. It rapidly decrease in the downstream and

maximum value is near the end of the stenosis. This is in agreement with Young's model [19].

Figure 7 represents the behaviour of axial velocity with the stenosis height for different values of permeability. It has been observed that the axial velocity possesses reverse behaviour on the either side of the centreline of the artery. This figure also shows an increase in axial velocity with the increase of permeability.

Figure 8 represents the behaviour of axial velocity with the magnetic field for a different tapering angle. It has been observed that with the increase of magnetic field the axial velocity shows a reverse behaviour. It is observed that with the increase of magnetic field, the curves representing the axial flow velocity do shift towards the origin for a converging region, while they shift away from the origin for a nontapered and diverging tapered artery.

Figure 9 represents the behaviour of axial velocity for different values of artery inclination and shows the remarkable changes with the inclination of artery. It has been observed that with the increase of inclination of artery, the curve representing the axial flow velocities does shift towards the origin.

Figure 10 represents that the axial velocity variation with the inclination of magnetic field.It has been observed that with the increase of inclination of magnetic field, the curve does shift towards the origin.

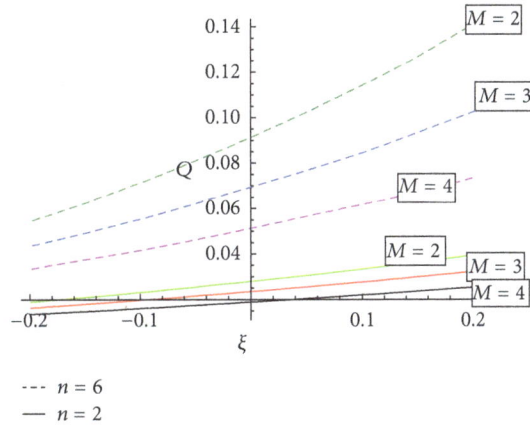

FIGURE 11: Variation of volumetric flow rate with the tapering angle, for different values of M and for fixed values of Re $= 0.1$, Fr $= 0.1$, $\alpha = \pi/3$, and $\kappa = 2$.

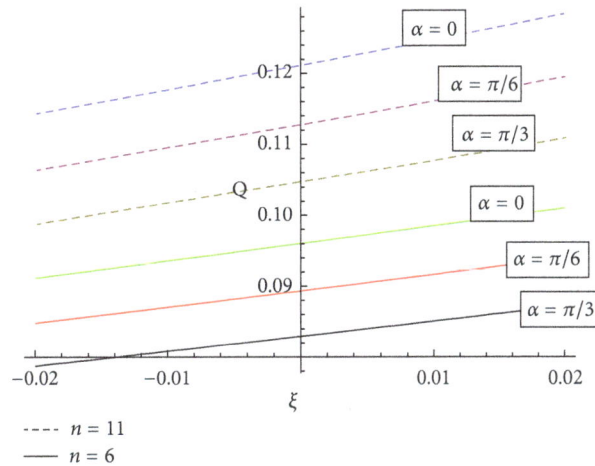

FIGURE 12: Variation of volumetric flow rate with the angle of inclination α of artery for converging, diverging region, and nontapered region.

Figure 11 reveals that the variation of volumetric flow rate for an inclined artery will decrease with the Hartmann number for the fixed values of Reynolds number Re $= 0.1$, Froude number Fr $= 0.1$, artery inclination $= \pi/3$, and permeability parameter $\kappa = 2$, but it will be greater in diverging region as compared to converging region.

Figure 12 reveals the information about that variation of volumetric flow rate with the angle of inclination "α" of artery for converging, diverging region, and nontapered region. It shows that with the increase of "α" the volumetric flow rate decreases.

Figure 13 reveals that the variation of volumetric flow rate with the angle of magnetic field θ will increase for all converging ($\xi < 0$), diverging region ($\xi > 0$), and nontapered region ($\xi = 0$).

5. Conclusion

The present study deals with the analysis of flow characteristics of the blood flowing through an inclined tapered porous artery with mild stenosis under the influence of an

inclined magnetic field. This investigation can play a vital role in the determination of axial velocity, shear stress, and fluid acceleration in particular situations. Since this study has been carried out for a situation when the human body is subjected to an external magnetic field, it bears the promise of significant application in magnetic or electromagnetic therapy, which has gained enough popularity. The study is also useful for evaluating the role of porosity. The analysis is carried out by employing appropriate analytical methods, and some important predictions have been made basing upon the study. The main findings of the present mathematical analysis are as follows.

(i) The behaviour of wall shear stress is inversely proportional to the permeability, and with the increase of shape parameter, the wall shear stress increases. Wall shear stress increases with an increase of tapering angle (ϕ) which is in agreement with Chaturani and Prahlad's model [20].

(ii) Shearing stress "τ" rapid decrease in the downstream and maximum value is near the end of the stenosis. This is in agreement with Young's model [19].

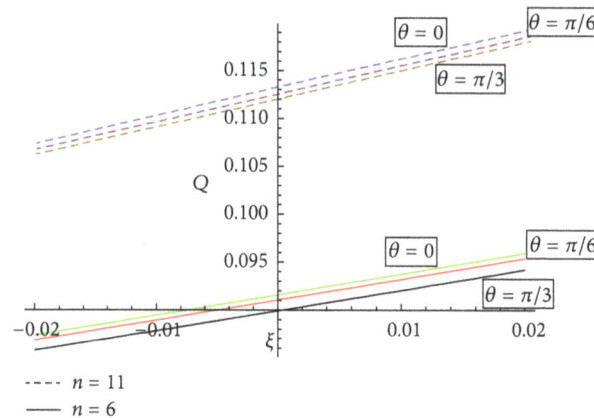

FIGURE 13: Variation of volumetric flow rate with the angle of magnetic field θ for converging, diverging region, and nontapered region.

(iii) Concerning the behaviour of axial velocity with the stenosis height for different values of permeability, it has been observed that the axial velocity possesses reverse behaviour on either side of the centreline of the artery.

(iv) With the increase of magnetic field, the axial velocity shows a reverse behaviour. It is observed that with the increase of magnetic field, the curves representing the axial flow velocity do shift towards the origin for a converging region, while they shift away from the origin for a nontapered and diverging tapered artery.

(v) Axial velocity shows the remarkable changes with the inclination of artery. It has been observed that with the increase of inclination of artery, the curve representing the axial flow velocities does shift towards the origin.

(vi) The volumetric flow rate increases with the increase of shape parameter n and decreases with the increase of Hartmann number. Also, Figure 11 shows an increasing behaviour of volumetric flow rate from converging to diverging region.

(vii) The variation of volumetric flow rate for an inclined artery will be greater in diverging region as compared to converging region.

(viii) Volumetric flow rate varies inversely with the angle of inclination "α" of artery for converging, diverging region, and nontapered region.

(ix) With the angle of inclination of magnetic field θ, the volumetric flow rate will increase and will be more for diverging region as compared to converging region.

Conflict of Interests

The author declares that there is no conflict of interests regarding the publication of this paper.

Acknowledgments

The author would like to thank the college authorities for their constant support and encouragement and referee for his valuable suggestions that lead to the improvement of this paper.

References

[1] H. Darcy, *The Flow of Fluids Through Porous Media*, Mc-Graw Hill, New York, NY, USA, 1937.

[2] H. C. Brinkman, "A calculation of the viscous force exerted by a flowing fluid on a dense swarm of particles," *Applied Scientific Research*, vol. 1, no. 1, pp. 27–34, 1949.

[3] A. Kolin, "Electromagnetic flow meter: principle of method and its applications to blood flow measurement," in *Proceedings of the Society for Experimental Biology and Medicine*, vol. 35, pp. 53–56, 1936.

[4] E. M. Korchevskii and L. S. Marochnik, "Magnetohydrodynamic version of movement of blood," *Biophysics*, vol. 10, no. 2, pp. 411–414, 1965.

[5] V. A. Vardanyan, "Effect of a magnetic field on blood flow," *Biophysics*, vol. 18, no. 3, pp. 515–521, 1973.

[6] V. K. Sud, P. K. Suri, and R. K. Mishra, "Effect of magnetic field on oscillating blood flow in arteries," *Studia Biophysica*, vol. 46, no. 3, pp. 163–171, 1974.

[7] P. K. Suri and R. Suri Pushpa, "Blood flow in a branched arteries," *Indian Journal of Pure and Applied Mathematics*, vol. 12, pp. 907–918, 1981.

[8] U. S. Chakraborty, D. Biswas, and M. Paul, "Suspension model blood flow through an inclined tube with an axially nonsymmetrical stenosis," *Korea Australia Rheology Journal*, vol. 23, no. 1, pp. 25–32, 2011.

[9] I. M. Eldesoky, "Slip effects on the unsteady MHD pulsatile Blood flow through porous medium in an artery under the effect of body acceleration," *International Journal of Mathematics and Mathematical Sciences*, vol. 2012, Article ID 860239, 26 pages, 2012.

[10] R. Bali and U. Awasthi, "Mathematical model of blood flow in the small blood vessel in presence of magnetic field," *Applied Mathematics*, vol. 2, pp. 264–269, 2011.

[11] M. Jain, G. C. Sharma, and R. Singh, "Mathematical modelling of blood flow in a stenosed artery under MHD effect through

Analysis of Flow Characteristics of the Blood Flowing through an Inclined Tapered Porous Artery with Mild Stenosis
under the Influence of an Inclined Magnetic Field

17

porous medium," *International Journal of Engineering, Transactions B*, vol. 23, no. 3-4, pp. 243–251, 2010.

[12] P. K. Mandal, "An unsteady analysis of non-Newtonian blood flow through tapered arteries with a stenosis," *International Journal of Non-Linear Mechanics*, vol. 40, no. 1, pp. 151–164, 2005.

[13] K. S. Mekheimer and M. A. E. Kot, "The micropolar fluid model for blood flow through a tapered artery with a stenosis," *Acta Mechanica Sinica*, vol. 24, no. 6, pp. 637–644, 2008.

[14] G. Ramamurty and B. Shanker, "Magnetohydrodynamic effects on blood flow through a porous channel," *Medical, Bioengineering and Computing*, vol. 32, no. 6, pp. 655–659, 1994.

[15] L. M. Srivastava, "Flow of couple stress fluid through stenotic blood vessels," *Journal of Biomechanics*, vol. 18, no. 7, pp. 479–485, 1985.

[16] D. Tripathi, "A mathematical model for blood flow through an inclined artery under the influence of an inclined magnetic field," *Journal of Mechanics in Medicine and Biology*, vol. 12, pp. 1–18, 2012.

[17] E. E. Tzirtzilakis, "A mathematical model for blood flow in magnetic field," *Physics of Fluids*, vol. 17, no. 7, 15 pages, 2005.

[18] N. Verma and R. S. Parihar, "Effect of Magnetohydrodynamics and hematocrit on blood flow in an artery with a multiple mild stenosis," *International Journal of Applied Mathematics and Computation*, vol. 1, no. 1, pp. 30–46, 2009.

[19] D. F. Young, "Fluid mechanics of arterial stenosis," *Journal of Biomechanical Engineering ASME*, vol. 101, pp. 157–175, 1968.

[20] P. Chaturani and R. N. Pralhad, "Blood flow in tapered tubes with biorheological applications," *Biorheology*, vol. 22, no. 4, pp. 303–314, 1985.

3

The Diamagnetic Susceptibility of the Tubulin Dimer

Wim Bras,[1] **James Torbet,**[1] **Gregory P. Diakun,**[2]
Geert L. J. A. Rikken,[3] **and J. Fernando Diaz**[4]

[1] *Netherlands Organisation for Scientific Research, Dutch-Belgian Beamlines, European Synchrotron Radiation Facility, BP 220, 38043 Grenoble, France*
[2] *Science and Technology Facility Council (STFC), Daresbury Laboratory, Cheshire WA4 4AD, UK*
[3] *National Centre for Scientific Research (CNRS), National High Magnetic Field Laboratory, 143 Avenue de Rangueil, 31400 Toulouse, France*
[4] *CIB Centro de Investigaciones Biológicas, Ramiro de Maeztu 9, 28040 Madrid, Spain*

Correspondence should be addressed to Wim Bras; wim.bras@esrf.eu

Academic Editor: Jianwei Shuai

An approximate value of the diamagnetic anisotropy of the tubulin dimer, $\Delta\chi_{\mathrm{dimer}}$, has been determined assuming axial symmetry and that only the α-helices and β-sheets contribute to the anisotropy. Two approaches have been utilized: (a) using the value for the $\Delta\chi_\alpha$ for an α-helical peptide bond given by Pauling (1979) and (b) using the previously determined anisotropy of fibrinogen as a calibration standard. The $\Delta\chi_{\mathrm{dimer}} \approx 4 \times 10^{-27}$ JT^{-2} obtained from these measurements are similar to within 20%. Although Cotton-Mouton measurements alone cannot be used to estimate $\Delta\chi$ directly, the value we measured, CM$_{\mathrm{dimer}}$ = (1.41 ± 0.03) × 10^{-8} T^{-2}cm^{2}mg^{-1}, is consistent with the above estimate for $\Delta\chi_{\mathrm{dimer}}$. The method utilized for the determination of the tubulin dimer diamagnetic susceptibility is applicable to other proteins and macromolecular assemblies as well.

1. Introduction

Microtubules, MT, are elongated macromolecular structures composed of protofilaments, of tubulin dimers assembled in long hollow tubes. The tubulin dimer is kidney shaped and has an approximate molecular weight of 55 kDa. The assembled microtubule is on average composed of 13 protofilaments which make up a hollow tube with a maximum diameter of 24.6±0.6 nm [1]. The length of this assembly *in vivo* is variable but can reach several microns.

The aligning effects of magnetic fields on macromolecular microtubules and other rigid fibrillar molecular assemblies are well known [1–3]. The driving force for magnetic orientation is the diamagnetic anisotropy of the tubulin subunits (dimers) combined with the large shape anisotropy and stiffness of the microtubules. Diamagnetic anisotropy originates from the anisotropic nature of chemical bonds and is more pronounced in resonance structures such as aromatic groups, peptide bonds, or double and triple carbon bonds [4].

A single peptide bond has a very weak diamagnetic anisotropy but when many of these bonds are linked together with a fixed and uniform orientation, as in α-helices and β-sheets, a relatively strong overall anisotropy can result [5].

The anisotropy of a single tubulin dimer is rather small, and so far the only effects of magnetic fields that have been observed are in concentrated assembled microtubule solutions [1, 3, 6] and microtubule containing structures [7–10]. Even for these systems, high magnetic fields (>8 Tesla) were required, although Vassilev et al. [2] reported high orientation in a significantly lower magnetic field for an individual microtubule in a diluted solution. An assessment of the value of the tubulin dimer anisotropy would be interesting from a fundamental point of view.

The molecular structure of the tubulin dimer has been elucidated by cryoelectron microscopy [11], and, in principle, this information could be used to determine the diamagnetic anisotropy by vectorially adding all the diamagnetically anisotropic components. However, assuming the aromatic amino acids have no overall preferred orientation, it should be possible to calculate the value of the diamagnetic anisotropy, to a good approximation, by considering only the contributions from the α-helices and β-sheets. Validation of this method by a direct cross correlation with experimentally obtained data is not possible since there is no method to measure the diamagnetic anisotropy directly. Therefore an indirect method has to be used.

In order to obtain experimental information about how a dilute solution of dimers responds to an applied magnetic field, Cotton-Mouton experiments can be performed [12]. For axial symmetric molecules and assuming a relatively low degree of orientation (<5% full alignment), the magnetically induced birefringence, Δn, increases linearly with the square of the applied field, B^2. The Cotton-Mouton constant, C_{CM}, is obtained from the slope; thus $C_{CM} = \Delta n / \lambda B^2$, where λ is the wavelength of the laser used in the birefringence experiments. The C_{CM} depends on the number of particles per unit volume and the product of their optical, $\Delta \alpha$, and diamagnetic, $\Delta \chi$, anisotropies; thus

$$C_{CM} = \frac{\Delta n}{\lambda B^2} \propto N \frac{\Delta \chi \Delta \alpha}{n_0 \, 15kT} \frac{1}{\lambda}, \qquad (1)$$

where k is the Boltzman constant and T is the absolute temperature of the solution.

$\Delta \chi$ depends on the internal structure of the molecule while $\Delta \alpha$ has potentially two components, one due to shape, form anisotropy, and the other, intrinsic anisotropy, due to internal structure.

Previous attempts to perform Cotton-Mouton experiments on tubulin dimers have been unsuccessful due to the low signal strength. Therefore, an improved protocol had to be designed with special care being taken to optimize the optical components of the equipment and stabilize the temperature of the solution. Experiments were performed on both dimers and double dimers, formed by the addition of Mg^{2+} ions.

2. Materials and Methods

2.1. Magnetic Birefringence and Cotton-Mouton Experiments. Magnetic birefringence [4] and Cotton-Mouton experiments, where the birefringence of the solution is measured as function of a steadily increasing magnetic field [12], were performed in the Grenoble High Magnetic Field Laboratory (Grenoble, France) using the optical equipment mounted on magnet M2. This magnet has a horizontal bore and can reach fields of upward to 17 Tesla [13]. For both experiments, the same optical equipment was used. In the Cotton-Mouton experiments the sample was maintained at constant temperature (4°C) and the magnetic field was ramped up and down at a constant dB/dt rate. All experiments were repeated a minimum of 6 times.

The effects that we were able to measure were rather weak and therefore care had to be taken to reduce the background contributions to the birefringence signal from the windows of the sample holder and from the solvent. The latter condition considerably reduces the possible choice

of buffer solutions since one should avoid all buffers that cause optical Schlieren effects, that is, all buffer solutions with increased viscosity. In these experiments we found that only a tris(hydroxylmethyl)aminomethane (Tris) buffer did not introduce an unacceptable background.

The samples were placed in a thermostated holder and loaded in the magnet after which the magnetic field was ramped to 17 T.

2.2. Biochemistry. For the Cotton-Mouton experiments analytical ultracentrifugation was performed on the samples in order to determine the particle weight/size distribution. For these experiments, the tubulin is prepared in 20 mM Tris buffer with 0.1 mM guanosine triphosphate (GTP) and at a pH = 7.5. The samples were diluted to 2, 4, and 8 mg/mL. To one aliquot, 4 mM Mg^{2+} was added in order to induce the formation of oligomers and none to the other.

The sedimentation velocities were determined using a Beckman Optima XL-I ultracentrifuge (Beckman Coulter, Brea, USA) equipped with an interference optical detection system that allowed us to monitor the sedimentation of tubulin at high (0.1 millimolar) concentrations of nucleotide. Samples were studied at a speed of 40,000 rpm and 20°C by using an An50Ti eight-hole rotor and double-sector center-pieces (Beckman Coulter, USA). Differential sedimentation-coefficient distributions, $c(s)$, were calculated by least-squares boundary modelling of sedimentation-velocity data, using the program SEDFIT [14].

The samples that were prepared without the addition of Mg^{2+} contained mainly (75 ± 5%) tubulin dimers, that is, the normal unit in which tubulin appears in vitro. A small amount of protein consisted of double or triple dimers (15 ± 5% and 7 ± 3%, resp.). This was independent of the protein concentration.

The samples to which Mg^{2+} was added had a more complex and concentration-dependent composition. At the lower concentration, 30 wt% of the material consisted of aggregates of several dimers. The remaining 70 wt% was composed of dimers and double dimers. At a concentration of 2 mg/mL the sample consisted of equal fractions of dimers and double dimers. The fraction of double dimers increased regularly as function of concentration such that at 8 mg/mL the fraction of material that was not assembled in larger aggregates (90 wt%) was solely composed of double dimers.

3. Results and Discussion

3.1. Cotton-Mouton Experiments on Dimers and Oligomers. In Figure 1, we show the Cotton-Mouton results obtained on a dimer solution of 4 mg/mL. This was the lowest concentration from which reproducible results could be obtained.

The dimer concentrations used were sufficiently dilute so that interdimer interactions could be ignored. This assumption is confirmed by the observation that the Cotton-Mouton constant normalised to the dimer concentration is independent of concentration (see Figure 2).

From these measurements we determined the value of the Cotton-Mouton constant for the dimer normalized to concentration to be $CM_{dimer} = (1.41 \pm 0.03) \times 10^{-8} \, T^{-2} cm^2 mg^{-1}$.

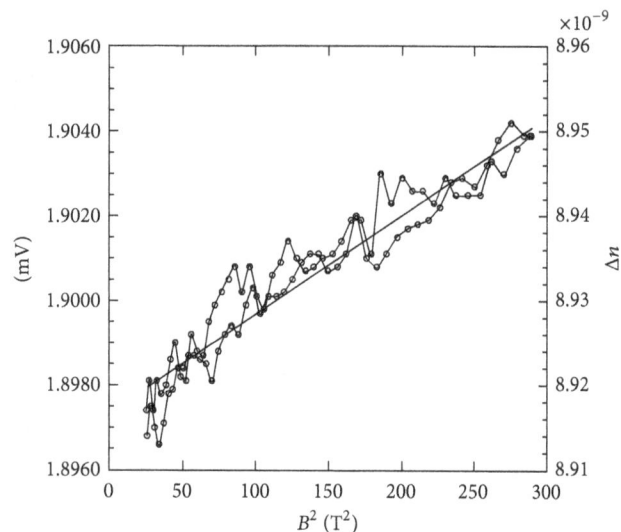

FIGURE 1: Representative example of Cotton-Mouton measurements on a dimer solution, 4 mg/mL, in Tris buffer. The magnetic field ramp rate was $\Delta B/\Delta T = 50$ (T min^{-1}). The results of the up and down sweep of the magnetic field are shown. A linear fit of the data (solid line) was used to obtain the Cotton-Mouton constant. Δn is also expressed in mV to emphasize that these experiments are on the limit of sensitivity of the instrument. The statistical errors can be estimated from the variations of the data with respect to the linear fit.

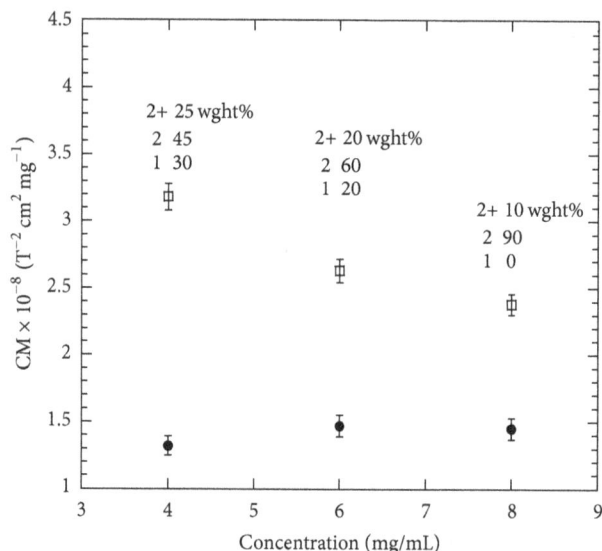

FIGURE 2: Cotton-Mouton constants, normalized to a concentration of 1 mg/mL, determined for both dimeric and double dimer samples as function of concentration. The estimated error margin is discussed in the text. The dimer solutions contain only a small, concentration independent fraction of larger aggregates. For the double dimer solutions, the composition is indicated in the figure. "2+" indicates the weight percentage of larger aggregates, "2" the weight percentage of double, dimers and "1" the amount of single dimers.

This value is 11 times smaller than that of fibrinogen [15] but when normalized to molecular weight, M_r, this difference is reduced by a third to 3.7 (M_r fibrinogen = 3 × M_r tubulin dimer). This comparison shows that tubulin dimers are significantly less anisotropic that fibrinogen molecules.

In order to have a cross check on the reliability of the data, oligomers were formed by adding MgCl$_2$ and GTP. Mass determination via ultracentrifuge showed that the oligomers predominantly consist of double-dimers plus single dimers with a small proportion of larger oligomers (see Figure 2). The relative proportion of the different species was found to vary with concentration, and this is probably the reason for the nonlinear increase in the Cotton-Mouton constant as the concentration rises (see Figure 2).

The average Cotton-Mouton constant of the oligomer solution with the largest proportion of double dimers CM$_{\text{dimer}}$ = $(2.73 \pm 0.03) \times 10^{-8}$ T^{-2}cm^{2}mg^{-1} is about twice that obtained from single dimer solutions. This was expected as an approximate doubling of the Cotton-Mouton constant is the maximum that can be expected when dimers are transformed into double dimers. The anisotropy of double dimers depends on the relative orientation of the constituent single dimers and is maximal when the long axes of the single dimers are parallel to each other as in this configuration their anisotropies add together to a good approximation. Any other relative orientation of single dimers would result in lower anisotropy. It is expected that the long axes will indeed be nearly parallel since this is the normal arrangement when the dimers are incorporated in the protofilaments and the assembled microtubules. However, a small deviation angle

between the long axes is feasible since the bond between dimers is not completely rigid.

Unfortunately, the magnetic fields available today are nowhere near strong enough to give rise to complete orientation. This limits further data interpretation since without the saturation value in birefringence it is not feasible to determine the optical anisotropy and hence the diamagnetic anisotropy from the CM constant [16]. One has to revert to comparisons with known materials.

3.2. Calculations of the Diamagnetic Moment. By idealizing the tubulin dimer as being axially symmetric, it is possible to obtain an approximate estimate of its diamagnetic anisotropy, $\Delta\chi_{\text{dimer}}$, by summing the contributions from the α-helices and β-sheets. In this we assume that the other potential sources of diamagnetic anisotropy, principally the aromatic amino acids, have no net preferred orientation.

Firstly, each α-helix and β-pleated sheet was identified, the number of peptide bonds in each, (N_α, N_β), was totaled, and the angle ($\theta_\alpha, \theta_\beta$) between the tubulin symmetry axis and the long axis of these secondary structural elements was estimated. The orientation factors (f_α, f_β) with ($f_{\alpha,\beta} = 1.5 \times \cos^2\theta_{\alpha,\beta} - 0.5$) could thus be obtained for each group. From this information $\Delta\chi_{\text{dimer}}$ was estimated using the following equation adapted from Torbet and Maret [16]:

$$\Delta\chi_{\text{TD}} = \sum_i (f_\alpha N_\alpha \Delta\chi_\alpha)_i + (f_\beta N_\beta \Delta\chi_\beta)_i. \tag{2}$$

$\Delta\chi_\alpha$ and $\Delta\chi_\beta$ are the diamagnetic anisotropies of a single peptide bond in either an α-helical or a β-pleated sheet conformation. Due to the difference in conformation, $\Delta\chi_\beta$ is only 25% of $\Delta\chi_\alpha$ (3) so α-helices are in general expected to make a greater contribution to the total $\Delta\chi_{\text{dimer}}$ than β-pleated sheets and we can simplify (2) to

$$\Delta\chi_{\text{TD}} = \Delta\chi_\alpha \sum_i (f_\alpha N_\alpha)_i + 0.25(f_\beta N_\beta)_i. \qquad (3)$$

In this way we estimate $\Delta\chi_{\text{dimer}} = 83.5\Delta\chi_\alpha$ with β-sheets contributing only 15% to the total anisotropy. Unfortunately the value of $\Delta\chi_\alpha$ is not accurately known. Two different estimates for $\Delta\chi_\alpha$ have been published, Pauling [17] gives 4.45×10^{-29} JT^{-2} while Worcester's [5] estimate is 60% greater. The former value is probably more reliable as it is closer to experimental estimates [18, 19] and is consistent with magnetic birefringence measurements made on two filamentous phages [16]. Thus, using Pauling's value we obtain $\Delta\chi_{\text{dimer}} = 3.7 \times 10^{-27}$ JT^{-2}.

The value of $\Delta\chi_{\text{dimer}}$ can also be estimated using fibrinogen for calibration as follows. The diamagnetic anisotropy can be calculated using the Cotton-Mouton constant and the birefringence at complete orientation or saturation orientation, Δn_{sat}, using the relationship [16] $\Delta\chi = 15\lambda kTC_{\text{CM}}/\Delta n_{\text{sat}}$. This equation cannot be exploited with tubulin as we do not have completely aligned samples. However, by putting the published values [15] for the Cotton-Mouton constant of fibrinogen and the saturation birefringence of fibrin into the latter equation we obtain 5×10^{-26} JT^{-2} for the $\Delta\chi$ of a single fibrinogen molecule. Again assuming aromatic residues make no net contribution, this anisotropy is due to the axially aligned α-helices constituting about 30% of the molecule (i.e., 930 residues). The $\Delta\chi$ of fibrinogen is thus equal to $930 \times \Delta\chi_\alpha$ as shown above. By comparison with fibrinogen, $\Delta\chi_{\text{dimer}} = 4.5 \times 10^{-27}$ JT^{-2} (i.e., $5 \times 10^{-26} \times 83.8/930$ JT^{-2}). This value is 20% larger than that calculated above using the Pauling value for $\Delta\chi_\alpha$.

The $\Delta\chi$ of fibrinogen is 3.7 times larger than that of $\Delta\chi_{\text{dimer}}$ normalized to molecular weight. As reported above the C_{CM} for fibrinogen is also 3.7 times larger than that of tubulin. This supports our estimate for $\Delta\chi_{\text{dimer}}$; however, this conclusion assumes that the diamagnetic and optical anisotropies are linked by the same proportionality in fibrinogen and tubulin. This is probably approximately true for the intrinsic component of the optical anisotropy because, like the diamagnetic anisotropy, it depends on the anisotropic mobility of electrons in the molecule [20]. But the optical anisotropy can also have a form component which might be relatively different for fibrinogen and tubulin. While the Cotton-Mouton measurements are supportive of our estimate for $\Delta\chi_{\text{dimer}}$, they do not constitute conclusive evidence.

$\Delta\chi_{\text{dimer}}$ can be used to obtain an estimate of the minimum number of tubulin dimers, acting cooperatively, required to attain a highly oriented system. It is known [21] that for a diamagnetically anisotropic object to attain better than 80% maximum orientation in a magnetic field, B, then $\Delta\chi B^2 > 20kT$. In a very strong magnetic field of 10 Tesla at 20°C the minimum number of tubulin dimers required, N_d, is

estimated to be in excess of 2×10^5 ($M_r > 2 \times 10^{10}$ daltons) using the lower value for $\Delta\chi_{\text{dimer}}$ calculated above. As diamagnetic anisotropy is additive to a good approximation, $\Delta\chi = N_d \times \Delta\chi_{\text{dimer}}$ for dimers arranged in parallel. Thus, if a single microtubule is undergoing orientation, in the absence of interaction with its neighbors and assuming a dimer length of 8 nm along the protofilament direction and an average of 13 dimer protofilaments in the tubulin wall, it would need to have a length in excess of $(2 \times 10^5 \times 8\,\text{nm})/13 \approx 12\,\mu\text{m}$, or alternatively a number of smaller microtubules orienting cooperatively could give rise to the same result. It should be pointed out that in concentrated solutions of rigid molecules like microtubules [22] the magnetic field might not be the only cause of alignment. Above a certain concentration, phase separation and subsequent formation of oriented domains could occur [23]. However, the directors of these domains will still display random orientation. The magnetic field will force these directors to become aligned with the magnetic field. Further research on this has been done but falls outside the scope of this paper.

4. Conclusions

We have been able to measure the Cotton-Mouton constant of tubulin dimers with a reasonable accuracy. The absence of a reliable value for the optical polarizability prevents a direct determination of the diamagnetic susceptibility but by comparing the Cotton-Mouton constants of the tubulin dimer with fibrinogen, for which the susceptibility is known, we can make a reasonable estimate. This estimate of the susceptibility corresponds well with the calculated contributions of the α-helices and β-sheets to the diamagnetic susceptibility. If the crystallographic structure is known these, calculations are relatively simple and can be carried out for other proteins as well. The cross correlations that can be made between the experimental results and the simple calculations validate the calculation method for the tubulin dimer and show that this method can be used as an initial assessment of the diamagnetic susceptibility of proteins for which no other data is available. The method that we have used to determine the tubulin dimer diamagnetic susceptibility can be used, for other, proteins and macromolecular assemblies for which this information is not readily available. This can become relevant in for instance medically applied Magnetic Resonance Imaging (MRI) where the applied fields keep increasing.

Conflict of Interests

The authors declare that there is no conflict of interests regarding the publication of this paper.

Acknowledgments

The Grenoble High Magnetic Field Laboratory is gratefully acknowledged for allowing the authors access to their facilities. Professor P. Christianen (Nijmegen High Magnetic Field

Laboratory) is thanked for his suggestions which the authors feel have improved the paper considerably.

References

[1] W. Bras, G. P. Diakun, J. F. Díaz et al., "The susceptibility of pure tubulin to high magnetic fields: a magnetic birefringence and X-ray fiber diffraction study," *Biophysical Journal*, vol. 74, no. 3, pp. 1509–1521, 1998.

[2] P. M. Vassilev, R. T. Dronzine, M. P. Vassileva, and G. A. Georgiev, "Parallel arrays of microtubules formed in electric and magnetic fields," *Bioscience Reports*, vol. 2, no. 12, pp. 1025–1029, 1982.

[3] J. Torbet, "Using magnetic orientation to study structure and assembly," *Trends in Biochemical Sciences C*, vol. 12, pp. 327–330, 1987.

[4] G. Maret and K. Dransfeld, *Strong and Ultrastrong Magnetic Fields and their Applications*, Springer, 1985.

[5] D. L. Worcester, "Structural origins of diamagnetic anisotropy in proteins," *Proceedings of the National Academy of Sciences of the United States of America*, vol. 75, no. 11, pp. 5475–5477, 1978.

[6] U. Raviv, D. J. Needleman, K. K. Ewert, and C. R. Safinya, "Hierarchical bionanotubes formed by the self assembly of microtubules with cationic membranes or polypeptides," *Journal of Applied Crystallography*, vol. 40, no. 1, pp. s83–s87, 2007.

[7] J. M. Denegre, J. M. Valles Jr., K. Lin, W. B. Jordan, and K. L. Mowry, "Cleavage planes in frog eggs are altered by strong magnetic fields," *Proceedings of the National Academy of Sciences of the United States of America*, vol. 95, no. 25, pp. 14729–14732, 1998.

[8] R. Emura, N. Ashida, T. Higashi, and T. Takeuchi, "Orientation of bull sperms in static magnetic fields," *Bioelectromagnetics*, vol. 22, no. 1, pp. 60–65, 2001.

[9] J. M. Valles Jr., "Model of magnetic field-induced mitotic apparatus reorientation in frog eggs," *Biophysical Journal*, vol. 82, no. 3, pp. 1260–1265, 2002.

[10] K. Guevorkian and J. M. Valles Jr., "Aligning Paramecium caudatum with static magnetic fields," *Biophysical Journal*, vol. 90, no. 8, pp. 3004–3011, 2006.

[11] E. Nogales, M. Whittaker, R. A. Milligan, and K. H. Downing, "High-resolution model of the microtubule," *Cell*, vol. 96, no. 1, pp. 79–88, 1999.

[12] P. G. Vulfson, *Molecular Magnetochemistry*, Gordon and Breach, 1998.

[13] P. Rub and G. Maret, "New 18-T resistive magnet with radial bores," *IEEE Transactions on Magnetics*, vol. 30, no. 4, pp. 2158–2161, 1994.

[14] P. Schuck and P. Rossmanith, "Determination of the sedimentation coefficient distribution by least-squares boundary modeling," *Biopolymers*, vol. 54, no. 5, pp. 328–341, 2000.

[15] J. M. Freyssinet, J. Torbet, G. Hudry Clergeon, and G. Maret, "Fibrinogen and fibrin structure and fibrin formation measured by using magnetic orientation," *Proceedings of the National Academy of Sciences of the United States of America I*, vol. 80, no. 6, pp. 1616–1620, 1983.

[16] J. Torbet and G. Maret, "High-field magnetic birefringence study of the structure of rodlike phages Pf1 and fd in solution," *Biopolymers*, vol. 20, no. 12, pp. 2657–2669, 1981.

[17] L. Pauling, "Diamagnetic anisotropy of the peptide group," *Proceedings of the National Academy of Sciences of the United States of America*, vol. 76, no. 5, pp. 2293–2294, 1979.

[18] K. Tohyama and N. Miyata, "Diamagnetic susceptibility and its anisotropy of poly-γ-ethyl-L-glutamate," *Journal of the Physical Society of Japan*, vol. 34, no. 6, article 1699, 1973.

[19] N. S. Murthy, J. R. Knox, and E. T. Samulski, "Order parameter measurements in polypeptide liquid crystals," *The Journal of Chemical Physics*, vol. 65, no. 11, pp. 4835–4839, 1976.

[20] R. Oldenbourg and T. Ruiz, "Birefringence of macromolecules. Wiener's theory revisited, with applications to DNA and tobacco mosaic virus," *Biophysical Journal*, vol. 56, no. 1, pp. 195–205, 1989.

[21] J. Torbet, "Internal structural anisotropy of spherical viruses studied with magnetic birefringence," *The EMBO Journal*, vol. 2, no. 1, pp. 63–66, 1983.

[22] M. G. L. van den Heuvel, S. Bolhuis, and C. Dekker, "Persistence length measurements from stochastic single-microtubule trajectories," *Nano Letters*, vol. 7, no. 10, pp. 3138–3144, 2007.

[23] L. Onsager, "The effects of shape on the interaction of colloidal particles," *Annals of the New York Academy of Sciences: Molecular Interaction*, vol. 51, pp. 627–659, 1949.

4

Comparative Trace Elemental Analysis in Cancerous and Noncancerous Human Tissues Using PIXE

Stephen Juma Mulware

Ion Beam Modification and Analysis Laboratory, Physics Department, University of North Texas, 1155 Union Circle, No. 311427, Denton, TX 76203, USA

Correspondence should be addressed to Stephen Juma Mulware; stephenmulware@my.unt.edu

Academic Editor: Janos K. Lanyi

The effect of high or low levels of trace metals in human tissues has been studied widely. There have been detectable significant variations in the concentrations of trace metals in normal and cancerous tissues suggesting that these variations could be a causative factor to various cancers. Even though essential trace metals play an important role such as stabilizers, enzyme cofactors, elements of structure, and essential elements for normal hormonal functions, their imbalanced toxic effects contribute to the rate of the reactive oxygen species (ROS) and formation of complexities in the body cells which may lead to DNA damage. The induction of oxidative-induced DNA damage by ROS may lead to isolated base lesions or single-strand breaks, complex lesions like double-strand breaks, and some oxidative generated clustered DNA lesions (OCDLs) which are linked to cell apoptosis and mutagenesis. The difference in published works on the level of variations of trace metals in different cancer tissues can be attributed to the accuracy of the analytical techniques, sample preparation methods, and inability of taking uniform samples from the affected tissues. This paper reviews comparative trace elemental concentrations of cancerous and noncancerous tissues using PIXE that has been reported in the published literature.

1. Introduction

Studies have shown that the imbalance in the composition of trace metals which are generally recognized to be essential to normal human homeostasis besides accumulation of potentially toxic and nonessential trace metals may cause disease. The significance of the essential trace metals is indisputable due to their positive roles when in specific concentration ranges while on the other hand displaying toxic effects in relatively high or low concentration levels. Physiochemical properties of trace metals govern their uptake, intracellular distribution, and the binding of the metal compounds in biological systems. In spite of diverse physiochemical properties of metals compounds, there are three main predominant mechanisms related to metal genotoxicity. The first is the interference with cellular redox regulation and production of oxidative stress which does cause oxidative DNA damage or trigger signaling cascades that may lead to stimulation of malignant growth [1]. Second is the ability to inhibit

the major DNA repairs mechanisms which may result in genomic instability and accumulation of critical mutations. And third is the deregulation of cell proliferation by induction of signaling pathways or ability to inactivate the growth controls such as tumor suppressor genes [2].

The essential trace metals have four main functions which include (i) stabilizers, (ii) elements of structure, (iii) essential elements for hormonal function, and (iv) cofactors in enzymes. Inadequate or lack of trace elements will affect the structure alone or will affect structural function due to lack of stabilization, change of charge properties, and allosteric configuration [3]. The deficiency of trace elements as enzyme cofactors is expected to expose the individual to carcinogenic stress [4]. Trace metals form core structures of superoxide dismutase (SOD) which are a group of metalloenzymes (containing Fe, Mn, or Cu and Zn) that catalyze the disproportionate of superoxide free radical ($O_2^{\bullet-}$) to form hydrogen peroxide and dioxide, hence breaking down the toxic reactive oxygen species (ROS) radical. These enzymes have been considered as the defense system against the cytotoxic superoxide free radical. Cu, Zn-SOD which is a prototypical dinuclear metalloproteinase is an important antioxidant enzyme for cellular protection from ROS and several proteins involved in DNA repair [5]. Metallothionein (MT) is a family of cysteine-rich, low-molecular-weight (MW ranging from 500 to 14000 Da) proteins which have the capacity to bind both physiological (such as zinc, copper, and selenium) and xenobiotic (such as cadmium, mercury, silver, and arsenic) heavy metals through the thiol group of its cysteine residues. Cysteine residues from MTs can capture harmful oxidant radicals like the superoxide and hydroxyl radicals, hence there important role in controlling oxidative stress that would lead to cell mutations. Due to MT's important role in transcription factor regulation, problems with MT function or expression may lead to malignant transformation of cells and even cancer. Research has found increased expression of MTs in some cancers of the breast, colon, kidney, liver, skin (melanoma), lung, nasopharynx, ovary, prostate, mouth, salivary gland, testes, thyroid, and urinary bladder; it has also found lower levels of MT expression in liver adenocarcinoma and hepatocellular carcinoma [4].

More than other 30 enzymes including ceruloplasmin, cytochrome oxidase, lysine oxidase, dopamine-hydroxylase, ascorbate oxidase, and tyrosinase among others in the body have copper as main building block. Some of these enzymes are involved in the synthesis of the main component of connective tissues called collagen [4]. On the other hand, iron plays a significant role in oxygen transportation, xenobiotic metabolism, and oxidative phosphorylation and being a prosthetic group in many enzymes; however, its excessive content in organism's binding capacity can be toxic even leading to cancer development. Most trace metals which occur in various oxidation states including Fe, Cu, Co, Cr, Hg, and V can generate ROS, a property which explains their carcinogenic effects [6–8].

Carcinogenesis is considered to occur in four stages: initiation, promotion, progression, and metastasis. The mechanism of metal-induced carcinogenesis is believed to be involved in all stages of cancer development. Consequently, the roles of trace metals in cancer development and inhibition have a complex character and have raised many questions because of their essential and toxic effects on people's health. In the last few decades, metals such as cadmium, nickel, arsenic, beryllium, and chromium (VI) have been recognized as human or animal carcinogens by the International Agency for Research on Cancer [9, 10]. The carcinogenic capability of these metals depends mainly on factors such as oxidation states and chemical structures. The oxidative concept in metal carcinogenesis proposes that complexes formed by these metals, in vivo, in the vicinity of DNA, catalyze redox reactions, which in turn oxidize DNA. The most significant effect of reactive oxygen species (ROS) in the carcinogenesis progression is DNA damage, which results in isolated base lesions or single-strand breaks, complex lesions like double strand breaks as well as some oxidatively generated clustered DNA lesions (OCDLs) and the sister chromatid exchange [11, 12]. It has been estimated that approximately 2×10^4 DNA

damaging events occur in every cell per day; a major portion of these occur via reactive oxygen species (ROS).

The oxidatively induced DNA damage associated with ROS is apurinic or apyrimidinic (abasic) DNA sites, oxidized purines and pyrimidines, and single- and double-strand DNA breaks. Even though the lesions are not ultimately lethal to the cell, they are considered highly mutagenic. The creation of an altered base or base loss is not expected to significantly destabilize the DNA molecule; however, a localized perturbation of the stacking forces, hydrogen bonds, and interaction with water molecules and positive metal ions like Na+ surrounding the DNA double helix is expected to occur [13, 14]. These perturbations generally change the DNA at the lesion site. Peroxyl radicals in the presence of oxygen are also believed to cause lipid peroxidation DNA damage and carcinogenesis [15]. Through Fenton-type reaction process, Fe^{2+} may reduce hydrogen peroxide creating hydroxyl radicals which attack DNA inducing base lesions and single-strand breaks which may lead to double-strand breaks in return. ROS can also attack various important cellular proteins that are essential for DNA repair thus affecting their binding to the DNA substrates [16]. These cumulative effects of ROS which in most cases are endogenously generated by trace metal presence in the body cells are a key element of carcinogenic process. Even though the increase in oxidative DNA lesions has been frequently attributed to metal exposures, the molecular mechanism leading to tumor formation after such exposures is still not well understood.

2. Sampling and Sample Preparation Techniques

Proper sampling is very important to ensure accuracy and reliability of the results. Trace metal concentrations can be heterogeneously distributed in some tissues. As has been reported in many studies, the inhomogeneity of the trace metal concentrations depends on an individual's age, metal type, and tissue specimen under investigation [14]. The samples should be homogeneous as much as possible, and all handling and preparation devices and reagents must be checked carefully to avoid or reduce contamination as much as possible if reliable results are to be achieved. For effective PIXE analysis, the samples must be dry; however, the process must be carefully done to reduce tissue disturbance from its original physiological state. Much attention should be taken during sample extraction if cancerous and noncancerous tissues are examined since the distribution of malignant molecules is generally heterogeneous. For example, levels of Fe, Zn, and K in healthy breast tissues extracted near tumor affected areas are found to be higher than those obtained from healthy breasts tissues [4]. This has been attributed to the physiological process leading to the accumulation of trace elements in tumors, and increased enzymatic activities may have affected the composition of healthy tissue in the margin surrounding lesions [17]. Additionally, the increased levels of trace elements in cancerous tissues could be due to fibroglandular specimens as compared to noncancerous breast tissues.

The samples are collected by autopsy or by punch biopsy from normal and cancerous sites of deidentified patients. The samples are then quickly frozen in 2-methylbutane cooled in liquid nitrogen (-176°C). Rapid freezing reduces ice crystal formation and minimizes morphological damage to the samples. Frozen sections may be used for a variety of procedures, including immunochemistry, enzymatic detection, and in situ hybridization. Cryosectioning is a critical step as temperature gradients and sample freezing out processes may occur when sample is positioned in mounting medium that can facilitate cell disruption and ion displacement. During sectioning frozen tissue, the sample must be firmly attached to the microtome stage or chuck. Saline or water will hold the sample when frozen; however, embedding media are preferred since they provide a supportive and protective aid to sectioning. The most commonly used embedding medium is OCT, an aqueous solution of glycols and resins which provides an inert matrix for sectioning. 30% bovine albumin and von Apathy's gum syrup have been used as alternative to OCT. Once the frozen tissue is firmly in place in the cryostat (-40°C), the microtome lock is released and advanced or retracted to the chuck position until the knife edge just touches the block. The superficial surface of the sample is roughly trimmed in small steps (15–25 μm) until an even, full face is achieved. The tissue debris is removed from the knife with a soft brush or tissue. Once the right cutting thickness is set and the cryostat and/or specimen allowed reaching the optimum cutting temperature, sections are cut using a slow even motion, except for hard tissue which requires a firmer stroke. The section should glide smoothly under the antiroll plate. Cutting is best done using a firm, steady motion and gentle pressure. Sections of 10–14 μm thickness are cut and removed from the knife with a saline moistened brush. Alternatively, with unfixed tissue and using a cooled dry knife, the section may be picked up directly onto a glass slide where they will thaw on contact and adhere to the slide surface.

The sections are then freeze-dried. The freeze-drying process is used to remove water from the frozen tissue by sublimation of the solid phase at a low temperature and under vacuum. Sublimation can only occur when the partial pressure of the water vapor of the ice exceeds that of the atmosphere. In practice, the process of drying a frozen sample takes place under a vacuum of 10^{-3} Torr or greater and at a temperature difference sufficient to heat the ice crystals in the sample and provide energy for sublimation to water vapor. The transfer of water molecules from ice to vapor, however, removes heat from the environment causing a drop in temperature and a reduction in the rate of sublimation. A constant temperature must therefore be maintained in the sample so that sublimation proceeds at a rate equivalent to the heat input (-30 to -40°C). The samples are allowed to freeze-dry to about 1-2 μm range and then mounted onto sample holders ready for irradiation. The sections can also be mounted on a foil of polycarbonate or silicon nitride adequately thin to enable high-resolution imaging using transmitted ions (STIM).

Other sample preparation techniques used in some of the studies involve freeze-drying the collected tissues and then converting into fine powder by pounding in an agate mortar. A fraction of this powdered material is dried overnight and mixed with a known quantity of internal standard like yttrium, and this mixture is again dried overnight. A known quantity of this powder is pressed into pellets, and these are used as targets.

3. PIXE as Analytical Technique

In the recent years, most scientists have adopted PIXE as an analytical method of choice for trace elemental analysis in biological samples. This is due to its ability to produce more reliable results that can consistently be compared to other studies. PIXE is a noninvasive analytical technique that offers the most detection limits of parts per million (ppm) levels of sensitivity. This opens up new fields in biological microanalysis where the measurement of important trace elements such as calcium, iron, zinc, copper, and selenium among others now becomes possible on a microscale level. It is not an overstatement that microbeam PIXE is the only technique that offers ppm elemental sensitivities with high quantitative accuracy at spatial resolutions smaller than cell dimensions [18, 19]. The measurement of trace elements in individual cells is a field that is open to nuclear microscopy scientists and is expected to continue to grow significantly especially with the spatial resolution of nuclear microprobes reaching the 100 nm level as in most microprobe laboratories today. In PIXE, the interaction of MeV protons with a target leads to the ionization of the target atoms resulting into emission of X-rays with energies that provide characteristic spectral signatures of the target atoms. The well-characterized nature of these interactions alongside the ability to determine the proton energy loss and X-ray absorption within the target and the detector makes it possible to directly model X-ray production and calculate the X-ray yields for a given target composition. The accurate charge measurement using a charge integrator and possible secondary electron suppressor are necessary to ensure accuracy of the PIXE results. The suitability of PIXE method is further enhanced by the high cross-sections for PIXE X-ray production and low levels of continuum background [20–22]. Figure 1 shows the schematic presentation of beam scanning and focusing for PIXE measurements in a typical microprobe.

4. Trace Metal Concentrations in Human Tissues

There have been a lot of studies and large number of data on the levels of trace elements in human cancerous and noncancerous tissues obtained using different analytical techniques as mentioned earlier. Due to their vital role in metabolism of trace metals, liver, lung, and kidney have been more frequently studied. With the rise of other types of cancer, more recently, much interest has been laid on the study of other organs including the breast, prostate, penis, thyroid, and stomach. We will concentrate on the studies done using PIXE as analytical techniques. Data from

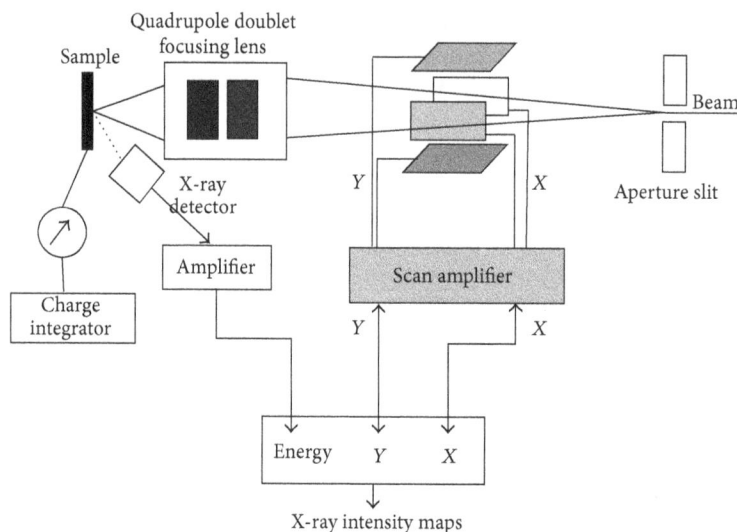

FIGURE 1: Schematic presentation of beam scanning and focusing for PIXE measurements in a typical microprobe setup.

the existing literature are reported in Table 1. Figures 2(a), 2(b), 2(c), and 2(d) show the bar charts of the log of concentrations of the trace elements for the breast, kidney, stomach, and testis tissues, respectively, which shows the correlation of the trace metal concentration of the cancerous and normal tissues of the selected organs.

4.1. Breast Tissue. Breast tissues were the most commonly studied sample compared to the others. One study found that the concentrations of the elements Cl, Ca, Cr, Fe, Cu, Zn, As, Se, BR, and Sr were significantly higher in cancerous tissues than normal tissues and that the concentrations of the elements K, Ti, and Ni had no significant difference from those of normal tissues [24]. These elevated levels were found to be similar to other previous studies that used other analytical techniques [17, 25–27]. The same study reported that the elevated level of chromium showed that it is a carcinogen as has been established. Chromium toxicity is commonly associated with exposure to hexavalent chromium compounds rather than to the low toxic trivalent chromium compounds [24]. Chromium (VI) has a characteristic ease of absorption by the body cells, and once in the cell, it is reduced to the trivalent state which produces genotoxic effects [28]. Its exposure can cause abnormal phenotypes due to generation of ROS and other several DNA lesions which in effect lead to DNA damage. Similarly, chromium has been found to repress p53 protein which is a tumor suppressor protein. Several studies reported that the inactivation of this protein is associated with various types of human cancers. P53 is involved in various biological processes including regulation of genes involved in the cell cycle, cell growth arrest after DNA damage, and apoptosis [29]. Due to the significance of cell cycle arrest and apoptosis, which are regulated by p53, their inactivation or alteration by reactive chromium intermediates such as Cr(V) and Cr(VI) can enhance cancer development [30]. Other carcinogen trace metals also include iron and copper. Fe catalyses hydrogen

peroxide conversion to free radical ions which attack cellular membranes causing DNA strand breaks, inactivating enzymes, depolymerizing polysaccharides, and also initiating lipid peroxidation. Several studies have also established that Fe promotes inflammation and increases cancer cell growth [31, 32].

Due to its ability to change between its two oxidation statuses Cu^+ and Cu^{2+}, Cu has been found to cause generation of ROS which produce hydroxyl radicals that modify proteins, lipids, and nucleic acids eventually causing DNA damage [33]. Similarly, copper ions can aid carcinogenesis due to their role in angiogenesis which is a vital process in tumor progression [24].

Another study using the blood serum from breast cancer patients reported significant reduction in concentrations of Ti, Cr, Mn, Ni, Zn, and Se in the sera of breast cancer patients, which might be caused by cancer [34]. Previously, other studies have shown that there is a correlation between the increase in tissue levels of certain trace elements and their consequent decrease in the serum of breast cancer patients [24]. Studies have suggested that the ratio of Cu and Zn can be used for diagnostic or prognostic value in cancer. Similarly, the ratio of concentration of one metal in a tumor tissue to that of the same metal in nontumor tissue can be used for cancer diagnosis. For instance, one study found that the ratio of mean tumor to mean healthy concentration for Fe was 1.6 for paired samples and 2.7 for nonpaired samples, while that for Cu was found to be 3.1 and 3.6 for paired sample and nonpaired sample, respectively [17, 35].

4.2. Thyroid. Thyroid tissues have also been widely studied for trace metal. One study observed higher levels of Ti, V, Cr, Mn, Co, Fe, and Sr in cancerous tissues compared to the normal thyroid tissues [36]. While comparing their observation with another study which also studied trace elements in thyroid, both studies established consistent concentrations

TABLE 1: Trace metal levels reported in the literature for cancerous and noncancerous human tissues using PIXE. The values are in μg/g.

Tissue	Tissue type	Cl	K	Ca	Ti	V	Cr	Mn	Fe	Co	Ni	Cu	Zn	As	Sr	I	Hg	Se	Br	Rb	Pb	Reference
Breast	Normal	3999	1381	157	14.3	—	31.9	18	299	—	7.4	42	56	2.57	6.7	—	—	0.66	9	12	—	[24]
	Cancerous	6815	1550	480	15.9	—	52.7	17.2	376	—	8.56	60.7	126	4.12	13.7	—	—	1.32	16	22	—	
Breast (serum)	Normal	—	—	—	416	32	18.6	29.8	291	2.9	7.2	24.6	38.5	0.9	—	—	—	2.5	46	—	—	[34]
	Cancerous	—	—	—	238	19	10	16	355	0.9	5.3	32.3	13	0.69	—	—	—	1.5	23	—	—	
Thyroid	Normal	200	11.6	688	18	4.5	6	17	569	11	8.4	54.9	149	125	11.7	229	98	—	—	—	17	
	Adenoma	215	18	577	15.9	4.2	36.6	14.4	525	11.6	10.6	6	68.7	17	6.8	222	19.8	—	—	—	11.6	[36]
	Cancerous	234	159	393	103	20	29.8	46.6	1397	17.9	7.7	12.9	71.8	2.8	—	—	7.7	—	—	—	17.8	
Kidney	Normal	2911	172	562	22	—	9.3	20.4	325	7.8	9.7	9.1	66	—	—	—	—	1.8	9.8	—	24.7	[40]
	Cancerous	2857	106	298	29	—	8.2	20.9	305	11.6	7.6	8.2	78	1.1	—	—	—	—	8.2	—	24.6	
Stomach	Normal	5295	387	647	31	—	7.3	28.6	2408	19.4	10.5	63.5	818	—	—	—	—	—	—	—	8.8	[40]
	Cancerous	2518	256	433	15	—	12.6	15.1	684	15.6	60	21.2	229	1.71	—	—	—	—	3.1	—	8.1	
Penis	Normal	3583	38.3	877	15.2	—	4.6	10.3	650	25	—	7.5	58.2	2.1	22	—	—	—	2	—	4.9	[45]
	Carcinoma	2133	57.8	815	16.8	—	2.6	9.2	464	11	—	79.6	148	5.6	16	—	—	—	1.5	—	5.2	
Testis	Normal	3301	69	3400	238	—	102	15.6	426.4	3.25	—	3.9	91.7	—	—	—	—	—	—	—	—	[45]
	Carcinoma	2887	131	2726	257	—	125	15.2	391.9	0.9	—	14.9	34	—	—	—	—	—	—	—	—	

of K, Ca, and I, while variations were observed in the concentration of Cu, Zn, and Fe in cancerous thyroid compared to noncancerous tissues [36, 37]. Zn concentration in adenoma and cancerous thyroid was found to be lower than that in normal tissues, an observation consistent with other findings in the study of kidney and hair samples, respectively [38, 39]. Zn deficiency has been linked to severe deficiency in immune function and disruption in T-Cell function which is directly related to pathogenic carcinoma. Zn deficiency also causes inactivation of p53, a tumor suppressor protein, which has been associated with many cancers, including thyroid. Since iodine plays an important role in the functioning of thyroid glands, through the production of thyroid hormones (thyroxin and triiodothyronine) which are essential for cellular oxidation, growth, reproduction, and the activity of the central and autonomic nervous system, its low level can be attributed to cancer as observed by Reddy and his group in thyroid carcinoma tissues [36]. The malignant tumor tissues contain areas of necrosis and some hemorrhage caused by capillary ruptures. The rupture produces accumulation of hemoglobin from red blood cells which fragment into Fe containing heme, a process that leads to observed increased level of Fe in carcinoma tissues.

4.3. Kidney.

Few studies have been done on kidney tissues. One such sturdy reported higher levels of Ti, Co, Zn, and Cd, and lower levels of K, Ca, Fe, Ni, and Se, respectively, in carcinoma kidney tissues compared to normal tissues [40]. The same study also observed that there was no variation of Cl, Cr, Mn, Cu, Br, and Pb between carcinoma tissues and the normal tissues. A different study, however, reported lower concentration of Cd, Cr, Ti, V, Cu, and Zn in cancerous kidney tissue than the noncancerous in the analysis, a fact this study attributed to the different metabolism and dynamics of cancer process compared to normal tissue [41]. Potassium

being a major intracellular cation was observed to be in low concentration in carcinoma tissue than in normal tissue. It is known that prolonged potassium deficiency (known as hypokalemia) can cause injury to the kidney.

4.4. Stomach.

With increasing cases of stomach cancer being reported, much attention has been turned to the study of stomach cancers. One such study reported higher concentrations of Cr, Ni, As, and Br and lower concentrations of Cl, K, Ca, Ti, Mn, Fe, Co, Cu, and Zn in stomach cancer tissues compared to noncancer tissues, respectively [40]. The high concentration of Cr in stomach cancer tissues supports its carcinogenic property [42]. Chromate ion enters the cells by sulfate uptake and is then reduced to Cr(III) through Cr(V)-glutathione intermediate species which binds with DNA producing a kinetically inert and damaging lesion. The same study observed that the low level of Fe in the stomach cancer tissue could be due to lack of HCl which may have resulted from the carcinoma nature of the stomach. Ni concentration was also much higher in the stomach cancer tissue, supporting its carcinogenic property as was reported in other studies [43, 44].

4.5. Penis and Testis.

High concentration of Cu, Zn, and As and low concentration of Cl, Fe, and Co in the carcinoma tissue of penis compared to the normal tissue, respectively, have been reported [45]. The same study also reported high concentrations of K, Cr, and Cu and low concentrations of Cl, Ca, Ti, and Mn in carcinoma tissue of testis compared to normal tissue, respectively. The observed high concentration of Cu in the carcinoma tissue of both organs underlines the important role Cu plays in cancer promotion through inflammation and angiogenesis. A clinical study demonstrated that cancer can be fought by reduction of body copper levels, since it is the common denominator of angiogenesis [45].

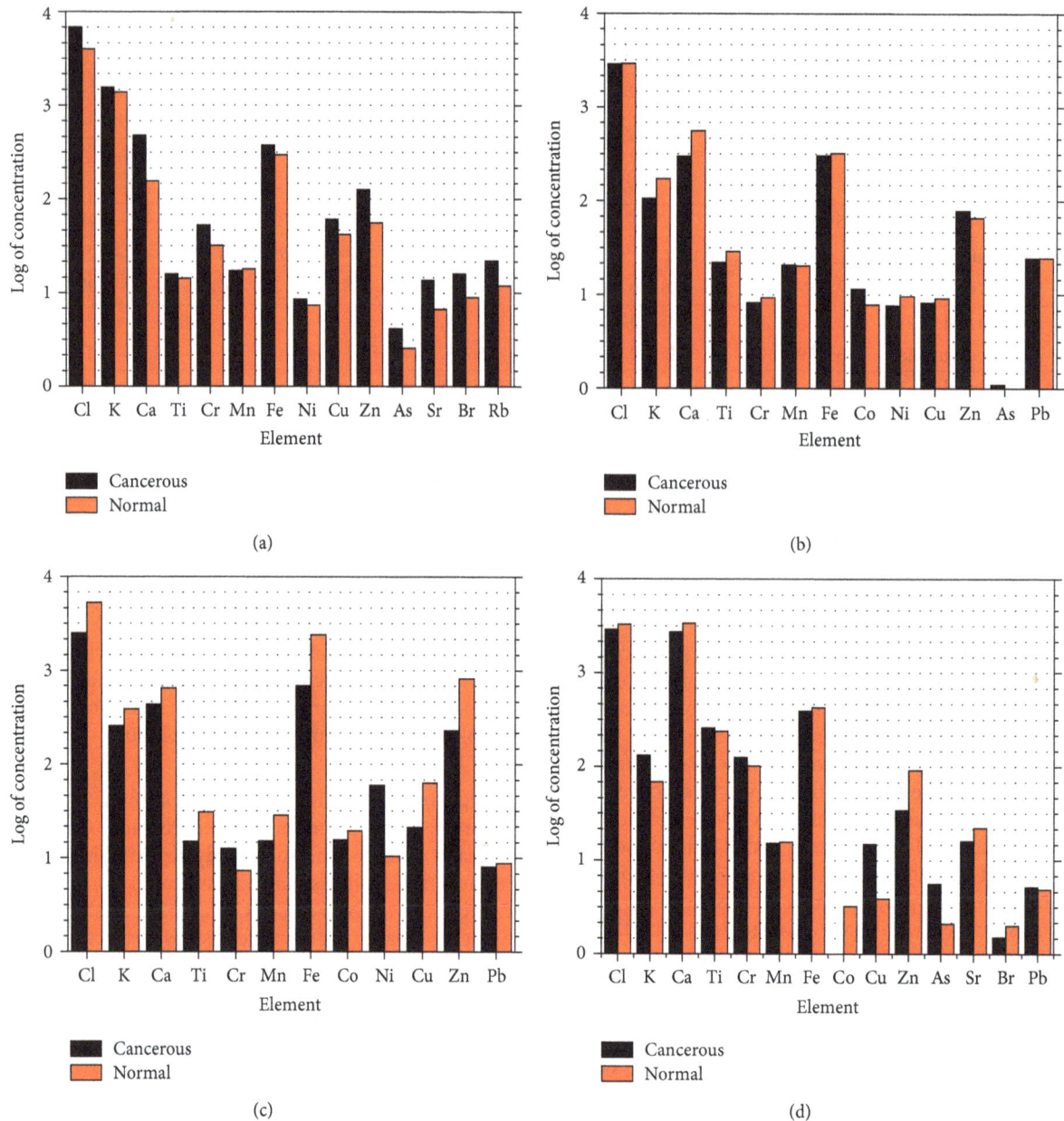

FIGURE 2: (a), (b), (c), and (d) show the bar charts of the log of concentrations of the trace elements for the breast, kidney, stomach, and testis tissues, respectively, which shows the correlation of the trace metal concentration of the cancerous and normal tissues of the selected organs.

They observed that cancer cells are easily proliferated into tumors in a high copper concentration environment, and thus most cancers growths can be inhibited through copper deficiency.

The Zn concentration levels were higher in the tissue of penis which supports the observations made in earlier works indicating that Zn is carcinogenic and that it is involved in tumor growth and development of neoplastic transformation [27, 28]. Zn has been found to enhance the activity of telomerase enzyme as well as antagonizing the inhibitory effect of bisphosphonates on breast and prostate tumor cell invasion as well [23]. In their study, Reddy and his group reported low concentration of Zn in testicular cancer and

high Zn levels in penis cancer, an observation that suggests different roles played by Zn in different cancerous organs [40].

5. Discussion

The results shown in Table 1 demonstrate that there are differences and variations in trace metal concentrations from one organ to the other depending on their cancerous or non-cancerous states. However, what is clear is that PIXE analysis can be effectively used to analyze trace metal concentrations of cancerous and noncancerous tissues. The variations of the trace metals can be attributed to the analysis technique used, sample sampling, and preparation technique including sample condition (wet/dry) and the reliability and accuracy level of the analytical technique. It is important to note that the PIXE technique would be very challenging for wet samples, while on the other hand drying the sample may also affect its physiology and hence the eventual results.

As analyzed for different organs above, the role of individual trace metals varies from one organ to the other. For example, one study reported low concentration of Zn in testicular cancer and high Zn levels in penis cancer [40]. Whether low or high, each different trace metal has specific effect on the carcinogenicity on the affected organ. The deficiency or excess of certain trace elements is correlated to the carcinogenesis of that organ. Other trace metals also act as inhibitors of cancer in different organs. For example, low levels of Se in the carcinoma tissue of kidney and low levels of Mn in carcinoma tissue of stomach, which supported the claims that they inhibit cancer growth in kidney and stomach, respectively, were reported by one group [36].

PIXE technique has this far proven to be a very sensitive technique for trace metal analysis especially of biological samples. When sampling and handling of samples during preparation are properly done, this analytical technique promises to be one of the best since a wide range of trace metals can easily be qualitatively and quantitatively detected. The use of a more sensitive detector like HPGe-detector can also improve the analysis especially of low-energy X-rays from lighter elements.

6. Conclusion

From the observations made in this paper, PIXE analysis can be successfully used to determine trace elemental concentrations of cancerous and noncancerous human tissues. The low or high levels of the trace metals in cancerous or noncancerous tissues can be used to determine the carcinogenic role of the deficiency or excess of each trace metal in specific organs. These concentration levels need to be closely monitored in further prognostic studies. The results also need to be verified by other advanced diagnostic techniques like high throughput sequencing, gene expression sequencing, and proteomics for conclusive determination to be made. For result reliability and correct assessment of the role of each trace metal in regard to carcinogenesis, the role in initiation, promotion, progression, or inhibition of cancer in various organs, there is need for acquisition of more data from several investigations. Similarly, there is a need to carry out the investigations from different regions using differentials like gender and age dietary habits among other variables.

References

[1] B. Ames and K. Shigenaga, "Oxidative stress is a major contribution to aging," *Annals of the New York Academy of Sciences*, vol. 663, pp. 85–96, 1992.

[2] D. Beyersmann and A. Hartwig, "Carcinogenic metal compounds: recent insight into molecular and cellular mechanisms," *Archives of Toxicology*, vol. 82, no. 8, pp. 493–512, 2008.

[3] E. L. Feinendegen and K. Kasperek, "Medical aspects of trace element research," in *Trace Element Analytical Chemistry in Medicine and Biology*, P. Bratter and P. Schramel, Eds., pp. 1–36, Walter de Gruyter, Berlin, Germany, 1980.

[4] M. Yaman, "Comprehensive comparison of trace metal concentrations in cancerous and non-cancerous human tissues," *Current Medicinal Chemistry*, vol. 13, no. 21, pp. 2513–2525, 2006.

[5] D. S. Auld, "Zinc coordination sphere in biochemical zinc sites," *BioMetals*, vol. 14, no. 3-4, pp. 271–313, 2001.

[6] P. Kovacic and J. D. Jacintho, "Mechanisms of carcinogenesis: focus on oxidative stress and electron transfer," *Current Medicinal Chemistry*, vol. 8, no. 7, pp. 773–796, 2001.

[7] H. Merzenich, A. Hartwig, W. Ahrens et al., "Biomonitoring on carcinogenic metals and oxidative DNA damage in a cross-sectional study," *Cancer Epidemiology Biomarkers and Prevention*, vol. 10, no. 5, pp. 515–522, 2001.

[8] Y. Wang, J. Fang, S. S. Leonard, and K. M. K. Rao, "Cadmium inhibits the electron transfer chain and induces reactive oxygen species," *Free Radical Biology and Medicine*, vol. 36, no. 11, pp. 1434–1443, 2004.

[9] International Agency for Research on Cancer, *IARC Monographs on the Evaluation of Carcinogenic. Risks of Chemicals to Humans*, vol. 1-2, supplement 7, IARC Scientific Publications, Lyon, France, 1984.

[10] International Agency for Research on Cancer, *IARC Be, Cd, Hg, and Exposures in the Glass Manufacturing Industry, IARC Monographs on the Evaluation of Carcinogenic Risks of Chemicals to Humans*, vol. 58, IARC Scientific Publications, Lyon, France, 1993.

[11] T. B. Kryston, A. B. Georgiev, P. Pissis, and A. G. Georgakilas, "Role of oxidative stress and DNA damage in human carcinogenesis," *Mutation Research—Fundamental and Molecular Mechanisms of Mutagenesis*, vol. 711, no. 1-2, pp. 193–201, 2011.

[12] M. Valko, H. Morris, and M. T. D. Cronin, "Metals, toxicity and oxidative stress," *Current Medicinal Chemistry*, vol. 12, no. 10, pp. 1161–1208, 2005.

[13] A. A. Konsta, E. E. Visvardis, K. S. Haveles, A. G. Georgakilas, and E. G. Sideris, "Detecting radiation-induced DNA damage: from changes in dielectric properties to programmed cell death," *Journal of Non-Crystalline Solids*, vol. 305, no. 1–3, pp. 295–302, 2002.

[14] V. K. Shukla, T. K. Adukia, S. P. Singh, C. P. Mishra, and R. N. Mishra, "Micronutrients, antioxidants, and carcinoma of the gallbladder," *Journal of Surgical Oncology*, vol. 84, no. 1, pp. 31–35, 2003.

[15] P. Lim, G. E. Wuenschell, V. Holland et al., "Peroxyl radical mediated oxidative DNA base damage: implications for lipid peroxidation induced mutagenesis," *Biochemistry*, vol. 43, no. 49, pp. 15339–15348, 2004.

[16] N. Gillard, M. Begusova, B. Castaing, and M. Spotheim-Maurizot, "Radiation affects binding of Fpg repair protein to an abasic site containing DNA," *Radiation Research*, vol. 162, no. 5, pp. 566–571, 2004.

[17] K. Geraki, M. J. Farquharson, D. A. Bradley, and R. P. Hugtenburg, "A synchrotron XRF study on trace elements and potassium in breast tissue," *Nuclear Instruments and Methods in Physics Research B*, vol. 213, pp. 564–568, 2004.

[18] D. R. Cousens, S. H. Ryan, and W. L. Griffin, "Self absorption and secondary fluorescence corrections of PIXE yields from

multilayered targets," in *Proceedings of 5th Australian Conference on Nuclear Techniques of Analysis*, pp. 58–60, Lucas Heights, Sydney, 1987.

[19] J. A. Maxwell, W. J. Teesdale, and J. I. Campbell, "The Guelph PIXE software package," *Nuclear Instruments and Methods in Physics Research B*, vol. 43, pp. 218–230, 1989.

[20] C. G. Ryan, *Quantitative Trace Element Imaging Using PIXE and the Nuclear Microprobe*, John Wiley & Sons, North Ryde, Australia, 2001.

[21] C. G. Ryan, D. R. Cousens, S. H. Sie, W. L. Griffin, G. F. Suter, and E. Clayton, "Quantitative pixe microanalysis of geological matemal using the CSIRO proton microprobe," *Nuclear Instruments and Methods in Physics Research B*, vol. 47, no. 1, pp. 55–71, 1990.

[22] S. E. Johansson and J. L. Campbell, *PIXE-A Novel Technique for Elemental Analysis*, John Wiley & Sons, Chichester, UK, 1988.

[23] S. Boissier, M. Ferreras, O. Peyruchaud et al., "Bisphosphonates inhibit breast and prostate carcinoma cell invasion, an early event in the formation of bone metastases," *Cancer Research*, vol. 60, no. 11, pp. 2949–2954, 2000.

[24] G. J. N. Raju, P. Sarita, M. R. Kumar et al., "Trace elemental correlation study in malignant and normal breast tissue by PIXE technique," *Nuclear Instruments and Methods in Physics Research B*, vol. 247, no. 2, pp. 361–367, 2006.

[25] K. Geraki, M. J. Farquharson, and D. A. Bradley, "X-ray fluorescence and energy dispersive x-ray diffraction for the quantification of elemental concentrations in breast tissue," *Physics in Medicine and Biology*, vol. 49, no. 1, pp. 99–110, 2004.

[26] H. W. Kuo, S. F. Chen, C. C. Wu, D. R. Chen, and J. H. Lee, "Serum and tissue trace elements in patients with breast cancer in Taiwan," *Biological Trace Element Research*, vol. 89, no. 1, pp. 1–11, 2002.

[27] K. H. Ng, D. A. Bradley, and L. M. Looi, "Elevated trace element concentrations in malignant breast tissues," *British Journal of Radiology*, vol. 70, pp. 375–382, 1997.

[28] M. Sugiyama, "Role of physiological antioxidants in chromium(VI)-induced cellular injury," *Free Radical Biology & Medicine*, vol. 12, no. 5, pp. 397–407, 1992.

[29] S. de Flora, M. Bagnasco, D. Serra, and P. Zanacchi, "Genotoxicity of chromium compounds. A review," *Mutation Research*, vol. 238, no. 2, pp. 99–172, 1990.

[30] J. Ye, X. Zhang, H. A. Young, Y. Mao, and X. Shi, "Chromium (IV)-induced nuclear factor-κB activation in intact cells via free radical reaction," *Carcinogenesis*, vol. 16, pp. 2401–2405, 1995.

[31] L. L. Miller, S. C. Miller, S. Y. Torti, Y. Tsuji, and F. M. Torti, "Iron-independent induction of ferritin H chain by tumor necrosis factor," *Proceedings of the National Academy of the United States of America*, vol. 88, no. 11, pp. 4946–4950, 1991.

[32] E. D. Weinberg, "Iron loading and disease surveillance," *Emerging Infectious Diseases*, vol. 5, no. 3, pp. 346–352, 1999.

[33] D. Armendariz and A. D. Vulpe, "11th International Symposium on Trace Elements in Man and Animals Abstracts," *The Journal of Nutrition*, vol. 133, no. 5, pp. 203E–282E, 2003.

[34] P. Sarita, G. J. Raju, A. S. Pradeep, T. R. Rutray, B. S. Reddy, and V. Vijayan, "Analysis of trace elements in blood sera of breast cancer patients by particle induced X-ray emission," *Journal of Radioanalytical and Nuclear Chemistry*, vol. 294, no. 3, pp. 355–361, 2012.

[35] K. Geraki, M. J. Farquharson, and D. A. Bradley, "Concentrations of Fe, Cu and Zn in breast tissue: a synchrotron XRF study," *Physics in Medicine and Biology*, vol. 47, no. 13, pp. 2327–2339, 2002.

[36] S. B. Reddy, M. J. Charles, M. R. Kumar et al., "Trace elemental analysis of adenoma and carcinoma thyroid by PIXE method," *Nuclear Instruments and Methods in Physics Research B*, vol. 196, no. 3-4, pp. 333–339, 2002.

[37] K. Maeda, Y. Yokode, Y. Sasa, H. Kusuyama, and A. M. Uda, "Multielemental analysis of human thyroid glands using particle induced X-ray emission (PIXE)," *Nuclear Instruments and Methods in Physics Research B*, vol. 22, no. 1–3, pp. 188–190, 1987.

[38] M. Uda, K. Maeda, Y. Yokode Y, Sasa, and H. Kusuyama, "An attempt to diagnose cancer by PIXE," *Nuclear Instruments and Methods in Physics Research B*, vol. 22, no. 1 3, pp. 184–187, 1987.

[39] X. Zeng, M. Yao, M. Mu et al., "An attempt to diagnose cancer by PIXE," *Nuclear Instruments and Methods in Physics Research B*, vol. 22, no. 1–3, pp. 184–187, 1987.

[40] S. B. Reddy, M. J. Charles, G. J. N. Raju et al., "Trace elemental analysis of carcinoma kidney and stomach by PIXE method," *Nuclear Instruments and Methods in Physics Research B*, vol. 207, no. 3, pp. 345–355, 2003.

[41] W. M. Kwiatek, T. Drewniak, M. Gajda, M. Gałka, A. L. Hanson, and T. Cichocki, "Preliminary study on the distribution of selected elements in cancerous and non-cancerous kidney tissues," *Journal of Trace Elements in Medicine and Biology*, vol. 16, no. 3, pp. 155–160, 2002.

[42] I. Bertini, H. B. Gray, S. J. Lippard, and J. S. Valentine, *Bioinorganic Chemistry*, Viva Books Private, 1989.

[43] H. Sigel, Ed., *Carcinogenicity and Metal Ions*, vol. 10 of *Metal Ions in Biological Systems*, Marcel Dekker, New York, NY, USA, 1980.

[44] A. M. Standeven and K. E. Wetterhahn, "Chromium(VI) toxicity: uptake, reduction, and DNA damage," *Journal of the American College of Toxicology*, vol. 8, no. 7, pp. 1275–1283, 1989.

[45] G. J. N. Raju, M. J. Charles, S. B. Reddy et al., "Trace elemental analysis in cancer-afflicted tissues of penis and testis by PIXE technique," *Nuclear Instruments and Methods in Physics Research B*, vol. 229, no. 3-4, pp. 457–464, 2005.

5

A Tree-Like Model for Brain Growth and Structure

Benjamin C. Yan[1,2] and Johnson F. Yan[1]

[1] *Yan Research, P.O. Box 4115, Federal Way, WA 98063, USA*
[2] *University of Illinois College of Medicine, 190 Medical Sciences Building, MC-714, 506 Mathews Avenue, Urbana, IL 61801, USA*

Correspondence should be addressed to Johnson F. Yan; jfyan131@gmail.com

Academic Editor: Giuseppe Chirico

The Flory-Stockmayer theory for the polycondensation of branched polymers, modified for finite systems beyond the gel point, is applied to the connection (synapses) of neurons, which can be considered highly branched "monomeric" units. Initially, the process is a linear growth and tree-like branching between dendrites and axons of nonself-neurons. After the gel point and at the maximum "tree" size, the tree-like model prescribes, on average, one pair of twin synapses per neuron. About 13% of neurons, "unconnected" to the maximum tree, migrate to the surface to form cortical layers. The number of synapses in each neuron may reach 10000, indicating a tremendous amount of flexible, redundant, and neuroplastic loop-forming linkages which can be preserved or pruned by experience and learning.

1. Introduction

The molecular weight distribution (MWD) of polycondensation of branched-chain monomers of the type RA_f has been derived classically by Flory [1] and generalized by Stockmayer [2]. Here, f is the number of functional groups, or "functionality" of group A. Mathematically, this widely quoted distribution function has been treated as power-series distribution and compound distribution that provides a simple concept; that is, single-parameter expressions of number- and weight-average "degrees of polymerization" (DP) are sufficient to generate the entire MWD for branched polymers [3]. Moreover, using a cascade formulation involving functionals and probability generating functions (PGF), this distribution can be extended to finite systems [4].

Here, the previously derived properties of this finite distribution are applied to synapse formation in the brain. A neuron has multiple dendritic processes and an axon, which can also be branched. Neurons are generally three or more orders of magnitude greater in size than molecular units. However, the functionality of a neuron may be 10^3 times larger than that of typical branched molecules ($f = 10000$). This large functionality also means there is great accessibility to the connection sites, and the long flexible axons offer a favorable condition for connection between

neurons. The "tree-like," or "ring-free," assumption in the Flory-Stockmayer theory can be satisfied by the initial linkage of head-to-tail linear chains and followed by a "tree-like branching." A neuron itself can be considered a small tree. Similarly, the peripheral nervous system (PNS) also resembles a tree made of the nerve bundles, which can be as large as 1.5 meters. The Finite Flory-Stockmayer theory (FFST) deals with numbers of highly branched repeat units and their association-dissociation mechanism. Therefore it is applicable to the statistical treatment of brain growth, neuronal connectivity, and information transmission. Linear growth and subsequent tree-like branching allow two simple equations to be applied up to a maximum tree size constrained by the FFST.

In practical, real and natural systems, lignin in wood is a tree-like molecule [5], and a major class of corals is also tree-like [6]. Thus the applicability of the F-S theory in molecular, cellular or animal size scales is equally sound if the repeating branched units can be properly identified.

Early brain growth is very robust and rapid. In the human brain the crosslinking between neurons and the growth of the linked neurons (polyneurons) proceed with a rapid rate. Because of their highly branched dendrites and axons, there are three types of crosslinking: axodendritic (A-D), axosomatic (A-S), and axoaxonic (A-A) synapses, in the order of decreasing abundance. In addition, there are at least three forms of crosslinks. The first are end-links which are links between the tips of an axon and a dendritic spine. The second form is a split-end link, which is shaped like the Greek letter Ω on top of V (the shape of a pair of dendritic spines), designated Ω/V-linking. The latter conformation represents a pair of twin synapses. The third form is the most abundant "X" shaped, mostly A-D linkages.

Structural and functional features of neurons set the limits and modes of various crosslinks. For example, there is no dendrite to dendrite (D-D) synapse because dendrites have been shown to demonstrate self-avoidance even in fruit fly brains [7]. With respect to a growing polyneuron tree, the "ring-free" assumption is also a self-avoidance rule. A conceptual illustration furnishes a visual model of a pair of Ω/V-twin synapses by removing the axonic "spaghetti maze" for a clearer view [8].

The rate of growth and rate of crosslinking in the brain occur exponentially in the months before and after birth in the human brain [9] and in the cortex of laboratory animals [10]. Initial linear growth and tree-like branching appear to be the most efficient path-way to grow and expand rapidly in a confined or open space [11].

The following is an ideal human brain model with neurons assumed to be homogeneous, with identical size, shape, and number of crosslinking sites.

There are two basic assumptions in the FFST. First, all linkages (synapses) are formed with equal probability, or the same "extent of reaction." Second, no ring is formed between branches of a growing chain. The second assumption is also termed the "tree-like model" [4]. The brain growth process is discussed with limiting forms of equations derived previously, and upper bounds of the variables and parameters are strictly constrained.

2. Limiting Equations and Numerical Examples

Symbols and abbreviations derived previously [4].

N = system size = number of neurons (in a brain or a "module").

f = functionality = number of connection sites per monomer (neuron) = 10000 or in thousands [11].

k = number of generations in the "transformer" probability generating functions (PGF) in the cascade formulation [4].

DP = degree of polymerization = number of connected neurons in a growing chain = x = 1, 2, 3, ..., N.

α = extent of reaction or fraction of connected synapse sites.

$\beta = (f - 1)\alpha$ which is approximately $f\alpha$ for large f = crosslinking index.

$\langle x_n \rangle$ = number-average DP.

$\langle x_w \rangle$ = weight-average DP.

$\beta_c = 1$ is value of β at gel point.

$\beta_m = 2$, maximum value of β allowed in tree-like model.

g_m = gel fraction at β_m.

At large f values, most of the finite system equations derived previously [4] become independent of f. The FFST is represented by the two averages:

$$\langle x_n \rangle = \left(1 - \frac{f\alpha}{2}\right)^{-1} \tag{1a}$$

$$= \left(1 - \frac{\beta}{2}\right)^{-1} \quad \text{for large } f, \tag{1b}$$

$$\langle x_w \rangle (\beta) = 1 + f\alpha \left(1 + \beta + \beta^2 + \cdots + \beta^k\right). \tag{2a}$$

The last equation is derived and proven in Supplementary Material session available online at http://dx.doi.org/10.1155/2013/241612. In its application, (2a) can be written in a more compact form:

$$\langle x_w \rangle (\beta) = 1 + \frac{\beta \left(1 - \beta^{k+1}\right)}{1 - \beta}, \quad \beta > 1 \text{ or } \beta < 1. \tag{2b}$$

In particular,

$$\langle x_w \rangle (\beta_c) = 1 + (k + 1), \quad \beta_c = 1. \tag{3}$$

The maximum value for $\langle x_w \rangle$ is N, that is, a giant gel particle with N neurons all connected in this "polyneuron" tree. With this value for $\langle x_w \rangle$ at $\beta_m = 2$ in (2b), the result is simply

$$N = 2^{k+2}. \tag{4a}$$

Equation (4a) expresses the system (brain) size in terms of the number of cell division, as powers of 2. The maximum value of $\langle x_n \rangle$ is also N, when using a precise value [4] of $\beta_m = 2(1 - N^{-1})$.

The unconnected monomer neurons have a weight fraction of $w(1)$ which can be approximated with the weight fraction in an infinite system [4] as

$$w(1) = (1 - \alpha)^f = e^{-\beta} \quad \text{for large } f. \tag{5}$$

Equation (5) provides a physically measurable definition for β. Since this monomer fraction does not join the tree growth scheme, the maximum "tree-like" value for β is 2 and the attainable DP for the gel particle is $(1 - e^{-2})N$, which may be applied to the left-hand side of (4a)

$$\left(1 - e^{-2}\right) N = 2^{k+2}, \tag{6}$$

which serves as a correction for computing k. But this is only a minor correction; for example, for $N = 10^{11}$, $e^{-2} = 0.135$, (4a) gives $k = 34.54$ and (6) yields $k = 34.33$ (Cf. Table 1). The gel fraction $1 - e^{-2} = 0.865$ is denoted as g_m.

The case of $N = 10^{11}$ and $f = 10000$ in Table 1 is the high-functionality system of the human brain [11]. The pregel distribution or the distribution in the sol fraction in Table 1 is narrow, with a dispersion ratio or dispersity r (ratio of weight-average DP to number-average DP) very close to 1, even for a system of 10^{11} neurons. Equation (5) indicates that the sol fraction contains mainly unconnected monomer neurons. These free, unconnected mononeurons may form other classes of aggregate structures, such as cortex layers (horizontal sheets) and minicolumns (vertical sheets) [11], or they may join the gel in a nontree fashion, as β reaches hundreds or even thousands.

3. Postgel Relations with Extensive Ring Formation

As $\beta > 2$, ring formation in the gel particle cannot be avoided [1]. Indeed when $\beta = \ln(10^{11}) = 25$, only 1 in 10^{11} neurons remains in the sol fraction. In a human neuron with $f = 10^4$ available connection sites, β may reach a large value of 8000. In this case, there is a near-zero chance of finding a free neuron that remains unconnected. The large values of β are a measure of excessive numbers of rings or loops. Information transfer is efficient when it takes a tree-like path in such a sea of ring structures.

This theory thus provides a statistical, nongeometrical model for brain growth, neuron packing, and neuron firing. Geometrical description such as calculation with cell and branch volumes tends to ignore the fact that in chemical solutions or cellular suspensions, volumes of cellular components are not additive [11].

In the postgel range of $1 < \beta < 2$, tree-like growth reaches a maximum size determined by (6). At $\beta > 2$, monomer neuron fraction can be considered as the sol fraction s. Thus the approximation

$$s = w(1) = e^{-\beta} \qquad (7)$$

or the gel fraction

$$g = 1 - e^{-\beta} \qquad (7^*)$$

is a good approximation for large β. Indeed the plot of g versus β, in the entire range of $0 < \beta < f$, is remarkably similar to a "brain growth" curve [12]. This kind of "growth curves" can also be depicted as developmental curves of spine densities in the cortex of rat, mouse, and guinea-pig [13].

The fractions of free and unconnected neurons, $w(1)$, being $e^{-1}(= 0.368)$, and $e^{-2}(= 0.135)$ at $\beta = 1$ and 2, respectively, are in exact agreement with those obtained by using a simple binomial distribution for the probability of connecting neighboring cortical neurons at the binomial averages [10] of 1 and 2. This agreement is not surprising since the PGFs of the originating "root" (zero generation) and subsequent generations of k-fold compounding all take the binomial form (Supplementary Material).

The asterisk on (7^*) denotes that it is a relation that holds well beyond gelation in the presence of excessive rings. In fact, denoting the sol and gel properties by and after the gel point,

the size distribution becomes heterogeneous and splits into two very narrow peaks at $w(1)$ and around N, with [4]

$$\frac{1}{\langle x_n \rangle} = \frac{s}{\langle x_n' \rangle} + \frac{g}{\langle x_n'' \rangle}, \qquad (8^*)$$

$$\langle x_w \rangle = s \langle x_w' \rangle + g \langle x_w'' \rangle, \qquad (9^*)$$

$$\beta = s\beta' + g\beta''. \qquad (10^*)$$

The FFST defines an x-mer as having $(x - 1)$ linkages without any loop structure. The maximum tree size has a DP of N neurons with N, or precisely $(N - 1)$, linkages. At this maximum size, $\beta = 2$, and (2a) reduces to (4a) as a constraint set by the system size. In the entire range $0 < \beta < f$; however, β is also the number of linkages per neuron.

If N pairs of twin synapses are uniformly distributed along a linear N-meric chain, it would then serve as a backbone for the entire system of connected neurons. This scheme leaves all the loop-forming X-linked connections in the branched, nonlinear chains. However, uniform distribution appears unlikely because at $\beta = 2$, there is still a fraction of e^{-2} or 13.5% (by weight) of neurons remained unconnected at the maximum tree size. This is approximately the same fraction as that of human cortex [14].

In a given neuron, there may be 10000 synapses that are formed by 10000 incoming post-synapses and 10000 outgoing presynapses [11]. The total number of linkages is $N\beta$, and the total redundancy is defined as

$$D = N(\beta - 1) \quad \text{for } \beta > 2. \qquad (11)$$

For $\beta = 2$, there is a redundancy of $D = N$ linkages. Of these $2N$ linkages at the maximum tree size, there are N tree-like linkages paired by exactly N loop-forming linkages. This seems contradictory, because in a strict sense of the ring-free assumption in FFST, a Ω/V-linkage, or a pair of split-end linkages, is a smallest loop. Such a pair of twin synapses stabilizes a crosslink or the entire network. For $\beta = 8000$, almost all $8000N$ redundant linkages provide not only greater redundant security, but also greater neuroplasticity to the gel-like network structure of the brain. At the early explosive growth stage of synaptogenesis [9], end-linking or specially the Ω/V-linking may be a preferred mode for linear and tree-like branching chain growth (Table 1). In this sense, each tree-like linkage is considered as having a pair of twin synapses.

Redundant, loop-forming linkages do not contribute to the DP of a growing chain. Thus the maximum-sized gel remains at a DP of $g_m N$, instead of the N discussed above. The maximum number-average DP is also $g_m N$, rendering the dispersity at 1. Finally the unconnected or cortical fraction of neurons should also join this growing gel, approaching a final DP of N, where the tree-like equations (1a), (1b), and (2a) no longer hold, but (8^*)–(10^*) are still valid as $\beta \rightarrow \beta'' \rightarrow f$. This is a state in which neurons in sol-sol, sol-gel, and gel-gel are all linked. This is the complete gelation scheme proposed by Flory [1, 4], which differs from that of Stockmayer and others [4].

TABLE 1: Properties calculated from a tree-like model for branched "polyneurons" with low ($f = 3$) and high ($f = 10^4$) functionalities. At a pregel stage, $\beta = 0.1$ and at the gel point, $\beta = 1$. For $f = 3$, the accurate relation $\beta = (f - 1)\alpha$ is used, as in (1a) and other equations in [4].

			$\beta = 0.1$			$\beta = 1$		
f	N	k	$\langle x_n \rangle$	$\langle x_w \rangle$	r	$\langle x_n \rangle$	$\langle x_w \rangle$	r
3	10^4	25	1.111	1.167	1.05	4	40	10
10^4	10^{11}	34	1.053	1.111	1.05	2	36	18

4. Neuron Wiring in the Brain

The simplest strategy for neuron wiring is a snake-like strategy as depicted in Figure 1(a) of [11], in which a long wire (axon fiber) is connected to all 209 neurons in that figure. In a "10000 nearest neighbors" model, the long wire makes a single connection to each of the 10000 light bulbs tangentially, meaning only 1 point of contact at each bulb.

Since a snake does not bite its own tail, the snaking scheme is ring-free.

The wiring model proposed here for the connection of *non-cortical* neurons may be called concentric "hardball" and "softball" model, obtained by invoking the above tree-like definitions, numbers, and equations, plus 3 types and 3 forms of synapses, and at least 2 self-avoidance rules.

4.1. The Hardball Is Where Most of the Neuron Mass Concentrated. It is centered at the soma of a given neuron, with a radius R_1 encompassing the entire soma, plus axon and dendrite branches near the soma. All 3 cell components (S, A, and D) can provide postsynapses to an incoming presynaptic axon. All 3 types (A-S, A-A, and A-D) of synapses may be present inside or on the surface of this hardball. The hardball is a (relatively) solid, rigid, and stationary target for the incoming straight-shooting pre-synaptic axon tip [10].

Incoming axons can be envisioned as straight arrows, each with a long string attached to its end [10]. In the early stage of synaptogenesis, an axon arrow is shot toward the hardball target, most likely with 0 or 1 hit [10].

The random hits to the hardball can be described with a binomial distribution:

$$b(y; np) = \left(\frac{n!}{y!(n-y)!} \right) p^y (1-p)^{n-y}, \quad y = 0, 1, 2, \ldots, \tag{12}$$

where $b(y; np)$ is a binomial distribution function, with random variable y and parameter p. But n, the available target sites, is a large constant. The PGF and the mean are

$$\phi(\theta) = (1 - p + p\theta)^n, \tag{13}$$
$$\phi'(1) = np.$$

Note that the random variable y starts with 0, and θ is the dummy variable [4] of the PGF. At low np values, for example, $np = 1$, the distribution is broad, with twin peaks at $y = 0$ and $y = 1$. This means the most probable outcome is 0 or only 1

hit [10]. The cases for $np = 1$ and 2, with $b(0; p) = e^{-1}$ and e^{-2}, have been discussed above as the fractions of "unconnected neurons" at these averages.

The simple binomial PGF in (13) and its distribution are different from FFST size distribution in other aspects, besides the similarity in fractions e^{-1} and e^{-2}. The simple binomial distribution is for the random hits because it has nothing to do with size of aggregates or polyneurons since there is no gelation. This "zero-or-one" scheme, proposed for random hits on linking of mouse cortical neurons [10], actually fits the principle of cascade formulation on tree-like connection of non-cortical neurons [4], which is also employed in the present treatment of ring-free assumption up to the limit of maximum tree size.

A more recent estimate is a narrow peak at $np = 5$ for linking two neighboring cortical neurons [14]. This means the distribution is peaked around $y = 5$, but the $y = 0$ peak is no longer significant. That is, with an expected 5 linkages between two cortical neurons, there is practically no chance for a "no hit." These probabilistic arguments suggest that the connection of cortical neurons is much more orderly or structured than that in the initial bulk phase of tree-like connection.

Contrary to the simple binomial probability distribution, the branched-chain size distributions in Table 1, for $f = 3$ or more, are narrow at low β, broadening with increasing β values and reaching a maximum at gel point [4] when $\beta = 1$.

It can be further simplified by assuming that synapses formed onto the hardball are entirely of the twin synapses of the Ω/V form, for the following reasons. Firstly, the rough, spiny, and more abundant dendrite branches provide conditions required for twin synapse formation. Secondly, even with a smooth and less active surface on the soma, the A-S connections may stabilize themselves with such twin linkages. All Ω/V-linkages in the hardball are designed to provide a "skeleton" for a brain lasting for a life-time of the host animal [9].

4.2. The Softball Layer. In addition to the long, thin axons, the thick layer outside the hardball, of radius R_2 from the central soma, is mainly composed of flexible and distal portions of dendrite branches. Incoming axons shot through this flexible, ribbon-like cloud have nonstationary targets for contact. Connections to this softball region are made with nontree, redundant X-links. A given neuron has up to 10000 contact sites for 10000 axonic presynapses from other neurons. If the synaptic density (in numbers of synapses per unit volume) is uniform in both hardball and softball regions, then

$$\frac{\left(R_2^3 - R_1^3 \right)}{R_1^3} = \frac{(f-2)}{2}, \tag{14}$$

where $f = 10000$. And 2 is the number of synapses, of the Ω/V-form, connected to the smaller hardball. The result is simply $R_2 > 17R_1$, which is easily achievable to satisfy this large number [11] of f.

5. Conclusions

For comparison with other statistical models, the FFST provides tree-like parameters such as k (an abstract concept of k-fold compounding in PGF), and $k + 2$ (as weight-average DP and as number of generations in cell divisions) at various crosslinking stages. Crosslinking index is $\beta \ll 1$ for pregel linear growth; $\beta = 1$ at the gel point, and $\beta = 2$ at the maximum tree size. When $2 < \beta < f$ extensive ring formation occurs, as the observed range of number of synapses per neuron.

Human cortical neurons are more homogeneous, less dense in synapses, and more ordered in their connection than the well-studied much smaller mouse cortex [10, 11]. In the human cortex, these advantages translate into a greater neuroplasticity and much greater freedom in the information transmission process. Low synapse density in non-cortical regions means an early onset of tree-like branching, so that the massive, late-occurring X-links can have sufficient space for high neuroplasticity. In this sense, the tree-like model proposed here is behaving as a "supporting" system—much like that animal skeletons providing support, with muscles and organs filling in, for their host bodies.

The medium for neurons is well dispersed with glia cells and other mixture components. The brain growth in terms of brain volume or synapse density follows a general curve shape of gel development in (7^*). Another developmental β versus time curve, in months before and after birth, surges with a rapid rate to a maximum f at 8 postnatal months. Thereafter, it declines and levels off at $f/2$ and continues to a very mature age of 70 years [9]. These surge and decline do not affect the shape of a growth curve because the values of β are in thousands.

The statistical model for a maximum tree gel leads to a physical model wherein all neuron units are connected entirely by Ω/V-twin synapses in the hardball region, accounting for 3 types and 3 forms of observed synapses. In the outer softball region highly redundant, flexible, accessible, and neuroplasticity are the characters of the X-linked synapses. In a brain, twin synapses are present in a much smaller amount than the X-links, but the twins are obviously more visible and easily identified [8, 11, 13].

Perhaps the simplest brain is that of $C.$ $elegans$ [15], in which a large sensory neuron has only one pair of "horse-shoe" shaped dendrite branches. Thus, the prediction of "one linkage, twin synapses" from FFST holds even for this primitive animal. In higher animals, the observation of twin synapses [8, 15] indicates that they are relics of tree-like structure.

Acknowledgments

The authors thank the following friends and colleagues for their help in preparing the paper and for their fundamental and vigorous discussions on this tree-like model: Dimitris Argyropoulos, Albert Chang, George Kletecka, C. C. Wang, and Alexander K. Yan.

References

[1] P. J. Flory, *Principles of Polymer Chemistry*, chapter 9, Cornell University Press, Ithaca, NY, USA, 1953.

[2] W. H. Stockmayer, "Theory of molecular size distribution and gel formation in branched-chain polymers," *The Journal of Chemical Physics*, vol. 11, no. 2, pp. 45–55, 1943.

[3] J. F. Yan, "A new derivation of molecular size distribution in nonlinear polymers," *Macromolecules*, vol. 11, no. 4, pp. 648–649, 1978.

[4] J. F. Yan, "Gelation in finite polycondensation systems," *The Journal of Chemical Physics*, vol. 78, no. 11, pp. 6893–6896, 1983.

[5] J. F. Yan, "Kinetics of delignification: a molecular approach," *Science*, vol. 215, no. 4538, pp. 1390–1392, 1982.

[6] http://news.stanford.edu/news/2008/february20/coralsr-022008.html.

[7] P. Soba, S. Zhu, K. Emoto et al., "Drosophila sensory neurons require Dscam for dendritic self-avoidance and proper dendritic field organization," *Neuron*, vol. 54, no. 3, pp. 403–416, 2007.

[8] "Photograph by Graham Johnson," *Science*, vol. 309, p. 1990, 2005, http://www.sciencemag.org/site/feature/misc/webfeat/vis2005/show/images/slide1_large.jpg.

[9] P. R. Huttenlocher, "Synaptogenesis in human cerebral cortex," in *Human Behavior and the Developing Brain*, G. Dawson and K. W. Fisher, Eds., chapter 4, pp. 137–153, Guilford Press, New York, NY, USA, 1994.

[10] V. Braitenberg and A. Schuz, *Anatomay of the Cortex*, Springer, Berlin, Germany, 1991.

[11] J. M. J. Murre and D. P. F. Sturdy, "The connectivity of the brain: multi-level quantitative analysis," *Biological Cybernetics*, vol. 73, no. 6, pp. 529–545, 1995.

[12] K. L. Sakai, "Language acquisition and brain development," *Science*, vol. 310, no. 5749, pp. 815–819, 2005.

[13] F. Karube, Y. Kubota, and Y. Kawaguchi, "Axon branching and synaptic bouton phenotypes in GABAergic nonpyramidal cell subtypes," *Journal of Neuroscience*, vol. 24, no. 12, pp. 2853–2865, 2004.

[14] D. E. Feldman and M. Brecht, "Map plasticity in somatosensory cortex," *Science*, vol. 310, no. 5749, pp. 810–815, 2005.

[15] http://www.sfu.ca/biology/faculty/hutter/hutterlab/research/Ce_nervous_system.html.

6

Analysis of the REJ Module of Polycystin-1 Using Molecular Modeling and Force-Spectroscopy Techniques

Meixiang Xu,[1] Liang Ma,[1] Paul J. Bujalowski,[1,2] Feng Qian,[3]
R. Bryan Sutton,[4] and Andres F. Oberhauser[1,2,5]

[1] Department of Neuroscience and Cell Biology, University of Texas Medical Branch, Galveston, TX 77555, USA
[2] Department of Biochemistry and Molecular Biology, University of Texas Medical Branch, Galveston, TX 77555, USA
[3] Department of Medicine, Division of Nephrology, University of Maryland School of Medicine, Baltimore, MD 21201, USA
[4] Department of Cell Physiology and Molecular Biophysics, Texas Tech University Health Sciences Center, Lubbock, TX 79430, USA
[5] Sealy Center for Structural Biology and Molecular Biophysics, University of Texas Medical Branch, Galveston, TX 77555, USA

Correspondence should be addressed to R. Bryan Sutton; roger.b.sutton@ttuhsc.edu
and Andres F. Oberhauser; afoberha@utmb.edu

Academic Editor: P. Bryant Chase

Polycystin-1 is a large transmembrane protein, which, when mutated, causes autosomal dominant polycystic kidney disease, one of the most common life-threatening genetic diseases that is a leading cause of kidney failure. The REJ (receptor for egg lelly) module is a major component of PC1 ectodomain that extends to about 1000 amino acids. Many missense disease-causing mutations map to this module; however, very little is known about the structure or function of this region. We used a combination of homology molecular modeling, protein engineering, steered molecular dynamics (SMD) simulations, and single-molecule force spectroscopy (SMFS) to analyze the conformation and mechanical stability of the first ~420 amino acids of REJ. Homology molecular modeling analysis revealed that this region may contain structural elements that have an FNIII-like structure, which we named REJd1, REJd2, REJd3, and REJd4. We found that REJd1 has a higher mechanical stability than REJd2 (~190 pN and 60 pN, resp.). Our data suggest that the putative domains REJd3 and REJd4 likely do not form mechanically stable folds. Our experimental approach opens a new way to systematically study the effects of disease-causing mutations on the structure and mechanical properties of the REJ module of PC1.

1. Introduction

PC1 is a large transmembrane protein, which, when mutated, causes autosomal dominant polycystic kidney disease (ADPKD), one of the most common life-threatening genetic diseases that is a leading cause of kidney failure [1]. PC1 may have a role in sensing of flow [2, 3], pressure [4] and the regulation of the cell cycle [5] and cell polarity [6]. PC1 may sense signals from the primary cilia, neighboring cells, and extracellular matrix and transduces them into cellular responses that regulate proliferation, adhesion, and differentiation that are essential for the control of renal tubules and kidney morphogenesis [1, 3, 7, 8]. The predicted amino acid sequence of PC1 (Figure 1(a)) suggests that it is a large multidomain membrane protein with 11 transmembrane domains. Its N-terminal extracellular region contains 4 leucine-rich repeats ((LRR) 250 amino acid long), a C-type lectin domain ((CLD) 130 amino acid long), a low-density-lipoprotein-like domain (LDL-A domain), 16 Ig-like domains (PKD domains, each 90 amino acid) and a region that is homologous to a sea urchin protein called receptor for egg jelly (REJ) [9, 10]. The PKD domains in PC1 have a similar topology fibronectin type III (FNIII) domain found in other modular proteins with structural and mechanical roles (recently reviewed in [11]). PC1 interacts with polycystin-2 (PC2) in the primary cilia of renal epithelial cells which forms a mechanically sensitive ion channel complex. Bending of the cilia induces Ca^{2+} flow into the cells, mediated by

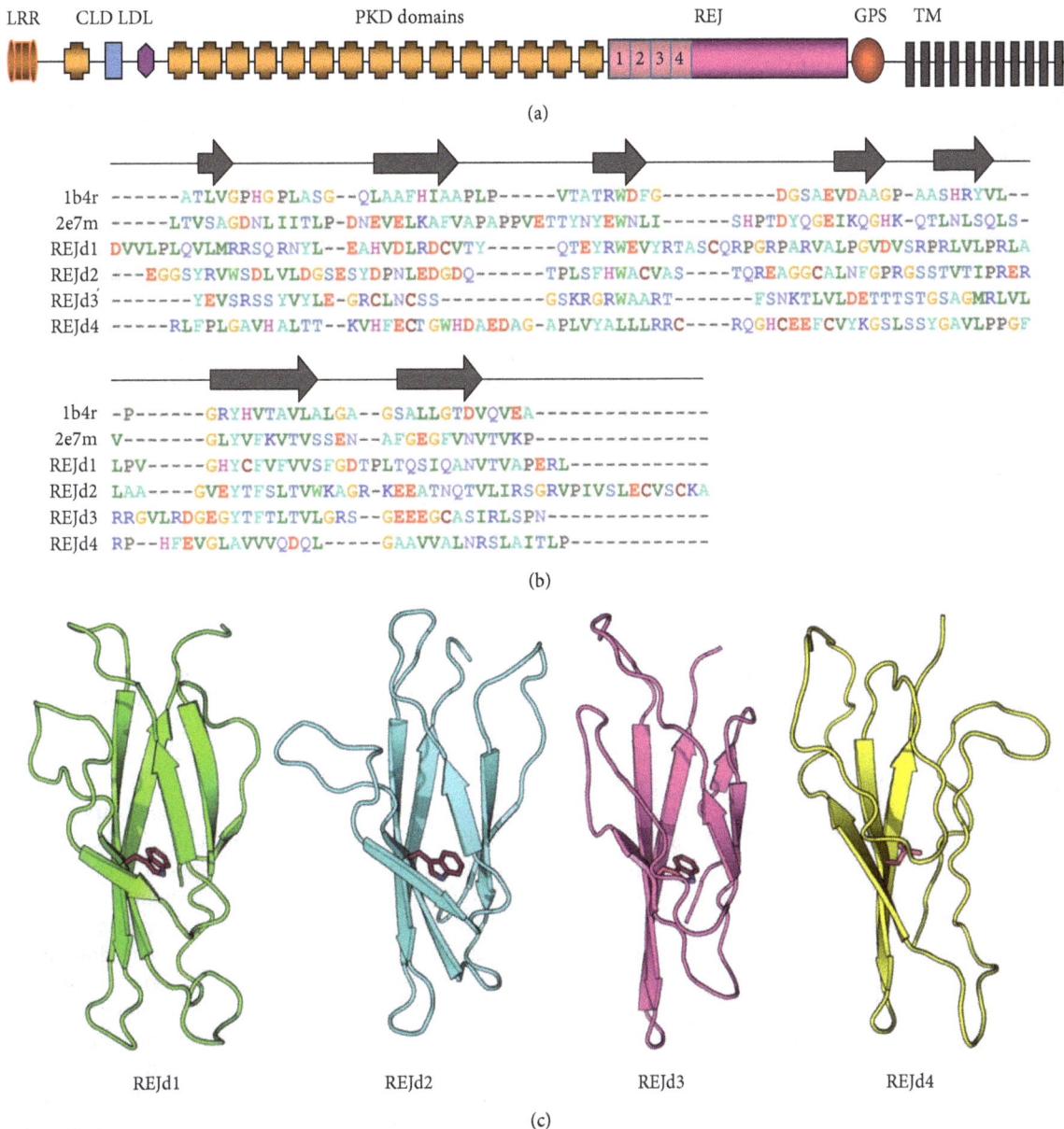

FIGURE 1: (a) Diagram of the predicted domain architecture of the extracellular region of PC1. The ectodomain has a large collection of domains: several leucine-rich repeats (LRR), a C-type lectin domain ((CLD) blue box), an low-density-lipoprotein-like domain ((LDL-A domain) purple octagon), 16 PKD domains (boxes in orange), and the 1000 aa long Receptor for Egg Jelly (REJ), in purple) region. GPS: G-protein coupled proteolytic site. TM: transmembrane domains. (b) Sequence alignments of the putative REJ domains with template structures of the human PKD domain no. 1 from polycystin-1 (1b4r) and the human PKD domain from protein KIAA0319 (2e7m). The arrows indicate beta-stranded secondary structure regions and are derived from the predicted secondary structure of 1b4r as calculated by the DSS algorithm in PyMol. The predicted secondary structure for 2e7m shares similar characteristics. The color of the various amino acids in the alignment reflects the chemical composition of the residues in the REJ fold, for example, red = acidic, blue = basic, and green = hydrophobic. (c) Homology models of putative FNIII domains within the REJ module of human PC1. The conserved Trp residue in REJd1, -d2, and -d3 is shown in purple.

the PC1-PC2 complex [2, 3, 12]. Mechanical signals are thus transduced into cellular responses that regulate proliferation, adhesion, and differentiation, essential for the control of renal tubules and kidney morphogenesis. Using SMFS, we and others have shown that the PC1 N-terminal extracellular region is highly extensible and that this extensibility is mainly caused by the unfolding and refolding of its PKD domains [13–15]. These force-driven reactions are likely to be important for cell elasticity and the regulation of cell signaling events mediated by PC1.

The REJ module is a major component of PC1 ectodomain that extends to about 1000 amino acids. A large number of mutations map on to this region. According to the Mayo Clinic PKD database, there are about 230 mutations including 80 missense mutations in the REJ region and of those about 65 missense mutations are predicted to be disease-causing mutations, highlighting the importance of this region for PC1 function. However, very little is known about the structure or function of this module. Recent evidence shows that PC1's ectodomain undergoes cleavage at the G-protein coupled proteolytic site (GPS), a process that requires the complete REJ region [16–18]. GPS cleavage is a process that is essential for kidney structure and function, as shown by the Pkd1$^{V/V}$ knock-in mouse [19], as well as by the fact that a number of mutations in the REJ indeed disrupt GPS cleavage [20, 21].

The REJ of PC1 shares similarity to the sea urchin sperm REJ proteins (such as SpREJ1, SpREJ2, and SpREJ3) and other members of the PC1 family (such as PKDREJ and PKD1L1) [22]. Initial secondary structure analysis predicted a total of four FNIII repeats in the first 400 amino acids of the REJ module of PC1 [23]. A later work concluded that the PC1 REJ module represents a novel sequence that contains no repeating motifs, and it does not show any homology to any known fold [9]. However, subsequent SMFS experiments indicated the existence of FNIII type of domains within the REJ module [14].

More recently, Schröder et al. used comprehensive sequence analysis together with CD spectroscopy and NMR techniques to analyze the first 425 amino acids of the REJ module [24]. They found that within this segment there are total of four predicted FNIII domain but only the first two domains could be expressed as soluble proteins, and only domain 2 was amenable for NMR analysis. Their data show that domain 2 has all the features of a bona-fide FNIII domain. The biophysical analysis of domain 1 was hindered because of partial aggregation. Domain 3 expressed well but in inclusion bodies and degraded quickly. Domain 4 expressed extremely poorly and in inclusion bodies.

In this work we used a different approach, where we combined homology modeling, protein engineering, and SMFS to systematically characterize the stability of the predicted four FNIII domains in the first 425 amino acids of the REJ module. After flanking the different putative FNIII sequences with titin I27 or MBP domains, we found that these constructs express well as soluble proteins in E. coli and were able to analyze their mechanical stability. We demonstrate that the REJ module contains several stable domains that are likely to have a fold similar to FNIII domains, confirming our previous predictions [14]. Our approach should make the analysis of the biophysical effects of mutations on the REJ module possible.

2. Materials and Methods

2.1. Homology Modeling. Multiple sequence alignment was performed with ClustalW (version 2.1, [25]) and visualized with JalView. We chose the best model based on the lowest calculated model energy values (molpdf) as reported by MODELLER (version 9.9, [26]) and low DOPE scores for each model [27]. Structures were rendered using PyMol

(http://www.pymol.org/). In the homology modeling analysis of the putative REJ domains we used template structures of the human PKD domains from polycystin-1 (1b4r) and from protein KIAA0319 (2e7m). Our assumption was that the REJ domains were a continuation of the PKD repeats from the more N-terminal domains. Since 1B4R and 2e7m shared at least some sequence similarity with the REJ regions, we picked these templates. We initially attempted to use both 1b4r and 2e7m to model all four REJ domains to strengthen the quality of the final model. However, due to the sequence degeneracy and the low overall sequence identity in each of the four sequences, the quality of the resulting homology models was poor as judged by the DOPE scores. We assumed that both the Trp residue in beta-strand B and the Tyr residue in beta-strand E make up essential elements as core residues. REJd1 and REJd2 could be most optimally aligned with 1b4r, based on local sequence homology with conserved secondary structure elements, while REJd3 and REJd4 could be aligned with 2e7m. The initial alignments against each of target structures (1b4r and 2e7m) were performed with Clustal; however, the alignments that were used for the homology models were manually adjusted to optimize the chemical nature of more conserved amino acids in each domain. Further, the perresidue DOPE analysis of REJd1 and REJd2 correlated best with using 1b4r as a template structure, while the perresidue DOPE analysis of REJd3 and REJd4 correlated best with 2e7m.

2.2. Construction, Expression, and Purification of REJ Segments for SMFS and CD Experiments. In order to characterize the mechanical properties of putative FNIII domains in the REJ module we made several protein constructs (Table 1) and expressed these in either E. coli or insect cells. Recombinant DNA techniques and multiple step cloning technique were used to construct different REJ segments heteropolyproteins [13, 14, 28, 29]. For E. coli expression system, REJ constructs were introduced into a modified pRSET A vector [28] or a p202 vector and expressed in E. coli BL21 or C41 strains. The p202 vector contains a maltose-binding protein (MBP) sequence upstream of the multicloning site to increase the solubility of the target protein. The proteins were purified by Ni-affinity chromatography as previously described [13–15, 30]. The proteins were kept in PBS containing 5 mM DTT (in order to prevent dimer formation since the I27 constructs have cysteine residues at the C-terminus to facilitate attachment to the gold coated AFM tip). For the insect expression system, REJ constructs were introduced into pVL1392 vector or pFastBac vector and expressed in insect cell Sf9, using the BaculoGold Transfection kit (BD Biosciences). All the constructs were cotransfected via Baculovirus Expression Vector System (BD Biosciences) into host cell Sf9 and cultured in Insect-Xpress w/L-Gln medium (LONZA Walkersville, Inc.) supplied with 5–10% FBS and penicillin/streptomycin. The infection and amplification protocols were based on the manual of BD BaculoGold Transfection kit. After 3 rounds of amplification of the recombinant baculoviruses, the cell pellets and supernatant were collected. The cell pellets were lyzed in insect cell lysis buffer (BD Biosciences) supplied with protease inhibitors (Roche) on ice bath for 30 min and

TABLE 1: REJ constructs used for SMFS experiments.

Protein construct	Amino acid (human PC1) Genbank no. L33243	Expression system	Remarks	Expression vector
REJd1	2151–2256	Insect cell Sf9	Very low expression/insoluble	pFastBac
REJd4	2468–2575	Insect cell Sf9	Very low expression/insoluble	pFastBac
REJd1-4	2151–2575	Insect cell Sf9	Very low expression/insoluble	pVL1392
MBP-REJd1-I27	2151–2256	E. coli	Good expression/soluble	p202
MBP-REJd1,2-I27	2151–2375	E. coli	Good expression/soluble	p202
MBP-REJd3,4-I27	2380–2575	E. coli	Very low expression/insoluble	p202
$(I27)_3$-REJd3,4-$(I27)_2$	2380–2575	E. coli/Sf9	Good expression/soluble	pRSETA/pVL1392
$(I27)_3$-REJd4-$(I27)_2$	2468–2575	E. coli/Sf9	Good expression/soluble	pRSETA/pVL1392

MBP: maltose-binding protein; I27: titin domain I27.

sonicated. The proteins were purified in native conditions with Ni-NTA resins and stored at 4°C for AFM studies.

We found that the REJd1-4, REJd1, and REJd4 recombinant constructs are expressed poorly as insoluble proteins in insect cells and bacteria. This is in agreement with a recent study that found that REJd1, -d2, -d3, and -d4 are very hard to express in E. coli [24]. In order to increase their solubility we flanked the REJ domains with maltose-binding protein (MBP) and titin I27 domains. We found that the MBP-REJd1-I27 and MBP-REJd1,2-I27 constructs are expressed as soluble proteins in E. coli. However, the MBP-REJd3,4-I27 is expressed poorly and mostly in inclusion bodies even at low induction temperatures (16°C). To further increase the solubility we flanked the REJd4 and REJd3,4 sequence with multiple titin I27 domains, $(I27)_3$-REJd4-$(I27)_2$ and $(I27)_3$-REJd3,4-$(I27)_2$. These constructs are expressed well as soluble proteins in both bacteria and insect cells. Our original plan was to characterize the secondary structure and thermodynamic stability of the different REJ proteins using far-UV CD and Equilibrium Denaturation techniques. However, we were unable to accomplish this goal for the following reasons: (i) we found that the native REJd1-4, REJd1, and REJd4 recombinant proteins are expressed poorly as insoluble proteins in both insect cells and in bacteria; (ii) in the MBP-REJd1-I27 and the MBP-REJd1,2-I27 constructs we included protease cleavage sites (TEV and thrombin) in between the MBP and REJd1 sequences. We found that, after cleavage, both proteins precipitated as an insoluble product; (iii) we were unable to make the $(I27)_3$-REJd1-$(I27)_2$ or $(I27)_3$-REJd2-$(I27)_2$ constructs; (iv) the only protein that expressed well enough for CD analysis was the $(I27)_3$-REJd4-$(I27)_2$ construct.

2.3. Single-Molecule Force Spectroscopy.

The mechanical properties of single proteins were studied using a home-built single-molecule atomic force microscope (AFM) as previously described in [31]. The spring constant of each individual cantilever (MLCT or Olympus OBL, Veeco Metrology Group) was calculated using the equipartition theorem [32]. In a typical experiment, a small aliquot of the purified proteins (~1–10 μL, 10–100 μg/mL) was allowed to adsorb onto a Ni-NTA coated glass coverslip [33, 34] for about 5 min and then rinsed with PBS. The pulling speed was in the range of 0.5–0.7 nm/ms. In single-molecule force-spectroscopy experiments the probability of picking up a protein is characteristically very low because the density of molecules has to be low enough to pull single molecules. Hence, in about 95% of the experiments, the approach of the AFM tip to the surface does not result in a contact with a protein [35, 36]. In addition the protein is contacted at random locations by the AFM tip and most does not show complete unfolding of the REJ protein construct. The AFM recordings traces were selected using the following criteria: (i) the trace should have clean initial force extension after retraction from the surface (i.e., little or no unspecific interactions); (ii) traces should have detachment forces higher than 200 pN to be sure that the protein is completely extended and unfolded. We chose the 200 pN threshold because most studied protein domains unfold at forces less than this force [37]. We found that typically about 1 in 500–1000 of force-extension traces fulfilled these criteria.

2.4. Contour Length Measurements.

The initial contour length of the folded protein (Lc) and the contour length increments (ΔLc) caused by domain unfolding were measured using the worm like chain (WLC) equation. The adjustable parameters of the WLC model are the persistence length, p and the contour length of the polymer [38, 39]. We measured Lc by manually fitting the first force peak of the sawtooth pattern to the WLC equation; the zero length point was defined as the point where the AFM cantilever tip contacts the coverslip. In a typical experiment, the cantilever tip is pressed into a layer of purified protein adsorbed onto a glass coverslip. Protein molecules are then stretched. Experimentally we find that the proximal region of the force-extension recording is frequently contaminated with nonspecific interactions due to entanglement with other protein molecules, making it difficult to get a clean estimation of zero-force-zero-length point. These nonspecific interactions can account to about 10–30 nm of the initial stretching region.

2.5. Circular Dichroism.

The far UV CD spectra of the titin I27 and the $I27_3$-REJd4-$I27_2$ polyprotein were recorded on a Jasco J-815 Spectropolarimeter. A 0.2 cm path length cuvette was used as the sample container. The protein concentration was 1 μM in 10 mM phosphate buffer. The data reported in Figure S1 corresponds to the average of 3 scans obtained at a scan rate of 50 nm/min in the range of 200–260 nm (see Suplemetary Material available online at http://dx.doi.org/10.1155/2013/525231). The secondary structure content was estimated using the CDNN program (version 2.0.3.188) [40].

2.6. Steered Molecular Dynamic (SMD) Simulations. We simulated the force-induced, linear unfolding of each putative REJ domain using Steered Molecular Dynamics as implemented in the GPU-accelerated version of NAMD [41, 42]. Coulombic forces were restricted using the switching function from 10 Å to a cutoff at 12 Å. The CHARMM22 force field was used throughout the simulations. Each of the REJ domain models was solvated in a water sphere with a boundary of 15 Å. The system was charge neutralized by adding Na^+ and Cl^-; the total ionic strength of the system was then adjusted to 0.150 M. The simulations of REJd1, -d2, -d3, and -d4 contained 9870, 8303, 13905, and 8608 atoms, respectively. Each system was then minimized to equilibrium using conjugate gradient minimization from an initial temperature of 298 K. This was followed by a 600 ps MD step to equilibrate the protein, water, and ions. For the SMD experiment, a spring constant of $10\,k_BT\,\text{Å}^{-2}$ was used. Simulated force was applied by fixing the C-terminal $C\alpha$ atom in the model and pulling the N-terminal $C\alpha$ SMD atom with constant velocity along a predetermined vector. The trajectories were recorded every 2 fs and then analyzed with VMD. The REJ domains were pulled at a constant velocity of $0.001\,\text{Å·ps}^{-1}$ and was followed for 150 Å. To validate the accuracy of our *in silico* experiments we carried out SMD simulations on titin I27. Our SMD results for I27 are very similar to those published previously [41, 43] (Figure S2).

3. Results and Discussion

3.1. Analysis of Potential FNIII Domains in the REJ Module Using Homology Modeling Techniques. Homology models for REJd1, REJd2, REJd3, and REJd4 (Figure 1(c)) were based on Clustal alignments of the primary sequences for predicted REJ domains with the primary sequences of the following template structures (Figure 1(b)): the human PKD domain no. 1 from PC1 (1b4r) and the human PKD domain from protein KIAA0319 (2e7m). The domain boundaries were based on Schröder et al. sequence analysis [24] and our Clustal multiple sequence alignment. The overall identity between each of the REJ domains and the template structures was low (~10% overall identity, ~27% similarity). Our method for homology model determination relies on finding periodicity within the primary sequence, that is, characteristic of beta-sheet structure. A similar technique was used to compute a homology model of the NS3 proteases of the Hepatitis C virus that have low sequence identity (~15%) [44]. For each homology model, 10 candidate structures were calculated. We chose the best model based on the lowest calculated model energy values as reported by MODELLER. Further, we assessed the perresidue DOPE score on the final model versus the template structure and refined any poorly scoring loop regions accordingly.

In order to assess the overall quality of the putative REJ domains structures with respect to well-determined structures we used the programs WHAT_CHECK [45] and ERRAT2 [46]. We found that REJd1 rated the highest of the four models on the ERRAT2 scale (quality factor = 72.7). In addition, the WHAT_CHECK packing quality scored best at $Z = -2.839$, while the RMSD of REJd1 versus the template structure (1b4r) was 3.66 Å. The REJd2 model scored lower on the ERRAT2 scale (35.2), but the RMSD versus the template structure was 2.52 Å. This could be indicative of a good alignment with the template structure, but the hydrophobic core of the REJd2 domain may not provide sufficient packing within the hydrophobic core of the domain that can be measured in other FNIII type domains. REJd3 and REJd4, on the other hand, score less well by these metrics. While the REJd3 matches well with its template, 2e7m (RMSD = 2.20 Å), the packing quality score is rated as "poor" (−4.3). REJd4 scored worse than the others. The putative REJd4 domain is the most divergent of the four; it does not have a well-defined core structure. The RMSD versus its template (2e7m) was 4.02 Å, while the quality factor was very low at 23.5. This is likely because REJd4 does not have a Trp in the core region where the Trp residue seems to be conserved in the REJ folds. Hence, our homology modeling analysis shows that the primary sequences for REJd1 and REJd2 are consistent with known FNIII domains, while the REJd3 domain may represent a partially structured domain, and REJd4 most likely lack a stable tertiary structure.

Based on this analysis we hypothesize that the putative REJd1 and REJd2 domains may have a fold similar to the FNIII domains and that REJd3 and REJd4 most likely lacks a stable tertiary structure. In order to test this hypothesis we used SMFS and SMD methods. These methods have been successfully used by a number of groups to obtain structural information, such as the mechanical stability (Ig and FNIII domains typically unfold at much higher forces than alpha-helical domains) and the increase in contour length upon unfolding (which is proportional to the number of residues that are exposed after unfolding) [47].

3.2. Mechanical Signatures of REJd1 and REJd2 Domains. We found that the MBP-REJd1-I27 and MBP-REJd1,2-I27 constructs are expressed as soluble proteins in *E. coli*. The advantage of using MBP and I27 proteins is that they provide unique mechanical unfolding fingerprints. Both MBP and I27 have been characterized using SMFS techniques [30, 48–51]. Stretching the construct containing MBP, titin I27 domain and sequences for REJd1,2 generated sawtooth patterns with distinctive force peaks and increases in contour lengths, ΔLc (Figure 2(a)). To determine the contribution of each domain to the unfolding pattern we analyzed the spacing between peaks in the unfolding patterns. We used the worm-like chain (WLC) model for polymer elasticity, which predicts the entropic restoring force generated upon the extension of a polymer [38, 39]. The thin lines in Figure 2(a) correspond to manual fits of the WLC equation to the curve that precedes each force peak. The I27 domains have been shown to unfold at forces of ~200 pN and produce an increase in contour length (ΔLc) of ~29 ± 8 nm upon unfolding [30]. On the other

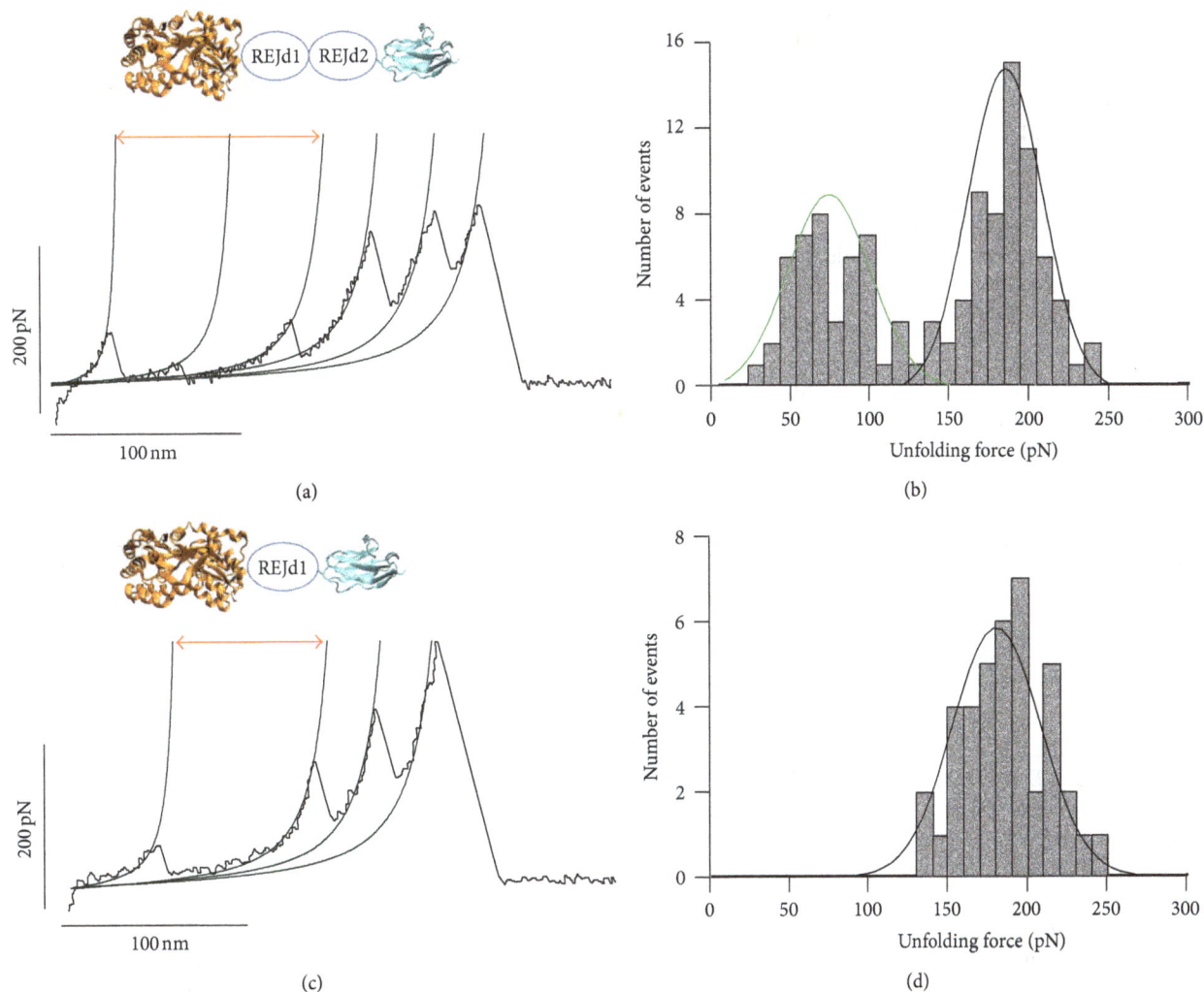

FIGURE 2: Analysis of the mechanical stability of putative REJ domains 1 and 2. (a) Typical unfolding pattern of the MBP-REJd1,2-I27 protein. The first two peaks in the force-extension curve correspond to the unfolding of MBP, with a total increase in ΔLc of about 100 nm (red double headed arrow) and an unfolding intermediate with a ΔLc of about 55 nm. We assign the third force peak to the unfolding of one of the REJ domains and the next two to the unfolding of the other REJ and I27 domains. (b) Unfolding force histogram for the I27 and REJ domains in the MBP-REJd1,2-I27 construct. In this histogram we did not include the unfolding of the MBP protein. The best fits to Gaussian distributions were obtained with the following parameters: 63 ± 37 pN ($n = 42$) and 190 ± 30 pN ($n = 61$; 47 traces). (c) Force-extension trace of the MBP-REJd1-I27 construct. This example shows the all-or-none unfolding of the MBP protein; in this example there is no unfolding intermediate. The increase in ΔLc is about 100 nm (red double headed arrow). The next two force peaks correspond to the unfolding of REJd1 and I27 domains. The black lines correspond to fits to the WLC equation using a ΔLc of 29 nm. (d) Unfolding force histogram for the I27 and REJ domains in the MBP-REJd1-I27 construct. In this histogram we did not include the unfolding of the MBP protein. There is a single distribution of force peaks with a mean of about 190 pN (188 ± 39 pN, $n = 40$; 21 traces).

hand, MBP is known to unfold at forces of about 70 pN with a total increase in ΔLc of ~100 nm upon unfolding [49, 50]; it was found that MBP can also unfold via a mechanically stable unfolding intermediate which contributes a ΔLc of ~50 nm upon unfolding [50]. Hence, we attribute the first 100 nm of the recording to the unfolding of the MBP protein. This trace also shows the unfolding intermediate. There are three force peaks before the detachment from the surface; these all show a ΔLc of about 29 nm. One of them has a peak force of about 70 pN. Given the construction of the protein this means that the REJd1 and REJd2 unfold at very different forces, one at about 70 pN and the other a force similar to the I27 domain

(~200 pN). Figure 2(b) shows an unfolding force histogram for the REJ and I27 domains. There are two clear populations, one unfolds at a low force of 63 ± 37 pN ($n = 42$) and the other at 190 ± 30 pN ($n = 61$).

Although we can confidently discriminate between one of the REJ domains and the I27 titin domain with these data, the identity of the individual REJ domains in these recordings cannot be established at this point. To unambiguously identify the force peaks from each REJ domain, we constructed a protein containing REJd1 with a flanking MBP and an I27 domain. Figure 2(c) shows a trace obtained after stretching the MBP-REJd1-I27 protein. This example shows the MBP protein unfolds in an all-or-none unfolding manner with no unfolding intermediate. The increase in ΔLc is about 100 nm. The next two peaks have unfolding forces of 148 pN and 204 pN, respectively. One of these events must correspond to the unfolding of the REJd1 domain. However, a precise assignment cannot be made since the unfolding force histogram shows only one distribution centered at ~190 pN (188 ± 39 pN, $n = 40$; Figure 2(d)).

Based on these data we can conclude that REJd1 unfolds at similar forces rather than the I27 domain (~200 pN), and REJd2 unfolds at a significantly lower force (~60 pN).

3.3. Mechanical Signatures of REJd3 and REJd4 Domains. In order to study the mechanical unfolding of putative REJ domains REJd3 and REJd4 we constructed a protein containing the REJd3 and REJd4 sequences plus MBP and I27 (MBP-REJd3,4-I27). However, we found that this is construct expressed poorly and mostly in inclusion bodies in *E. coli*. To further increase the solubility we used a chimeric I27 polyprotein approach that has been successfully used to study proteins that tend to aggregate such as alpha-synuclein, huntingtin polyQ, and tau proteins [52–54]. For this purpose we inserted the REJd3,4 sequence in between several I27 domains. We made two constructs in this way, (I27)$_3$-REJd4-(I27)$_2$ and (I27)$_3$-REJd3,4-(I27)$_2$. These are expressed well as soluble proteins in *E. coli*. Figure 3 shows examples obtained after stretching these constructs. To facilitate the analysis we selected traces that had five I27 unfolding peaks and had a clean initial force extension after retraction from the surface (i.e., little or no unspecific interactions). In the case of the (I27)$_3$-REJd4-(I27)$_2$ and (I27)$_3$-REJd3,4-(I27)$_2$. These recordings show only five unfolding peaks. There are two possible scenarios that the mechanical stabilities of REJd3 and REJd4 are much higher or lower (within the noise) than those for the I27 domain. It is unlikely that the mechanical stabilities of REJd3 and REJd4 exceed that of titin I27 because: (i) we observed no more than five force peaks that show the mechanical fingerprint of I27 domains (i.e., unfolding at ~200 pN and an interpeak spacing of ~28 nm), (ii) the detachment forces (last force peak) are >400 pN and all protein domains studied so far unfold at forces less than this force [37], and (iii) we typically observed a spacer before the unfolding of the I27 domains. In the example shown in Figure 3(a), the distance to the first I27 force peak is about 60 nm, and in Figure 3(b) this distance is about 85 nm. The spacers observed in Figure 3 are characteristically seen in domains that have a mostly disordered or random coil conformation, such as, for example, tropoelastin, the titin PEVK domain, or some neurotoxic proteins [52–55]. In addition Far-UV CD analysis is also consistent with REJd4 forming an unstructured random coil (Figure S1). The spectra for the (I27)$_3$-REJd4-(I27)$_2$ protein shows a significant higher random-coil content (43%) than that of a (I27)$_8$ protein (35%).

Hence, we conclude that domains REJd3 and REJd4 form mechanically weak structures (random-coil or unfolded conformation) that unfold at forces that are below the resolution of our AFM (<10 pN).

3.4. Steered Molecular Dynamic Simulations of REJ FNIII Domains. As our homology models of REJd1-d4 are based on preexisting structures, we sought to assay the biophysical correspondence between our *in silico* domains and experimental data. Steered molecular dynamic simulations (SMDs) analysis has been used in the past to study the mechanical unfolding pathways of FNIII domains [56–59]. For example the simulated force-extension curves for FNIII domain no. 10 of fibronectin show a single dominant force peak which corresponds to the rupture of the tertiary structure of FNIII [59]. Figure 4 shows constant velocity SMD simulations of the mechanical unfolding of the four REJ domains. The force-extension curves were obtained from SMD simulations by stretching domains between its C-terminus and its N-terminus at a pulling speed of 0.001 Å·ps^{-1}. The magnitude of the forces observed in the SMD simulations does not directly correspond to those measured with AFM. This is partially because the pulling speeds are several orders of different magnitude. However, the simulations are qualitatively consistent with AFM experiments [60–62]. To validate the accuracy of our *in silico* experiments we carried out SMD simulations on titin I27. Our SMD results for I27 are very similar to those published previously [41, 43] (Figure S2). Our simulations show that force-extension profiles of REJd1 and REJd2 are very similar and show a force peak of about 3000 pN at around 40 Å. The shaded area corresponds to the initial burst of force that is typical of other FNIII domains [56–59].

Both the REJd1 (black curve) and REJd2 (red curve) agree with experimental measurements of the REJd1 and REJd2 proteins. Further, the blue SMD force-curve (REJd4) agrees with our experimental data for REJd4; that is, it is likely not well folded. However, there are clearly limitations to the current computational techniques. While the primary sequence that we have assigned as REJd3 fits to an Ig-like model and it reacts like a properly folded FNIII-like domain in our SMD simulation, it is clearly intermediate between a folded and a nonfolded domain for reasons that SMD technique is not accurate enough to simulate.

4. Conclusions

The available evidence indicates that PC1 has a role in sensing of flow [2, 3], pressure [4], cell cycle [5], cell polarity regulation [6], and kidney development [63]. PC1 may sense signals from the primary cilia, neighboring cells, and extracellular matrix and transduces them into cellular responses that regulate proliferation, adhesion, and differentiation that are essential for the control of renal tubules and kidney morphogenesis [1, 3, 7, 8].

In this work we combined homology modeling, protein engineering SMD simulations, and SMFS to systematically characterize the mechanical stability of the predicted four FNIII domains in the REJ module of PC1. Of the 80 missense mutations in the REJ module about 20 disease-causing mutations map onto the region studied in this work. After flanking the different putative FNIII sequences with titin I27 or MBP domains we found that these constructs were expressed well as soluble proteins in *E. coli* and were able to analyze their mechanical stability. Our study provides

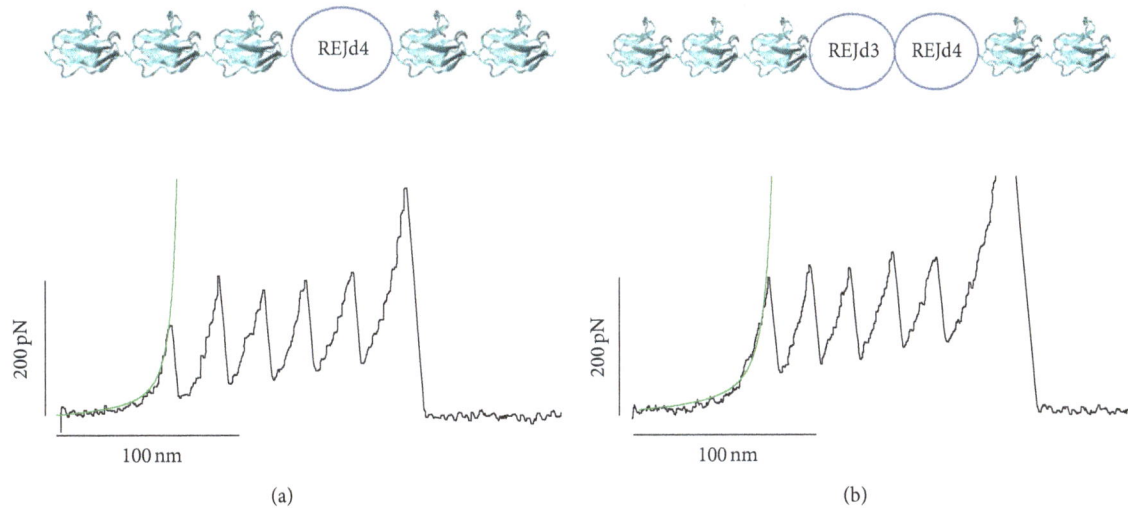

(a)   (b)

FIGURE 3: Analysis of the mechanical stability of putative REJ domains 3 and 4. (a) Example of unfolding pattern observed after stretching the $(I27)_3$-REJd4-$(I27)_2$ protein. (b) Example of unfolding pattern observed after stretching the $(I27)_3$-REJd3,4-$(I27)_2$ protein. In these two examples the unfolding of the five titin I27 domains (they unfold on average at 200 pN with an increase length of 28 nm) is preceded by a long spacer.

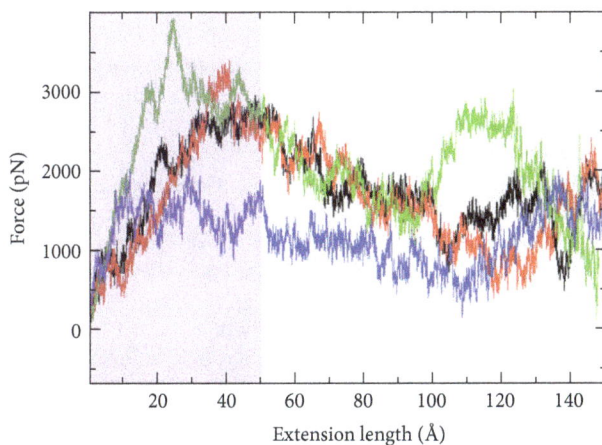

FIGURE 4: Constant velocity steered molecular dynamics simulations of the mechanical unfolding of REJ domains. Constant velocity steered molecular dynamics simulation of the mechanical unfolding of REJd1 (black), REJd2 (red), REJd3 (green), and REJd4 (blue). Force-extension curves were obtained from the SMD simulation of each REJ domain model by first fixing the C-terminal $C\alpha$ atom and then applying a constant force to the N-terminal $C\alpha$ atom along a predetermined vector. Forces (in pN) were recorded for each time step along the simulation. The shaded area corresponds to the initial burst of force that is typical of other FNIII domains.

direct mechanical force measurements of four putative FNIII domains in the REJ module and provides *in vitro* evidence that these domains have a different mechanical stability. Stretching a construct containing the REJd1 and REJd2 sequences generated mechanical fingerprints that correspond to the unfolding of the MBP and I27 domains as well as extra unfolding events that represent the unfolding of the REJd1 and -d2 domains. Our data show that REJd1 and REJd2

domains are mechanically stable and unfold at forces of about 200 pN and 60 pN, respectively. This range of values is consistent with those reported for the mechanical unfolding of FNIII domains [64, 65]. We found that constructs harboring the REJd3 and REJd4 sequences are expressed well as soluble proteins in *E. coli* or insect cells (Sf9); however they do not form mechanically stable folded domains. Our results do not exclude the possibility that REJd3 and REJd4 form stable folds when expressed in other cells, such as human kidney epithelial cells, or when expressed within the native full length extracellular region. It is also possible that these domains need molecular chaperones to acquire a proper stable fold. Hence, future experiments in more physiological settings would be necessary to resolve this important issue and facilitate the univocal characterization of these domains. The REJ module of PC1 could represent a continuation of the PKD domain structure known to exist in PC1. That is, the protein could possess a series of PKD domains that terminate in a series of homologous but mechanically weaker FNIII domains. It is possible that the primary sequence of the REJd3 and REJd4 repeating units degraded over evolution into structures that still retained the general character of the parental FNIII domain but do not possess the mechanical stability preset in the more N-terminal domains. We speculate that the REJd3 and REJd4 domains may function as entropic springs designed to adjust the length of the extracellular region of PC1 in response to mechanical shear stress. This spring-like behavior might also be important for the autoproteolysis in the GPS domain, a process that requires the complete REJ module.

We and others (e.g., [13, 66]) have shown that SMFS techniques can be used to accurately quantify the effects of disease causing mutations on single protein domains. Pathogenic missense mutations that target PC1 PKD domains were shown to result in a loss in mechanical stability which may lead to the abnormal mechanical function of PC1 [13]. Our SMFS results demonstrate a powerful experimental approach to study the domain architecture and stability of the REJ module and should pave the way to systematically characterize the effects of disease-causing mutations in the REJ module of human PC1.

Abbreviations

AFM: Atomic force microscopy
SMFS: Single-molecule force spectroscopy
SMD: Steered molecular dynamics
REJ: Receptor for egg jelly
PC1:   Polycystin-1.

Acknowledgments

This work was funded by NIH Grant R01DK073394, a PKD Foundation (Grant 116a2r), the John Sealy Memorial Endowment Fund for Biomedical Research, P30AG024832 (Pepper Grant) (to Andres F. Oberhauser), and NIH R01 DK062199 and P30 DK090868 (to Feng Qian). The authors thank Odutayo Odunuga for help with the construction and expression of some of the REJd3 and REJd4 constructs.

References

[1] P. C. Harris and V. E. Torres, "Polycystic kidney disease," *Annual Review of Medicine*, vol. 60, pp. 321–337, 2009.

[2] S. M. Nauli and J. Zhou, "Polycystins and mechanosensation in renal and nodal cilia," *BioEssays*, vol. 26, no. 8, pp. 844–856, 2004.

[3] S. M. Nauli, F. J. Alenghat, Y. Luo et al., "Polycystins 1 and 2 mediate mechanosensation in the primary cilium of kidney cells," *Nature Genetics*, vol. 33, no. 2, pp. 129–137, 2003.

[4] R. Sharif-Naeini, J. H. Folgering, D. Bichet et al., "Polycystin-1 and -2 dosage regulates pressure sensing," *Cell*, vol. 139, pp. 587–596, 2009.

[5] A. K. Bhunia, K. Piontek, A. Boletta et al., "PKD1 induces p21waf1 and regulation of the cell cycle via direct activation of the JAK-STAT signaling pathway in a process requiring PKD2," *Cell*, vol. 109, no. 2, pp. 157–168, 2002.

[6] H. Happe, E. de Heer, and D. J. Peters, "Polycystic kidney disease: the complexity of planar cell polarity and signaling during tissue regeneration and cyst formation," *Biochimica et Biophysica Acta*, vol. 1812, pp. 1249–1255, 2011.

[7] O. Ibraghimov-Beskrovnaya and N. Bukanov, "Polycystic kidney diseases: from molecular discoveries to targeted therapeutic strategies," *Cellular and Molecular Life Sciences*, vol. 65, no. 4, pp. 605–619, 2008.

[8] A. J. Streets, L. J. Newby, M. J. O'Hare, N. O. Bukanov, O. Ibraghimov-Beskrovnaya, and A. C. M. Ong, "Functional analysis of PKD1 transgenic lines reveals a direct role for polycystin-1 in mediating cell-cell adhesion," *Journal of the American Society of Nephrology*, vol. 14, no. 7, pp. 1804–1815, 2003.

[9] G. W. Moy, L. M. Mendoza, J. R. Schulz, W. J. Swanson, C. G. Glabe, and V. D. Vacquier, "The sea urchin sperm receptor for egg jelly is a modular protein with extensive homology to the human polycystic kidney disease protein, PKD1," *Journal of Cell Biology*, vol. 133, no. 4, pp. 809–817, 1996.

[10] R. Sandford, B. Sgotto, S. Aparicio et al., "Comparative analysis of the polycystic kidney disease 1 (PKD1) gene reveals an integral membrane glycoprotein with multiple evolutionary conserved domains," *Human Molecular Genetics*, vol. 6, no. 9, pp. 1483–1489, 1997.

[11] A. F. Oberhauser and M. Carrión-Vázquez, "Mechanical biochemistry of proteins one molecule at a time," *Journal of Biological Chemistry*, vol. 283, no. 11, pp. 6617–6621, 2008.

[12] F. J. Alenghat, S. M. Nauli, R. Kolb, J. Zhou, and D. E. Ingber, "Global cytoskeletal control of mechanotransduction in kidney epithelial cells," *Experimental Cell Research*, vol. 301, no. 1, pp. 23–30, 2004.

[13] L. Ma, M. Xu, J. R. Forman, J. Clarke, and A. F. Oberhauser, "Naturally occurring mutations alter the stability of polycystin-1 polycystic kidney disease (PKD) domains," *Journal of Biological Chemistry*, vol. 284, no. 47, pp. 32942–32949, 2009.

[14] F. Qian, W. Wei, G. Germino, and A. Oberhauser, "The nanomechanics of polycystin-1 extracellular region," *Journal of Biological Chemistry*, vol. 280, no. 49, pp. 40723–40730, 2005.

[15] J. R. Forman, S. Qamar, E. Paci, R. N. Sandford, and J. Clarke, "The remarkable mechanical strength of polycystin-1 supports a direct role in mechanotransduction," *Journal of Molecular Biology*, vol. 349, no. 4, pp. 861–871, 2005.

[16] S. Yu, K. Hackmann, J. Gao et al., "Essential role of cleavage of Polycystin-1 at G protein-coupled receptor proteolytic site for kidney tubular structure," *Proceedings of the National Academy of Sciences of the United States of America*, vol. 104, no. 47, pp. 18688–18693, 2007.

[17] W. Wei, K. Hackmann, H. Xu, G. Germino, and F. Qian, "Characterization of cis-autoproteolysis of polycystin-1, the product of human polycystic kidney disease 1 gene," *Journal of Biological Chemistry*, vol. 282, no. 30, pp. 21729–21737, 2007.

[18] F. Qian, A. Boletta, A. K. Bhunia et al., "Cleavage of polycystin-1 requires the receptor for egg jelly domain and is disrupted by human autosomal-dominant polycystic kidney disease 1-associated mutations," *Proceedings of the National Academy of Sciences of the United States of America*, vol. 99, no. 26, pp. 16981–16986, 2002.

[19] S. Yu, K. Hackmann, J. Gao et al., "Essential role of cleavage of Polycystin-1 at G protein-coupled receptor proteolytic site for kidney tubular structure," *Proceedings of the National Academy of Sciences of the United States of America*, vol. 104, no. 47, pp. 18688–18693, 2007.

[20] M. A. Garcia-Gonzalez, J. G. Jones, S. K. Allen et al., "Evaluating the clinical utility of a molecular genetic test for polycystic kidney disease," *Molecular Genetics and Metabolism*, vol. 92, no. 1-2, pp. 160–167, 2007.

[21] F. Qian, A. Boletta, A. K. Bhunia et al., "Cleavage of polycystin-1 requires the receptor for egg jelly domain and is disrupted by human autosomal-dominant polycystic kidney disease 1-associated mutations," *Proceedings of the National Academy of Sciences of the United States of America*, vol. 99, no. 26, pp. 16981–16986, 2002.

[22] H. J. Gunaratne, G. W. Moy, M. Kinukawa, S. Miyata, S. A. Mah, and V. D. Vacquier, "The 10 sea urchin receptor for egg jelly proteins (SpREJ) are members of the polycystic kidney disease-1 (PKD1) family," *BMC Genomics*, vol. 8, p. 235, 2007.

[23] J. Hughes, C. J. Ward, B. Peral et al., "The polycystic kidney disease 1 (PKD1) gene encodes a novel protein with multiple cell

recognition domains," *Nature Genetics*, vol. 10, no. 2, pp. 151–160, 1995.

[24] S. Schröder, F. Fraternali, X. Quan, D. Scott, F. Qian, and M. Pfuhl, "When a module is not a domain: the case of the REJ module and the redefinition of the architecture of polycystin-1," *Biochemical Journal*, vol. 435, no. 3, pp. 651–660, 2011.

[25] R. Chenna, H. Sugawara, T. Koike et al., "Multiple sequence alignment with the Clustal series of programs," *Nucleic Acids Research*, vol. 31, no. 13, pp. 3497–3500, 2003.

[26] N. Eswar, B. Webb, M. A. Marti-Renom et al., "Comparative protein structure modeling using MODELLER," *Current Protocols in Protein Science*, chapter 2:unit 2.9, 2007.

[27] D. Eramian, M. Y. Shen, D. Devos, F. Melo, A. Sali, and M. A. Marti-Renom, "A composite score for predicting errors in protein structure models," *Protein Science*, vol. 15, no. 7, pp. 1653–1666, 2006.

[28] A. Steward, J. L. Toca-Herrera, and J. Clarke, "Versatile cloning system for construction of multimeric proteins for use in atomic force microscopy," *Protein Science*, vol. 11, no. 9, pp. 2179–2183, 2002.

[29] K. L. Fuson, L. Ma, R. B. Sutton, and A. F. Oberhauser, "The c2 domains of human synaptotagmin 1 have distinct mechanical properties," *Biophysical Journal*, vol. 96, no. 3, pp. 1083–1090, 2009.

[30] M. Carrion-Vazquez, A. F. Oberhauser, S. B. Fowler et al., "Mechanical and chemical unfolding of a single protein: a comparison," *Proceedings of the National Academy of Sciences of the United States of America*, vol. 96, no. 7, pp. 3694–3699, 1999.

[31] M. Carrion-Vazquez, A. F. Oberhauser, T. E. Fisher, P. E. Marszalek, H. Li, and J. M. Fernandez, "Mechanical design of proteins studied by single-molecule force spectroscopy and protein engineering," *Progress in Biophysics and Molecular Biology*, vol. 74, no. 1-2, pp. 63–91, 2000.

[32] E. L. Florin, M. Rief, H. Lehmann et al., "Sensing specific molecular interactions with the atomic force microscope," *Biosensors and Bioelectronics*, vol. 10, no. 9-10, pp. 895–901, 1995.

[33] N. Sakaki, R. Shimo-Kon, K. Adachi et al., "One rotary mechanism for F1-ATPaSe over ATP concentrations from millimolar down to nanomolar," *Biophysical Journal*, vol. 88, no. 3, pp. 2047–2056, 2005.

[34] H. Itoh, A. Takahashi, K. Adachi et al., "Mechanically driven ATP synthesis by F1-ATPase," *Nature*, vol. 427, no. 6973, pp. 465–468, 2004.

[35] A. F. Oberhauser, P. K. Hansma, M. Carrion-Vazquez, and J. M. Fernandez, "Stepwise unfolding of titin under force-clamp atomic force microscopy," *Proceedings of the National Academy of Sciences of the United States of America*, vol. 98, no. 2, pp. 468–472, 2001.

[36] E. M. Puchner, A. Alexandrovich, A. L. Kho et al., "Mechanoenzymatics of titin kinase," *Proceedings of the National Academy of Sciences of the United States of America*, vol. 105, no. 36, pp. 13385–13390, 2008.

[37] J. I. Sulkowska and M. Cieplak, "Stretching to understand proteins—a survey of the protein data bank," *Biophysical Journal*, vol. 94, no. 1, pp. 6–13, 2008.

[38] J. F. Marko and E. D. Siggia, "Stretching DNA," *Macromolecules*, vol. 28, no. 26, pp. 8759–8770, 1995.

[39] C. Bustamante, J. F. Marko, E. D. Siggia, and S. Smith, "Entropic elasticity of λ-phage DNA," *Science*, vol. 265, no. 5178, pp. 1599–1600, 1994.

[40] G. Bohm, R. Muhr, and R. Jaenicke, "Quantitative analysis of protein far UV circular dichroism spectra by neural networks," *Protein Engineering*, vol. 5, no. 3, pp. 191–195, 1992.

[41] H. Lu, B. Isralewitz, A. Krammer, V. Vogel, and K. Schulten, "Unfolding of titin immunoglobulin domains by steered molecular dynamics simulation," *Biophysical Journal*, vol. 75, no. 2, pp. 662–671, 1998.

[42] J. C. Phillips, R. Braun, W. Wang et al., "Scalable molecular dynamics with NAMD," *Journal of Computational Chemistry*, vol. 26, no. 16, pp. 1781–1802, 2005.

[43] H. Lu and K. Schulten, "Steered molecular dynamics simulations of force-induced protein domain unfolding," *Proteins*, vol. 35, pp. 453–463, 1999.

[44] A. Tramontano, "Homology modeling with low sequence identity," *Methods*, vol. 14, no. 3, pp. 293–300, 1998.

[45] G. Vriend, "WHAT IF: a molecular modeling and drug design program," *Journal of Molecular Graphics*, vol. 8, no. 1, pp. 52–56, 1990.

[46] C. Colovos and T. O. Yeates, "Verification of protein structures: patterns of nonbonded atomic interactions," *Protein Science*, vol. 2, no. 9, pp. 1511–1519, 1993.

[47] A. F. Oberhauser and M. Carrión-Vázquez, "Mechanical biochemistry of proteins one molecule at a time," *Journal of Biological Chemistry*, vol. 283, no. 11, pp. 6617–6621, 2008.

[48] M. Bertz and M. Rief, "Ligand binding mechanics of maltose binding protein," *Journal of Molecular Biology*, vol. 393, no. 5, pp. 1097–1105, 2009.

[49] M. Bertz and M. Rief, "Mechanical unfoldons as building blocks of maltose-binding protein," *Journal of Molecular Biology*, vol. 378, no. 2, pp. 447–458, 2008.

[50] V. Aggarwal, S. R. Kulothungan, M. M. Balamurali, S. R. Saranya, R. Varadarajan, and S. R. Ainavarapu, "Ligand modulated parallel mechanical unfolding pathways of Maltose Binding Proteins (MBPs)," *The Journal of Biological Chemistry*, vol. 286, pp. e9–e10, 2011.

[51] J. Bravo, V. Villarreal, R. Hervas, and G. Urzaiz, "Using a communication model to collect measurement data through mobile devices," *Sensors*, vol. 12, pp. 9253–9272, 2012.

[52] M. Sandal, F. Valle, I. Tessari et al., "Conformational equilibria in monomeric alpha-synuclein at the single-molecule level," *PLoS Biology*, vol. 6, no. 1, article e6, 2008.

[53] L. Dougan, J. Li, C. L. Badilla, B. J. Berne, and J. M. Fernandez, "Single homopolypeptide chains collapse into mechanically rigid conformations," *Proceedings of the National Academy of Sciences of the United States of America*, vol. 106, no. 31, pp. 12605–12610, 2009.

[54] S. Wegmann, J. Schöler, C. A. Bippes, E. Mandelkow, and D. J. Muller, "Competing interactions stabilize pro- and anti-aggregant conformations of human Tau," *Journal of Biological Chemistry*, vol. 286, no. 23, pp. 20512–20524, 2011.

[55] R. Hervas, J. Oroz, A. Galera-Prat et al., "Common features at the start of the neurodegeneration cascade," *PLoS Biology*, vol. 10, Article ID e1001335, 2012.

[56] Q. Peng, S. Zhuang, M. Wang, Y. Cao, Y. Khor, and H. Li, "Mechanical design of the third FnIII domain of tenascin-C," *Journal of Molecular Biology*, vol. 386, no. 5, pp. 1327–1342, 2009.

[57] M. Gao, D. Craig, V. Vogel, and K. Schulten, "Identifying unfolding intermediates of FN-III10 by steered molecular dynamics," *Journal of Molecular Biology*, vol. 323, no. 5, pp. 939–950, 2002.

[58] B. Isralewitz, J. Baudry, J. Gullingsrud, D. Kosztin, and K. Schulten, "Steered molecular dynamics investigations of protein function," *Journal of Molecular Graphics and Modelling*, vol. 19, no. 1, pp. 13–25, 2001.

[59] A. Krammer, H. Lu, B. Isralewitz, K. Schulten, and V. Vogel, "Forced unfolding of the fibronectin type III module reveals a tensile molecular recognition switch," *Proceedings of the National Academy of Sciences of the United States of America*, vol. 96, no. 4, pp. 1351–1356, 1999.

[60] M. Sotomayor and K. Schulten, "Single-molecule experiments in vitro and in silico," *Science*, vol. 316, no. 5828, pp. 1144–1148, 2007.

[61] H. Lu, B. Isralewitz, A. Krammer, V. Vogel, and K. Schulten, "Unfolding of titin immunoglobulin domains by steered molecular dynamics simulation," *Biophysical Journal*, vol. 75, no. 2, pp. 662–671, 1998.

[62] P. E. Marszalek, H. Lu, H. Li et al., "Mechanical unfolding intermediates in titin modules," *Nature*, vol. 402, no. 6757, pp. 100–103, 1999.

[63] K. Piontek, L. F. Menezes, M. A. Garcia-Gonzalez, D. L. Huso, and G. G. Germino, "A critical developmental switch defines the kinetics of kidney cyst formation after loss of Pkd1," *Nature Medicine*, vol. 13, no. 12, pp. 1490–1495, 2007.

[64] A. F. Oberhauser, C. Badilla-Fernandez, M. Carrion-Vazquez, and J. M. Fernandez, "The mechanical hierarchies of fibronectin observed with single-molecule AFM," *Journal of Molecular Biology*, vol. 319, no. 2, pp. 433–447, 2002.

[65] M. Rief, M. Gautel, and H. E. Gaub, "Unfolding forces of titin and fibronectin domains directly measured by AFM," *Advances in Experimental Medicine and Biology*, vol. 481, pp. 129–141, 2000.

[66] B. R. Anderson, J. Bogomolovas, S. Labeit, and H. Granzier, "Single molecule force spectroscopy on titin implicates immunoglobulin domain stability as a cardiac disease mechanism," *The Journal of Biological Chemistry*, vol. 288, pp. 5303–5315, 2013.

7

Cell Matrix Remodeling Ability Shown by Image Spatial Correlation

Chi-Li Chiu,[1] Michelle A. Digman,[1,2] and Enrico Gratton[1,2]

[1] *Department of Developmental and Cell Biology, University of California, Irvine, CA 92697, USA*
[2] *Laboratory for Fluorescence Dynamics, Department of Biomedical Engineering, University of California, Irvine, CA 92697, USA*

Correspondence should be addressed to Enrico Gratton; egratton22@yahoo.com

Academic Editor: Jianwei Shuai

Extracellular matrix (ECM) remodeling is a critical step of many biological and pathological processes. However, most of the studies to date lack a quantitative method to measure ECM remodeling at a scale comparable to cell size. Here, we applied image spatial correlation to collagen second harmonic generation (SHG) images to quantitatively evaluate the degree of collagen remodeling by cells. We propose a simple statistical method based on spatial correlation functions to determine the size of high collagen density area around cells. We applied our method to measure collagen remodeling by two breast cancer cell lines (MDA-MB-231 and MCF-7), which display different degrees of invasiveness, and a fibroblast cell line (NIH/3T3). We found distinct collagen compaction levels of these three cell lines by applying the spatial correlation method, indicating different collagen remodeling ability. Furthermore, we quantitatively measured the effect of Latrunculin B and Marimastat on MDA-MB-231 cell line collagen remodeling ability and showed that significant collagen compaction level decreases with these treatments.

1. Introduction

Extracellular matrix (ECM) remodeling through cell-ECM interactions is a critical step of many biological and pathological processes such as embryonic development [1], angiogenesis [2], wound healing [3], and cancer cell metastasis [4]. For instance, when cancer cells move through a dense ECM, they can generate actomyosin forces that deform the collagen fibers to push the cell through the ECM [5]. Reciprocally, the physical properties of the ECM including matrix structure, mechanics, and dimensionality can profoundly influence cellular behavior. In the case of cancer ECM which is usually stiffer than normal tissues, the compromised tensional homeostasis affects cell phenotype, Rho-dependent cell contractility and oncogene-mediated transformation [6]. Clinically, the increased matrix stiffness and ECM remodeling were observed in premalignant tissue, and this increase was shown to contribute to malignant transformation [7].

Collagen, the most abundant protein in a mammalian body, is the major contributor to tissue mechanical properties. These mechanical properties have roots in collagen's microstructure, network organization, and orientation.

Collagen has been used in a number of studies that aim to quantify cell-mediated ECM remodeling process and its mechanics. The measurement scale of ECM remodeling ranges from whole gel contraction assay at centimeter scale [8–11] to collagen properties at micrometer scale such as fiber diameter, fiber length, and pore size [12]. Also the organization of the collagen matrix in the proximity of the cells has been a subject of several investigations [13, 14]. Although these methods provide ways to describe the collagen matrix property changes around the cells and at the microscopic level, they lack the information of heterogeneity due to ECM remodeling at the mesoscale that is comparable to cell size. Alternatively, Stevenson et al. [15] attempted to quantify the collagen matrix compaction level in the pericellular region by creating an intensity contour map of a confocal collagen reflection image. This method, however, does not allow an absolute quantitative comparison of different images due to the dependence of intensity range of each image.

Nonlinear microscopy techniques such as second harmonic generation (SHG) provide a noninvasive and label-free tool to image collagen fibers. Collagen has highly crystalline triple-helix structure that is not centrosymmetric, which makes it extremely bright in SHG. SHG does not involve the excitation of molecules; as a result, the molecules do not suffer the effects of phototoxicity or photobleaching. Since the first publication of SHG on collagen three decades ago [16], this technique has become a robust tool for imaging tissue structure in both *ex vivo* and *in vivo* preparations [4, 17–19].

In this study, we aimed to develop a quantitative method based on image spatial correlation of collagen SHG images to analyze collagen organization at mesoscale, that is, at the scale of 10–100 μm, comparable to cellular size. Image spatial correlation provides an unbiased way for estimating the size of spatial organization. By applying image spatial correlation to collagen SHG images with proper resolution, it is possible to extract the collagen mesoscale organization information. During cell-mediated ECM remodeling, cells may generate collagen-dense areas through cell-ECM interactions that modify the density of collagen in the regions around the cells. We applied this method to estimate the mesoscale changes of SHG high intensity area size due to the presence of cells in the collagen matrix few days after seeding the cells. This parameter is referred to as collagen compaction level in this report, the same as Stevenson et al.'s definition [15]. Relevant to our methodology, applying image correlation spectroscopy to SHG image has been described by Raub et al. [20], which focused on quantifying collagen matrix pore size and fiber dimension from collagen. In contrast to the emphasis of the microscopic scale in the previous report, here we applied the spatial correlation technique to analyze the mesoscale structure of collagen and showed that the SHG image of collagen combined with image correlation spectroscopy is an effective means to assess the mesoscale properties of an *in vitro* ECM that involves cellular remodeling. We also emphasize that the image correlation method used in this work provides unbiased quantitative evaluation of the mesoscale collagen organization. This is important for comparison of different cell types and for the assessment of the effect of drugs treatment in regard to the large scale organization of the ECM.

In this report, we compared the collagen matrix remodeling ability of two human breast cancer cell lines with distinct degrees of invasiveness, MDA-MB-231 and MCF-7, and fibroblast cell line NIH/3T3. Tumor cell invasiveness has been associated with increased contractile force generation [21], and ECM compaction level may reflect cell contractility. Our results suggested that cancer cells with high invasiveness have significantly more ability to remodel the collagen matrix, consistent with the previous report [21], while NIH/3T3 cell line moderately changed the collagen compaction level. Furthermore, the matrix remodeling ability of invasive cancer cell line can be altered through metalloproteinases (MMP) inhibitor or sequestering F-actin, suggesting that MMP and actin cytoskeleton affect collagen remodeling at the mesoscale.

2. Methods

2.1. Cell Culture in 3D Collagen Matrix.
MDA-MB-231, MCF-7, and NIH/3T3 cell lines (American Type Culture Collection HTB-26, HTB-22, and CL-1658, resp.) were obtained from Dr. Wen-Hua Lee's lab in UC Irvine. Cells were cultured in Dulbecco's modified Eagle's medium (DMEM) with high glucose (Sigma, St. Louis, MO) supplemented with 10% (v/v) fetal bovine serum (FBS) at 37°C in a 5% CO_2 humid incubator.

Type I collagen with original concentration of 3.75 mg/mL was purchased from BD Biosciences (Franklin Lakes, NJ). Collagen was diluted with 10X PBS and water to achieve final concentration of 1X PBS and 2 mg/mL collagen. NaOH was added to neutralize collagen solution before mixing with cells. Cells in serum-free DMEM were mixed with collagen solution. The final concentration of $5 * 10^4$ cells/mL cell-collagen mixture was added to 8-well Lab-Tek chambered coverglass with surface area 0.7 cm^2 per well (Thermo Scientific, Rochester, NY).

Collagen was polymerized at 20°C for 1 hr and then at 37°C for 20 minutes. Full medium was applied after polymerization. Medium was changed everyday after collagen polymerization for both collagen-only and cell-collagen matrix to ensure minimum variation caused by evaporation.

Matrices were kept at 37°C, 5% CO_2 incubator. Images were taken under room temperature after 1, 2, and 4 days of collagen polymerization.

2.2. Pharmacological Treatment.
MDA-MB-231 cells were cultured in type I collagen matrix and applied pharmacological treatment everyday when changing the medium. The treatment was applied to the cell-collagen matrix for 4 consecutive days before imaging. The F-actin sequestering drug Latrunculin B (Molecular Probes, Eugene, OR) was used at 1 μM final concentration. The matrix metalloproteinase inhibitor Marimastat (Tocris Bioscience, Bristol, United Kingdom) was used at 10 μM final concentration.

2.3. 2-Photon Microscopy.
To assess the degree of collagen matrix remodeling, collagen second harmonic generation (SHG) images were collected using LSM 710 (Carl Zeiss, Maple Grove, MN) with a 40 × 0.75 N.A. water immersion lens. The Mai-Tai laser (Newport, Irvine, CA) with emission at 900 nm was used for second harmonic generation. The collagen SHG propagating in backward direction was collected using a bandwidth filter (442–463 nm). Laser power and detector gain were fixed for all images taken.

We conducted 4 independent experiments for the samples containing collagen only, collagen with MCF cell line, and collagen with NIH cell line. We did 8 independent experiments for the collagen with MB231 cell line, and 3 independent experiments for each drug treatment. For each independent experiment, we used 3 different regions for analysis. Each measurement has 10 z-stacks.

For each measurement, as shown in Figure 1 upper right panel, z-stack SHG images with total of 10 slices across 200 μm depth were taken (each slice is 22 μm apart along the axial direction). Each x-y scan slice of the z-stack image was 1024 × 1024 pixels with pixel size of 0.8 μm and pixel dwell time of 50 μs/pixel. The lowest slice was taken right above the glass surface of culture dishes (slice 1 in the upper left panel of Figure 1) and went all the way up to 200 μm above (slice 10 in the upper left panel of Figure 1).

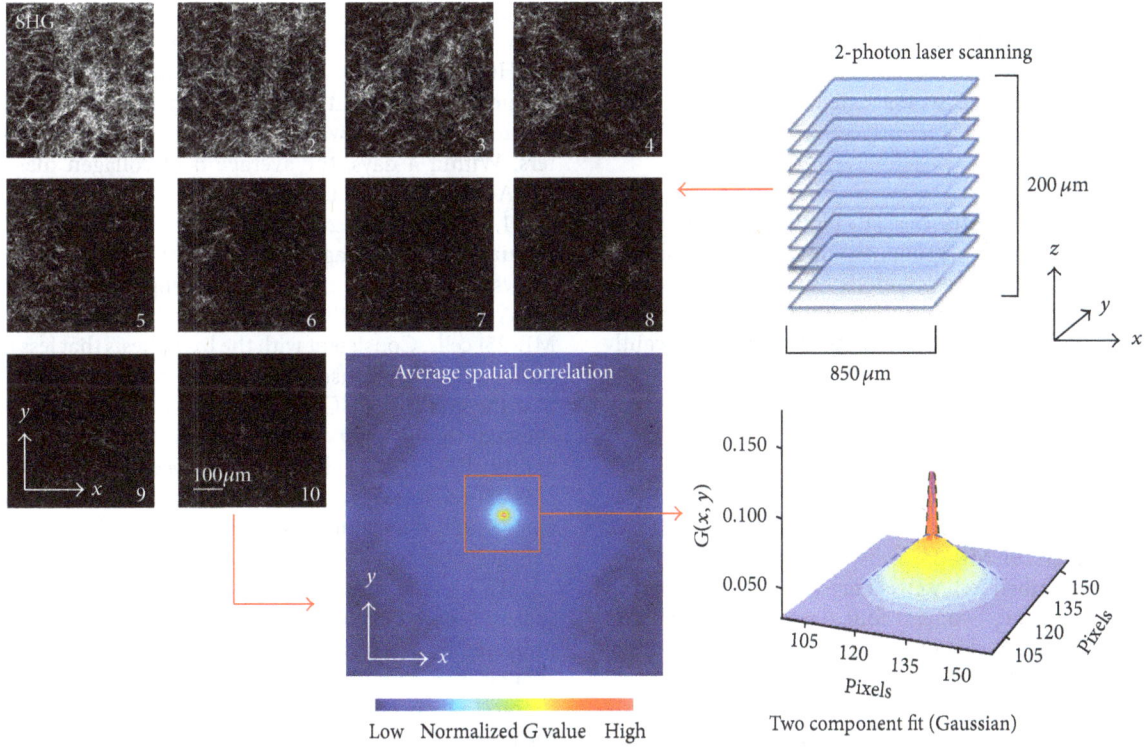

FIGURE 1: *Experimental workflow—Collagen remodeling analysis using image spatial correlation*. Upper right panel: collagen SHG images were acquired by 2-photon microscopy. 10 z-stack slices across 200 μm depth from culture dish surface were collected. Each slice covers 850 μm $\times$ 850 μm. Upper left panel: an example of collagen SHG images from close to the culture dish surface (slice 1) to 200 μm above (slice 10). The SHG intensity decreases with the increasing depth of acquisition. However, since image spatial correlation is not sensitive to average intensity change, image spatial correlation (1) was applied to each slice, and the results were averaged to produce one correlation image. Lower left panel: the spatial correlation image calculated from the SHG z-stack. The relatively symmetrical autocorrelation pattern indicates that with our sample preparation, collagen fibers were not preferentially aligned toward specific direction at the scale of the whole image. Lower right panel: the x and y cross sections from the spatial correlation image were fitted with two Gaussian components. One component has the standard deviation close to pixel size (green curve); the other component has broader standard deviation (blue curve), and its width varies based on the collagen compaction level.

2.4. Image Analysis—Image Spatial Correlation. The spatial correlation function below (1) was then applied to each x-y plane of SHG images:

$$G_s(\xi, \psi) = \frac{\langle I(x, y) I(x + \xi, y + \psi) \rangle_{x,y}}{\langle I(x, y) \rangle_{x,y}^2} - 1, \qquad (1)$$

where I is the intensity, ξ and ψ are the spatial increments in the x and y directions, respectively, and the angle bracket indicates the average over all the spatial locations in both x and y directions. Since smaller scale structures are not the subject of this study, the original 1024 $\times$ 1024 pixel images were binned 4 $\times$ 4, which reduce the images to 256 $\times$ 256 pixels. The image spatial correlation was computed for the binned 256 $\times$ 256 image. After binning, the correlation image corresponds to pixel size of 3.2 μm. More description of this method can be found at [20, 22]. The binning and spatial correlation analysis was done by SimFCS software developed at the Laboratory for Fluorescence Dynamics (available at http://www.lfd.uci.edu/).

2.5. Image Analysis—Estimation of the Collagen Compaction Level. In this study, we have found no specific directionality of collagen fibers and organization, which results in symmetrical spatial correlation images along both x axis and y axis (see, e.g., Figure 1, the lower left panel of the correlation image). Hence, the cross-sections of x-axis and y-axis were used for fitting.

A script was written in MATLAB (MathWorks, Natick, MA) to perform a fit of the cross-section of the image spatial correlation function. Figure 1 lower right panel shows that we need at least 2 Gaussian functions for the fitting, one very narrow that corresponds to the size of the pixel (the green dashed curve in Figure 1 lower right panel) and a broad Gaussian which corresponds to the low spatial frequency fluctuation of the collagen matrix (the blue dashed curve in Figure 1 lower right panel). The amplitude and standard deviation of the broad Gaussian (A, σ) were extracted. The average peak amplitude of low spatial frequency components is 4.5 (the average of A). For the images with A smaller than 0.001, the value of σ cannot be reliably measured. In the following, we mainly discuss the σ value, which reports the mesoscale collagen spatial compaction level. Note that in this repot we used the Gaussian fit not as a model of the spatial frequency fluctuations but as an algorithmic approach to estimate the spatial extent of the SHG image structure.

2.6. Data Plotting and Statistics. Data plotting was done using MATLAB. For each data group, the highest and lowest σ values were removed before analysis. The error bars on all graphs were one standard deviation. Two-sample t-test was used to compare data. We report the data as significantly different if the P value is smaller than 0.01.

3. Results

To evaluate the impact of ECM remodeling by MCF-7, NIH/3T3, and MDA-MB-231 cells, type I collagen matrices with cells embedded were imaged using two-photon microscopy after one, two, and four days of collagen polymerization. SHG z-stack images covered 200 μm depth (10 z slices in total) and 819 μm of both width and length were collected and analyzed as shown in Figure 1. Collagen matrix with no cells was prepared under the same polymerization and incubation conditions for baseline comparison.

Figures 2(a)–2(d) shows the collagen matrix SHG images of collagen alone, collagen with MCF-7, NIH/3T3, and MDA-MB-231 cell line, respectively. Contrary to the relatively homogenous collagen fiber distribution from collagen matrix without cell seeding shown in Figure 2(a), collagen organization heterogeneity can be clearly seen after one day of MDA-MB-231 cell seeding, and the collagen compaction level at some local regions increased over time (Figure 2(d)). On the other hand, the less-invasive breast cancer cell line MCF-7 and fibroblast cell line NIH/3T3 have more mild effects on collagen compaction level (Figures 2(b) and 2(c)).

The association of the high collagen matrix density region with the cell locations can be seen in Figure 2(e), where the collagen fibers (magenta) were densely packed around MDA-MB-231 cells (green). Our aim is to quantitatively assess the average size of these pericellular collagen compactions due to cell-ECM remodeling. As described in the method section, to capture this mesoscale collagen matrix feature, we calculated image spatial correlation of SHG images (Figure 1) to statistically estimate the average size of collagen compaction. The image spatial correlation gives rise to a sharp peak, which corresponds to the pixel size, and a broader distribution, which reflects the actual size of the structure. The lower right panel in Figure 1 represents an example of spatial correlation with this property. To identify the characteristic scale of these two distinct spatial frequency features, we used two Gaussian components fitting to the spatial correlation images. The standard deviation value of the sharp Gaussian of the SHG spatial correlation, as shown in Figure 3, was not affected by cell ECM remodeling. It coincides with the value of the pixel size and does not have significant change over time. In contrast, the standard deviation of low frequency component reflects the average size of the collagen compaction, and

the amplitude is a function of signal-to-background ratio and of the proportion of the broad Gaussian.

Here we define σ as the standard deviation of low spatial frequency component (i.e., the broader distribution), which in our experiment was in the order of tens of micrometers. Within 4 days, the average σ of collagen matrix with MDA-MB-231 cells increases significantly from 17 μm in day 1, 21 μm in day 2, to 39 μm after 4 days of incubation (Figure 3). The average σ of collagen matrix without cells shows much smaller change, increasing by only 1 μm after 4 days compared to ~20 μm increase in the presence of MDA-MB-231 cells. Consistent with the hypothesis that less invasive tumor cells may cause less collagen compaction, we found that the collagen SHG images with MCF-7 cells embedded showed significantly smaller σ. The collagen remodeling ability of NIH/3T3 cell line lies between MCF-7 and MDA-MB-231 cell line. However, unlike MDA-MB-231 cell line, the change of σ over time was minimal for NIH/3T3 embedded collagen matrix.

We exclude the possibility of the value of σ due to the space occupied by cells based on the following. (1) The size indicated by σ is larger than cell body; (2) cell size does not change over time, but collagen compaction level increases dramatically over time; and (3) the cell size is comparable for the three cell lines examined.

To further assess the method's applicability to evaluate collagen remodeling, we performed Latrunculin B treatment and Marimastat treatment to MDA-MB-231 cells cultured in the type I collagen matrix. Latrunculin B affects the actin polymerization degree, which leads to the altered mechanical properties of the cell, including decreased contractile force [23]. Marimastat is a broad-spectrum matrix metalloproteinase inhibitor that blocks the activity of MMP-1, 2, 3, 7, 9, and 12 [24] and has been shown to affect collagen remodeling *in vitro*. With either Latrunculin B or Marimastat treatment, the collagen compaction level by MDA-MB-231 cells was significantly hampered (Figure 5). The treatment with Marimastat decreased the magnitude of collagen remodeling from the average $\sigma = 37\,\mu$m to $\sigma = 29\,\mu$m, while Latrunculin B showed even more severe effect that leaded to σ similar to MCF-7 cell line ($\sigma = 21\,\mu$m).

For a structure with given geometry in an image, the structure size is linearly correlated to the standard deviation of spatial correlation as obtained from the Gaussian fit. The resulting spatial correlation will have a standard deviation that is half of the original structure size due to the properties of the correlation function. The amplitude, on the other hand, is proportional to the area of the structure and signal-to-background ratio. The result from collagen SHG images used in this study, thus, represents a weighted average of complex patterns, with two major components: the pixel-size correlation due to the small collagen fiber width and its random orientation (Figure 3) and the collagen compaction area created by cell-matrix interaction, which is highly variable (Figures 4 and 5). Here, we demonstrated that by using spatial image correlation it is possible to quantify and differentiate the impact of cells on collagen matrix under different conditions. The estimated collagen compaction structure size from MDA-MB-231 cell after 4 days in collagen is around 70 μm.

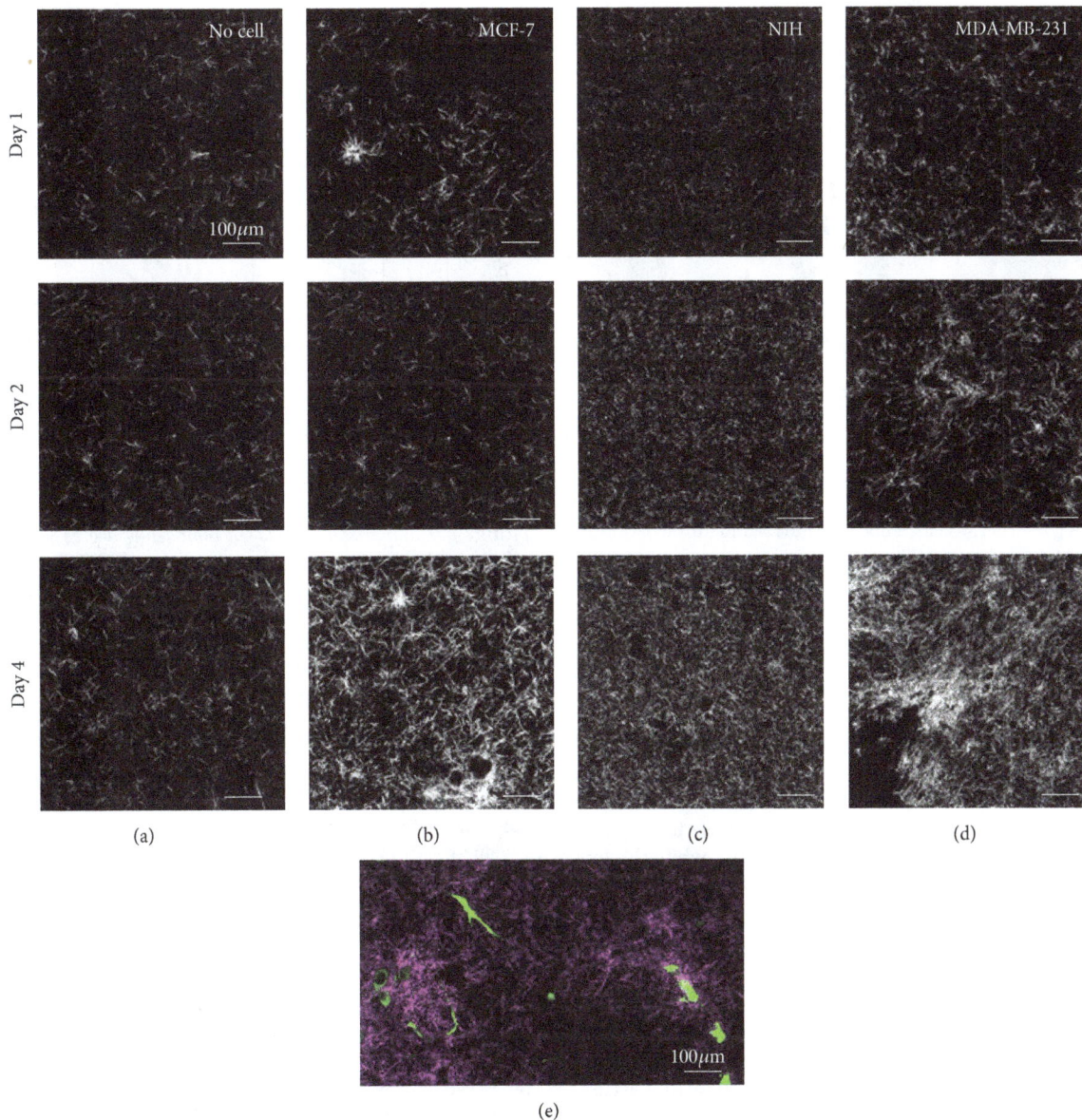

FIGURE 2: *Representative SHG images of collagen matrix with and without cells.* (a) Type I collagen matrix without cells after 1, 2, and 4 days of incubation. There is no recognizable collagen fiber distribution heterogeneity or collagen compaction. (b) Collagen matrix with MCF-7 cells after 1, 2, and 4 days of incubation. There is no distinguishable feature until day 4, in which collagen matrix with MCF-7 cells shows mild remodeling, although the majority of the collagen matrix was not compacted by cells. (c) Collagen matrix with NIH/3T3 cells after 1, 2, and 4 days of incubation. A higher degree of collagen compaction compared to the collagen matrix with MCF-7 cells can be seen since day 2. Higher collagen density was also observed, possibly due to collagen synthesis by the fibroblasts. (d) Collagen matrix with MDA-MB-231 cells after 1, 2, and 4 days of incubation. Collagen matrix shows significantly higher degree of remodeling compared to both MCF-7 and NIH/3T3 cell-embedded collagen matrices. (e) Actin-GFP labeled MDA-MB-231 cells cultured in type I collagen matrix for 4 days, showing that collagen matrix (magenta) has higher density close to cells (green). The higher density area around the cell is referred to as collagen compaction, and its size is quantitatively evaluated by image spatial correlation in this report.

Compared to an average cell body diameter of 20–30 μm, the collagen structure emerging after cell seeding may be much larger than the single cell.

4. Discussion

The interplay between the biophysical properties of the cell and ECM establishes a dynamic mechanical reciprocity between the cell and the ECM in which the cell's ability to exert contractile stresses against the ECM balances the elastic resistance of the ECM to deformation. In this sense, ECM is effectively a physical extension of the cell and cytoskeleton [25]. In this study, we described a method based on spatial correlation to quantify the degree of matrix remodeling at the tens of micron scale. We chose cell lines (MCF-7, NIH/3T3, and MDA-MB-231) that presumably could show a difference

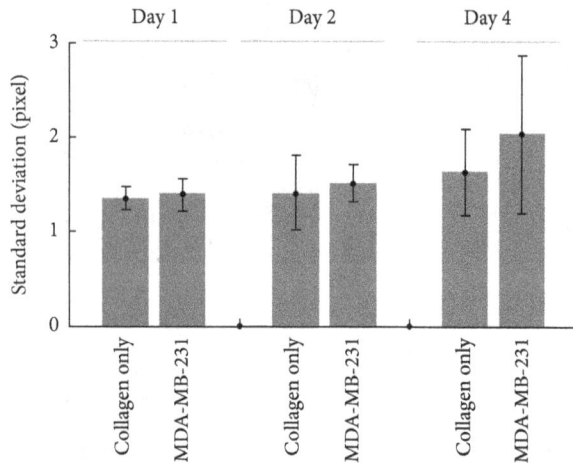

FIGURE 3: *The standard deviation of narrow peak is not significantly affected by collagen remodeling.* For the collagen matrix with or without MDA-MB-231 cells embedded, the standard deviation of the narrow peak from spatial correlation images (the green curve in Figure 1) did not show any significant difference, and its value was close to the pixel size.

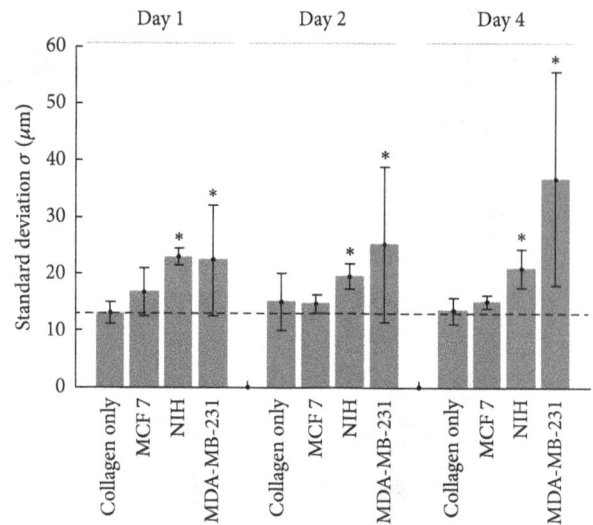

FIGURE 4: *Comparison of collagen remodeling degree with different cell lines.* The degree of collagen remodeling by MCF-7, NIH/3T3, and MDA-MB-231 cell lines over 4 days of incubation was evaluated by σ of the low spatial frequency component (the blue curve in Figure 1). Of all three cell lines, MDA-MB-231 showed the highest degree of collagen remodeling, and the collagen compaction degree increases over 4-day period. MCF-7 exhibits the lowest degree of collagen compaction and has the σ value similar to the collagen matrix without cells. Dashed line shows the average σ value of collagen matrix without cells at day 1. * indicates that the standard deviation is significantly different from the baseline control (collagen matrix without cells for each equivalent age of the sample, P value < 0.01 by t-test).

in regard to collagen compaction at the mesoscale. We were able to estimate the average collagen compaction area by MCF-7 cells, NIH/3T3 cells, and MDA-MB-231 cells. With the same initial type I collagen concentration, the more invasive cancer cell line, MDA-MB-231, was able to generate significantly larger collagen compaction areas, while the less invasive cell line (MCF-7) has a modest effect on the collagen matrix. The NIH/3T3 cell line exhibits the collagen remodeling ability between MCF-7 and MDA-MB-231 cell lines. This result is consistent with previous studies showing that tumor cell invasiveness is associated with increased contractile force generation [21], which affects the collagen remodeling ability. This notion was further supported by the significant decrease of σ when MDA-MB-231 cells were treated with Latrunculin B, which inhibits actin polymerization and decreases cellular contractility [23]. Blocking MMPs by Marimastat, although with smaller impact compared to the effect of actin cytoskeleton disruption, also showed significant alteration on the degree of collagen compaction. In fact, MMPs have been associated to tumor stroma and fibrosis *in vivo* [26, 27]. A recent research conducted by tracking bead displacement to assess ECM remodeling also showed the treatment of MMP inhibitor reduced the magnitude of ECM deformation [28], supporting our evaluation by spatial correlation method.

High breast tissue density due to increased type I collagen is one of the single largest risk factors for developing breast cancer [29]. This higher stiffness correlates with increased mammographic density and has been exploited to detect cancer [4, 30–32]. Matrix stiffness from type I collagen crosslinking has also been implicated as a contributor to the enhanced invasive behavior in tumor [33], enhanced cell growth and survival, and promoted migration. However, what causes the increase in tumor stromal stiffness and how stromal stiffness contributes to tumor progression is still not clear. In addition, the causality of matrix stiffness and tumor

invasiveness has yet to be determined. Our data indicate that tumor cells may be involved in ECM remodeling and contribute to the different degree of collagen compaction based on their invasiveness. This hypothesis is also supported by *in vivo* tumor collagen images which showed collagen stretching near invasive cancer cells [4], which is similar to what we observed *in vitro*.

In this report, we showed that the image spatial correlation method provides a measure of ECM spatial organization, in this example the collagen compaction by cells, at the scale comparable or larger than the cellular size. The advantages of the analysis method proposed in this report include (1) the simple statistical procedure that does not require image segmentation or other predefined parameters, which could be more subjective; (2) it does not depend on the intensity range of each image acquired, so the comparison between different images can be achieved without difficulty; and (3) the use of SHG to image collagen fibers that does not require extra labeling and could be directly applied to *in vivo* studies. Due to its simplicity of implementation, this method has the potential to not only facilitate *in vitro* ECM remodeling study or ECM assessment for tissue engineering purposes but can be also applied to *in vivo* research for determining regional collagen organization differences in tissues and tumor sites. By choosing the proper pixel size, it is possible to use this method to detect structures of different scales and to quantify

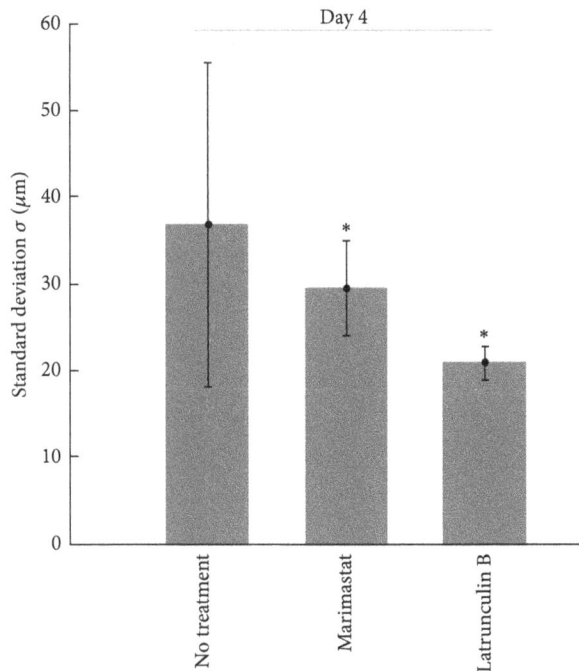

FIGURE 5: *MMP inhibitor and actin polymerization inhibitor affect MDA-MB-231 collagen remodeling ability*. MDA-MB-231 cells were cultured in collagen matrix and treated with Marimastat (MMP inhibitor) or Latrunculin B (actin polymerization inhibitor) for 4 consecutive days. Both treatments significantly compromised the collagen compaction ability of MDA-MB-231 cells. While the effect of Marimastat is milder, Latrunculin B treatment resulted in the compaction degree similar to collagen matrix with MCF-7 cells. * indicates the standard deviation is significantly different from the collagen matrix without cells (P value < 0.01 by t-test).

each component's significance. Provided there is enough axial resolution, image spatial correlation can be used in three-dimensions, making it possible to tell whether collagen exhibits different spatial organization along lateral and axial directions. Furthermore, although not shown in this report, the directional matrix alignment could be estimated by 2D Gaussian fitting to the spatial correlation image, which may increase the application of this method.

Conflict of Interests

The authors declare no conflict of interests.

Authors' Contribution

Chi-Li Chiu did the experiments and wrote the paper. Michelle A. Digman provided the samples and helped in analyzing the data. Enrico Gratton developed the software for the spatial correlation analysis and wrote part of the paper. All authors reviewed the paper.

Acknowledgments

The authors thank Jose S. Aguilar for helping in the sample preparation. Funding was provided by National Institutes of Health P50 GM076516. This project was supported by grants from the National Center for Research Resources (5P41RR003155-27) and the National Institute of General Medical Sciences (8 P41 GM103540-27).

References

[1] Z. Werb and J. R. Chin, "Extracellular matrix remodeling during morphogenesis," *Annals of the New York Academy of Sciences*, vol. 857, pp. 110–118, 1998.

[2] D. R. Senger and G. E. Davis, "Angiogenesis," *Cold Spring Harbor Perspectives in Biology*, vol. 3, no. 8, Article ID a005090, 2011.

[3] M. Abercrombie, M. H. Flint, and D. W. James, "Wound contraction in relation to collagen formation in scorbutic guinea-pigs," *Journal of Embryology and Experimental Morphology*, vol. 4, no. 2, pp. 167–175, 1956.

[4] P. P. Provenzano, K. W. Eliceiri, J. M. Campbell, D. R. Inman, J. G. White, and P. J. Keely, "Collagen reorganization at the tumor-stromal interface facilitates local invasion," *BMC Medicine*, vol. 4, article 38, 2006.

[5] J. B. Wyckoff, S. E. Pinner, S. Gschmeissner, J. S. Condeelis, and E. Sahai, "ROCK- and myosin-dependent matrix deformation enables protease-independent tumor-cell invasion in vivo," *Current Biology*, vol. 16, no. 15, pp. 1515–1523, 2006.

[6] M. J. Paszek, N. Zahir, K. R. Johnson et al., "Tensional homeostasis and the malignant phenotype," *Cancer Cell*, vol. 8, no. 3, pp. 241–254, 2005.

[7] K. R. Levental, H. Yu, L. Kass et al., "Matrix crosslinking forces tumor progression by enhancing integrin signaling," *Cell*, vol. 139, no. 5, pp. 891–906, 2009.

[8] Y. Noda, Y. Hata, T. Hisatomi et al., "Functional properties of hyalocytes under PDGF-rich conditions," *Investigative Ophthalmology and Visual Science*, vol. 45, no. 7, pp. 2107–2114, 2004.

[9] R.-I. Kohno, Y. Hata, S. Kawahara et al., "Possible contribution of hyalocytes to idiopathic epiretinal membrane formation and its contraction," *British Journal of Ophthalmology*, vol. 93, no. 8, pp. 1020–1026, 2009.

[10] M. Miura, Y. Hata, K. Hirayama et al., "Critical role of the Rho-kinase pathway in TGF-β2-dependent collagen gel contraction by retinal pigment epithelial cells," *Experimental Eye Research*, vol. 82, no. 5, pp. 849–859, 2006.

[11] C. M. Kraning-Rush, S. P. Carey, J. P. Califano, B. N. Smith, and C. A. Reinhart-King, "The role of the cytoskeleton in cellular force generation in 2D and 3D environments," *Physical Biology*, vol. 8, no. 1, Article ID 015009, 2011.

[12] D. Harjanto, J. S. Maffei, and M. H. Zaman, "Quantitative analysis of the effect of cancer invasiveness and collagen concentration on 3D matrix remodeling," *PLoS One*, vol. 6, no. 9, Article ID e24891, 2011.

[13] C. Bayan, J. M. Levitt, E. Miller, D. Kaplan, and I. Georgakoudi, "Fully automated, quantitative, noninvasive assessment of collagen fiber content and organization in thick collagen gels," *Journal of Applied Physics*, vol. 105, no. 10, Article ID 102042, 11 pages, 2009.

[14] A. M. Pena, D. Fagot, C. Olive et al., "Multiphoton microscopy of engineered dermal substitutes: assessment of 3-D collagen matrix remodeling induced by fibroblast contraction," *Journal of Biomedical Optics*, vol. 15, no. 5, Article ID 056018, 2010.

[15] M. D. Stevenson, A. L. Sieminski, C. M. McLeod, F. J. Byfield, V. H. Barocas, and K. J. Gooch, "Pericellular conditions regulate

extent of cell-mediated compaction of collagen gels," *Biophysical Journal*, vol. 99, no. 1, pp. 19–28, 2010.

[16] S. Roth and I. Freund, "Optical second-harmonic scattering in rat-tail tendon," *Biopolymers*, vol. 20, no. 6, pp. 1271–1290, 1981.

[17] E. Brown, T. McKee, E. DiTomaso et al., "Dynamic imaging of collagen and its modulation in tumors in vivo using second-harmonic generation," *Nature Medicine*, vol. 9, no. 6, pp. 796–800, 2003.

[18] A. Zoumi, A. Yeh, and B. J. Tromberg, "Imaging cells and extracellular matrix in vivo by using second-harmonic generation and two-photon excited fluorescence," *Proceedings of the National Academy of Sciences of the United States of America*, vol. 99, no. 17, pp. 11014–11019, 2002.

[19] M. Strupler, A.-M. Pena, M. Hernest et al., "Second harmonic imaging and scoring of collagen in fibrotic tissues," *Optics Express*, vol. 15, no. 7, pp. 4054–4065, 2007.

[20] C. B. Raub, J. Unruh, V. Suresh et al., "Image correlation spectroscopy of multiphoton images correlates with collagen mechanical properties," *Biophysical Journal*, vol. 94, no. 6, pp. 2361–2373, 2008.

[21] C. T. Mierke, D. Rösel, B. Fabry, and J. Brábek, "Contractile forces in tumor cell migration," *European Journal of Cell Biology*, vol. 87, no. 8-9, pp. 669–676, 2008.

[22] M. A. Digman and E. Gratton, "Analysis of diffusion and binding in cells using the RIGS approach," *Microscopy Research and Technique*, vol. 72, no. 4, pp. 323–332, 2009.

[23] T. Wakatsuki, B. Schwab, N. C. Thompson, and E. L. Elson, "Effects of cytochalasin D and latrunculin B on mechanical properties of cells," *Journal of Cell Science*, vol. 114, no. 5, pp. 1025–1036, 2001.

[24] R. Hoekstra, F. A. L. M. Eskens, and J. Verweij, "Matrix metalloproteinase inhibitors: current developments and future perspectives," *Oncologist*, vol. 6, no. 5, pp. 415–427, 2001.

[25] S. Kumar, I. Z. Maxwell, A. Heisterkamp et al., "Viscoelastic retraction of single living stress fibers and its impact on cell shape, cytoskeletal organization, and extracellular matrix mechanics," *Biophysical Journal*, vol. 90, no. 10, pp. 3762–3773, 2006.

[26] A. Y. Strongin, "Mislocalization and unconventional functions of cellular MMPs in cancer," *Cancer and Metastasis Reviews*, vol. 25, no. 1, pp. 87–98, 2006.

[27] Y. Y. Li, C. F. McTiernan, and A. M. Feldman, "Interplay of matrix metalloproteinases, tissue inhibitors of metalloproteinases and their regulators in cardiac matrix remodeling," *Cardiovascular Research*, vol. 46, no. 2, pp. 214–224, 2000.

[28] R. J. Bloom, J. P. George, A. Celedon, S. X. Sun, and D. Wirtz, "Mapping local matrix remodeling induced by a migrating tumor cell using three-dimensional multiple-particle tracking," *Biophysical Journal*, vol. 95, no. 8, pp. 4077–4088, 2008.

[29] N. F. Boyd, G. S. Dite, J. Stone et al., "Heritability of mammographic density, a risk factor for breast cancer," *The New England Journal of Medicine*, vol. 347, no. 12, pp. 886–894, 2002.

[30] D. T. Butcher, T. Alliston, and V. M. Weaver, "A tense situation: forcing tumour progression," *Nature Reviews Cancer*, vol. 9, no. 2, pp. 108–122, 2009.

[31] M. H. Zaman, L. M. Trapani, A. Siemeski et al., "Migration of tumor cells in 3D matrices is governed by matrix stiffness along with cell-matrix adhesion and proteolysis," *Proceedings of the National Academy of Sciences of the United States of America*, vol. 103, no. 29, pp. 10889–10894, 2006.

[32] R. Sinkus, J. Lorenzen, D. Schrader, M. Lorenzen, M. Dargatz, and D. Holz, "High-resolution tensor MR elastography for breast tumour detection," *Physics in Medicine and Biology*, vol. 45, no. 6, pp. 1649–1664, 2000.

[33] M. R. Ng and J. S. Brugge, "A stiff blow from the stroma: collagen crosslinking drives tumor progression," *Cancer Cell*, vol. 16, no. 6, pp. 455–457, 2009.

8

Reduced Dynamic Models in Epithelial Transport

Julio A. Hernández

Sección Biofísica, Facultad de Ciencias, Universidad de la República, Iguá esq. Mataojo, 11400 Montevideo, Uruguay

Correspondence should be addressed to Julio A. Hernández; jahern@fcien.edu.uy

Academic Editor: Andreas Herrmann

Most models developed to represent transport across epithelia assume that the cell interior constitutes a homogeneous compartment, characterized by a single concentration value of the transported species. This conception differs significantly from the current view, in which the cellular compartment is regarded as a highly crowded media of marked structural heterogeneity. Can the finding of relatively simple dynamic properties of transport processes in epithelia be compatible with this complex structural conception of the cell interior? The purpose of this work is to contribute with one simple theoretical approach to answer this question. For this, the techniques of model reduction are utilized to obtain a two-state reduced model from more complex linear models of transcellular transport with a larger number of intermediate states. In these complex models, each state corresponds to the solute concentration in an intermediate intracellular compartment. In addition, the numerical studies reveal that it is possible to approximate a general two-state model under conditions where strict reduction of the complex models cannot be performed. These results contribute with arguments to reconcile the current conception of the cell interior as a highly complex medium with the finding of relatively simple dynamic properties of transport across epithelial cells.

1. Introduction

The transport of water and solutes across epithelia is a relevant physiological property of higher organisms. To perform transport, epithelial cells develop a polarized distribution of membrane molecules, which localize at distinct apical and basolateral domains of the plasma membrane [1, 2]. The analysis and interpretation of quantitative data about solute and water transport across epithelia have constituted a major objective of cell physiologists. The majority of the models classically developed to represent solute transport across epithelia have considered that the interior of the epithelial cells constitutes a well-stirred, homogeneous compartment, characterized by a single value of concentration of the transported species [3–5]. This view implicitly assumes that the intracellular diffusion coefficient of the species remains constant and that diffusion occurs freely and rapidly, so that the intracellular solute concentrations attain a single equilibrium value at a faster time scale than the overall process. This conception differs markedly from the current view about the structural and functional characteristics of the cell interior. In this conception, the intracellular compartment is regarded as a highly crowded media of marked structural and

functional heterogeneity [6–8]. The effects of macromolecular crowding and structural organization on the activity of macromolecules and smaller dissolved species represent major topics for the understanding of the cellular behavior [9]. Realistic approaches to describe diffusion in cellular media require computational simulations that employ, for instance, Brownian dynamics [10, 11], finite-element methods [12, 13], or The Virtual Cell framework [14, 15].

Can the finding of relatively simple dynamic properties of transport processes in epithelia be compatible with the complex structural conception of the cell interior? The general objective of this work is to contribute with the basic aspects of one formal theoretical approach to answer this question. In particular, this study employs mathematical modeling to uncover properties that could be employed to measure structural cellular complexity. Since a detailed computer simulation of the solute movement throughout the intracellular medium would, although more realistic, not be easy to incorporate in a representation of the overall transport process, this study adopts a simpler approach which, nevertheless, may provide some basic conclusions. In this way, as an alternative to explicit computational simulations of the intracellular media, this study assumes that the unidirectional solute movement can approximately be represented by a discrete, multicompartment model. To be noted, discrete approaches to describe flow through nonhomogeneous media have been employed to understand the basic aspects of percolation [16]. Similarly, the simple approach adopted here represents an initial attempt to reconcile macroscopic physiological evidence with microscopic cellular complexity. The multicompartment representation permits to express the transition of the solute between adjacent intracellular compartments via kinetic expressions; the overall dynamics are, therefore, governed by a system of linear differential equations. Multicompartmental strategies have been utilized, for instance, to understand the role of diffusion in brain processes [17] and to describe sarcomeric calcium movement [18].

In essence, the findings of this theoretical study suggest that the basic processes of transcellular transport across an epithelial cell between the two extracellular compartments may be reduced to an equivalent two-state linear model. The strategy of model reduction represents an alternative to study discrete systems with a high degree of complexity, such as biochemical networks, and permits to derive models that retain some of the relevant system properties under specific conditions. In this respect, linear systems of a relatively large number of components can be handled in a rather straightforward fashion. Thus, in macromolecular systems the reduction of linear intermediate transitions of multistate diagrams to yield simpler models has provided a tool, for instance, to understand the finding of simple kinetic behaviors in complex membrane transport systems [19]. In epithelial transport, nonlinearity may emerge as a consequence of interactions between different transported species or from the existence of feedback mechanisms, such as those involved in crosstalk responses [20–22]. The loss of linearity underlies the emergence of more complex behaviors of multistate systems and their reduction may inevitably require the design of alternative computational strategies [23]. In the present work, only the basic aspects of transcellular transport across epithelial cells are considered, which permit to conform a linear model with an arbitrary number of intermediate intracellular states. In this study, techniques analogous to the ones utilized for linear macromolecular kinetics are employed to obtain reduced two-state models from the original multistate ones [19]. The numerical simulations performed here also permit to obtain the noteworthy result that, under conditions where strict model reduction does not occur, an equivalent pseudo-two-state dynamic model can nevertheless be approximated. These results contribute with some arguments to reconcile the current conception of the cell interior as a highly complex media with the finding of relatively simple dynamic properties of transport across epithelial cells.

2. Models of Transcellular Transport of Solutes across Epithelial Cells

One of the simplest models of transcellular transport of a solute across an epithelial cell (e.g., an intestinal cell) is

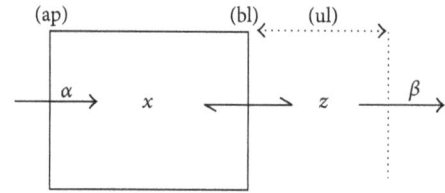

FIGURE 1: Scheme of an epithelial cell performing transcellular transport of a solute. The solute enters the cell at the apical membrane (ap) via active transport at rate α and exits at the basolateral membrane (bl) to an adjacent unstirred layer (ul), from which it is extracted at rate β. One of the simplest situations is represented, where the cell interior is assumed to be a homogeneous compartment characterized by a single value of solute concentration (x). The extracellular unstirred layer is also characterized by a single value of the solute concentration (z).

depicted in Figure 1. In this scheme, a solute (e.g., glucose) is being driven inside the cell via an active transport system of the apical membrane (e.g., the Na-glucose cotransporter). Under physiological conditions, this transport system is assumed to operate irreversibly at rate α. In this model, the strict homogeneous condition of the cell applies; that is, the solute concentration x is the same throughout the whole intracellular compartment. The solute is driven out of the cell at the basal domain via a passive, reversible transport system (e.g., the glucose transporter (GLUT2)). In the model considered (Figure 1), the solute accumulates in the unstirred layer adjacent to the basal membrane at a concentration z and exits this compartment at rate β. Models exploring the possible role of unstirred layers at the extracellular cell surface in transcellular transport have been developed, for instance, to explain contradictory data about solute and solvent coupling in epithelia [24]. It is not the objective of this work to contribute to the discussion of the importance of unstirred layers in explaining quantitative data about epithelial solute and water transport, a matter that has received attention in the past [25], but to consider plausible models of transcellular transport of a single dissolved solute for illustrative purposes. An alternative to the meaning of the intermediate state z is to assume that it directly corresponds to the solute concentration at the apical extracellular compartment. In this case, β would represent its rate of extraction from other tissues. The elementary dynamic model governing the transport process described in Figure 1 (Model I) is shown in more detail in Figure 2(a) and is given by

$$x' = \alpha - k_{1f}x + k_{1b}z,$$
$$z' = k_{1f}x - (k_{1b} + \beta)z,$$

(1)

where $x'(z')$ denotes the time derivative of $x(z)$ and, α, k_{1f}, k_{1b}, and β are (positive) rate parameters. The solution and basic properties of this model are given in Appendix A, solely as a reference to the studies performed in this work. It can be easily concluded from the study of the explicit solution (Appendix A) or from the stability analysis (not shown) that the steady state of this model represents an asymptotically

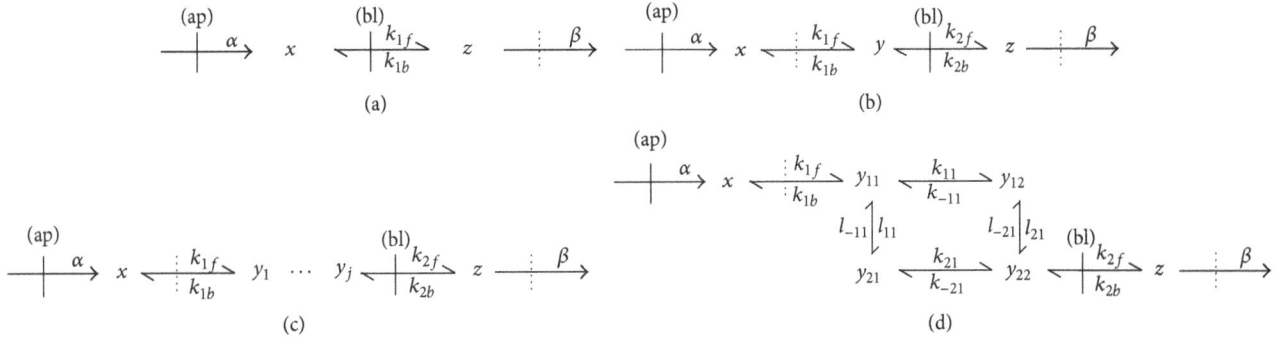

FIGURE 2: Different models of transcellular solute transport across an epithelial cell (cf. Figure 1). The dotted lines denote limits of unstirred layers (cf. Figure 1), the rest of the symbols as in Figure 1. The model in (a) (Model I) corresponds to the situation depicted in Figure 1. (b) shows a more complex model (Model II), where the solute distributes into two distinctive intracellular compartments (at concentrations x and y). In the general scheme of (c), the solute may distribute along a larger number of intracellular compartments (at concentrations x and y_1 to y_j). In particular, this work considers models having from two (Model III) to four (Model V) intermediate states "y." (d) corresponds to a simple case of bidimensional intracellular distribution of the solute (Model VI).

stable configuration, a characteristic property of integrated systems of membrane transport in general [26–29].

A somewhat more complex model can be obtained if one assumes, for instance, that an unstirred layer additionally exists at the intracellular surface of the apical membrane (Figure 2(b)). In this case, the solute accumulates at this layer at concentration x and then diffuses reversibly to the rest of the cell, where it achieves a uniform concentration y. As in the previous case, it is then transported reversibly to the extracellular space at the level of the basal domain. The corresponding dynamic model (Model II) is given by

$$x' = \alpha - k_{1f}x + k_{1b}y,$$

$$y' = k_{1f}x - \left(k_{1b} + k_{2f}\right)y + k_{2b}z, \qquad (2)$$

$$z' = k_{2f}y - \left(k_{2b} + \beta\right)z,$$

where α, k_{1f}, k_{1b}, k_{2f}, k_{2b}, and β are rate parameters. Assuming that y is a quasistationary intermediate, the model given by (2) can be reduced to a simple two-state model formally analogous to Model I (Appendix B).

As mentioned above (Section 1), the present view about the intracellular compartment is far from the homogeneous, dilute perspective classically invoked to perform quantitative interpretations of cellular transport properties. A more realistic conception of the cell interior implies a highly crowded, heterogeneous media where instant equilibration to a unique intracellular concentration of a specific species may not represent a realistic approximation. Figure 2(c) depicts a general model of unidirectional intracellular transport that assumes the existence of several intermediate internal compartments for the transported species. These successive compartments, extending to the rest of the cell starting from the unstirred layer at the intracellular apical domain, are characterized by specific concentrations (y_1 to y_j) of the transported species. The transitions between adjacent compartments are reversible and governed by first-order rate constants. In this work, we shall further consider models of the general type of Figure 2(c), ranging from two (i.e.,

y_1, y_2) to four ($y_1, \ldots, y_4$) intermediate states, to perform some numerical studies (see below). These models shall be designated as Models III to V, respectively. An extension of the unidimensional model to more realistic situations would consider the inclusion of a larger number of intermediate states. Still further complexity is attainable if one assumes two-dimensional distribution of the intracellular solute. As an example, Figure 2(d) shows a situation where the solute is distributed inside the cell, apart from the unstirred layer at the apical membrane, into the simplest two-dimensional network of intermediate states (Model VI). More complex configurations in the two- and even three-dimensional domains are certainly conceivable, but their analysis would require the employment of alternative procedures, such as Monte Carlo simulations. Appendix B illustrates the procedures of linear model reduction [19, 30] employing Models II and VI as examples. It is shown there that, under some conditions, complex models of the type of Models II to VI can be reduced to a simple two-state model qualitatively similar to Model I (A.1). As examples, (B.1) and (B.2) give the expressions obtained for the reduced rate constants r_{12} and r_{21} (A.1) for the cases of Models II and VI, respectively.

In order to illustrate the concepts introduced here, the next section contains numerical studies of some dynamic properties of the models. Of particular interest is the finding that, under conditions not permitting strict reduction, the models nevertheless exhibit a dynamic behavior approximately equivalent to a two-state dynamic model governed by the general equations (A.1).

3. Numerical Results and Discussion

In this section, numerical studies are performed to compare the dynamic behaviors between the original and the reduced models, for the different models considered and for different values of some of the parameters. In essence, the procedure followed here consists in simulating the time courses of the model dynamics in response to perturbations from the steady state. The results shown are not exhaustive and only

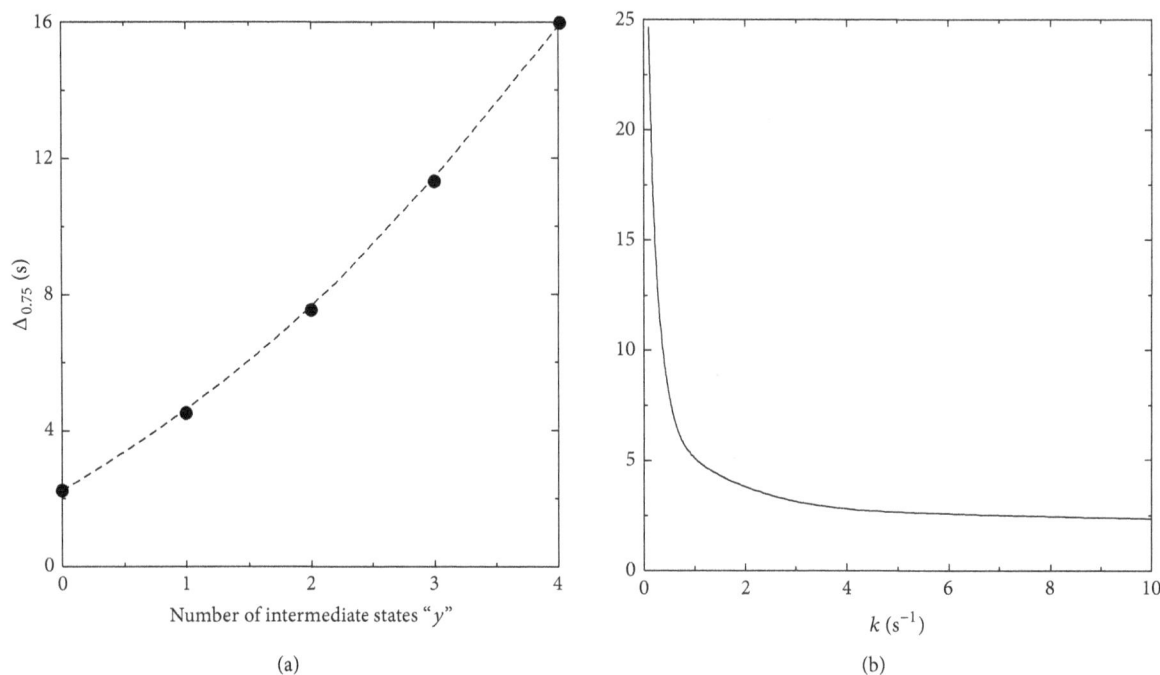

FIGURE 3: Plots of the time delay $\Delta_{0.75}$ versus (a) the number of intermediate states "y" in Models I to V for $k = 1$ and versus (b) k for the case of Model II ($\Delta_{0.75}$: time elapsed between $t = 0$ and t for $z/z^* = 0.75$; $k = k_{1f} = k_{2b}$, cf. Figure 2). The numerical integrations were performed employing the Runge-Kutta fourth-order method. For every run, for $t = 0$, $z/z^* = 0.5$. For every case, $\alpha = 10$ and the numerical value for the rest of the parameters was 1.

intended to illustrate some basic properties of the models. For this reason, the numerical values employed here for the rate constants are arbitrary and only results of the relative variations of the variable z with respect to the steady-state value (z^*) are shown. For the choice of the numerical values, the only restrictive condition assumed was that the parameter α should have a larger value than the rest of the parameters, since it represents the rate of active transport of the solute (cf. Figure 1).

The increasing complexity of the models (i.e., from Model I towards Model V, Figure 2) in turn determines modifications in properties that may have physiological significance, such as the time delay to achieve the steady state from an initial perturbed condition. Figure 3 shows the effect of increasing complexity and of the rate constants on the time delay to achieve the steady state. As expected, the increasing complexity (measured by the number of intermediate states "y," Figure 3(a)) determines an increase in the time delay, while the rise in some of the intermediate rate constants produces the opposite effect (Figure 3(b)). Since the dynamic behavior of the complex model may be indistinguishable from that of a two-state model (Figures 1 and 2(a)), either by satisfying the conditions of model reduction or by approximate behavior (see below), measurements of the actual values of the time delays, if possible, may provide clues to infer the degree of structural complexity of the cellular transport system.

Figures 4 and 5 show the dynamic responses of Models II and VI (Figure 2), respectively, to perturbations of the steady

state. The figures display the numerical integrations of the complete models, the corresponding reduced models ((A.1) and (B.1) for Model II and (A.1) and (B.2) for Model VI), and the approximations to the complete models (A.6). For the two models, the numerical integration of (A.1) yielded similar results to the direct numerical solution of (A.2). As can be seen in Figures 4 and 5, for parameter values satisfying the necessary reduction conditions (Appendix B), the strictly reduced models yield dynamic behaviors undistinguishable from the original ones (Figures 4(c) and 5(c)). The necessary conditions for model reduction may be somewhat unrealistic, however, since they imply the quasistationary hypothesis for the intermediate states (Appendix B). It is, therefore, a noteworthy result that, for values not satisfying the reduction conditions (Figures 4(a), 4(b), 5(a), and 5(b)), the numerical studies nevertheless permitted to approximate two-state models by the simple procedure of introducing an adjust factor Ψ to the time constants of the corresponding reduced models. This property, not further analyzed here, is possibly a consequence of the linear character of the model. As revealed by Figures 4(d) and 5(d), for large values of k (i.e., far from the reduction conditions) low values of Ψ are required to obtain a proper approximation to the original model behavior. The figures also show that, in order to obtain that approximation, Ψ tends to unity as k tends to zero. Thus, Ψ may represent a measurement of the degree of complexity of the original model, since its value depends on how distant the actual model dynamics are from the reduction conditions.

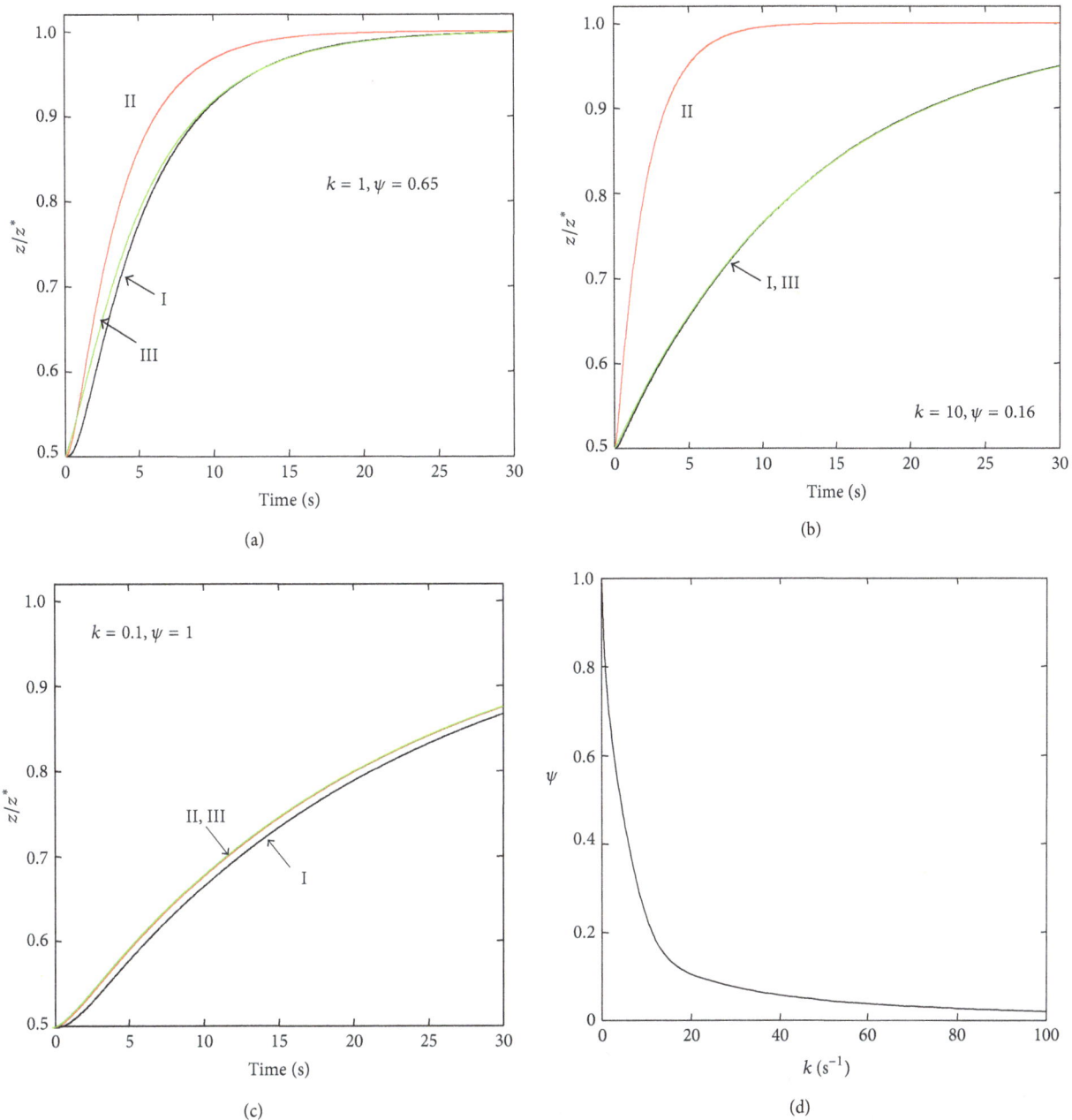

FIGURE 4: (a)–(c) Plots of z/z^* versus time for the case of Model II and for different values of k. Numerical integrations: Curve I, (2); Curve II, (A.1) and (B.1). Direct numerical solutions: curve superposed to Curve I, (A.2); Curve III, (A.6). (d) Plot of ψ versus k. Numerical methods: numerical values of the rest of the parameters and definition of k are the same as in Figure 3. For each value of k, the numerical value for the adjust factor ψ (Appendix A) was obtained by trial and error in order to attain the best approximation to Curve I. The curve ψ versus k (d) was obtained as the best fit to a sample of values of ψ for the corresponding values of k, throughout the whole range of values of k considered.

Similar results to the ones displayed in Figures 4 and 5 were obtained for Models III to V (not shown). In particular, in every case it was possible to empirically approximate a two-state model to the complete one when strict conditions for model reduction did not apply. Simulations performed for different values of the intermediate rate parameters (although conserving the rule that α should be larger than the rest of the parameters), also permitted to obtain reasonable approximations to Model II employing (A.6) (not shown). These results suggest that, at least for the case of some processes of epithelial transport of solutes, it is possible to describe these processes by a relatively simple model of the general type given by (A.1). However, the results of this work also suggest that, under these circumstances, it is not possible to conclude that the actual underlying process strictly corresponds to the simple model described in Figure 1, characterized by a unique intracellular concentration of the solute. The actual intracellular distribution of the solute may

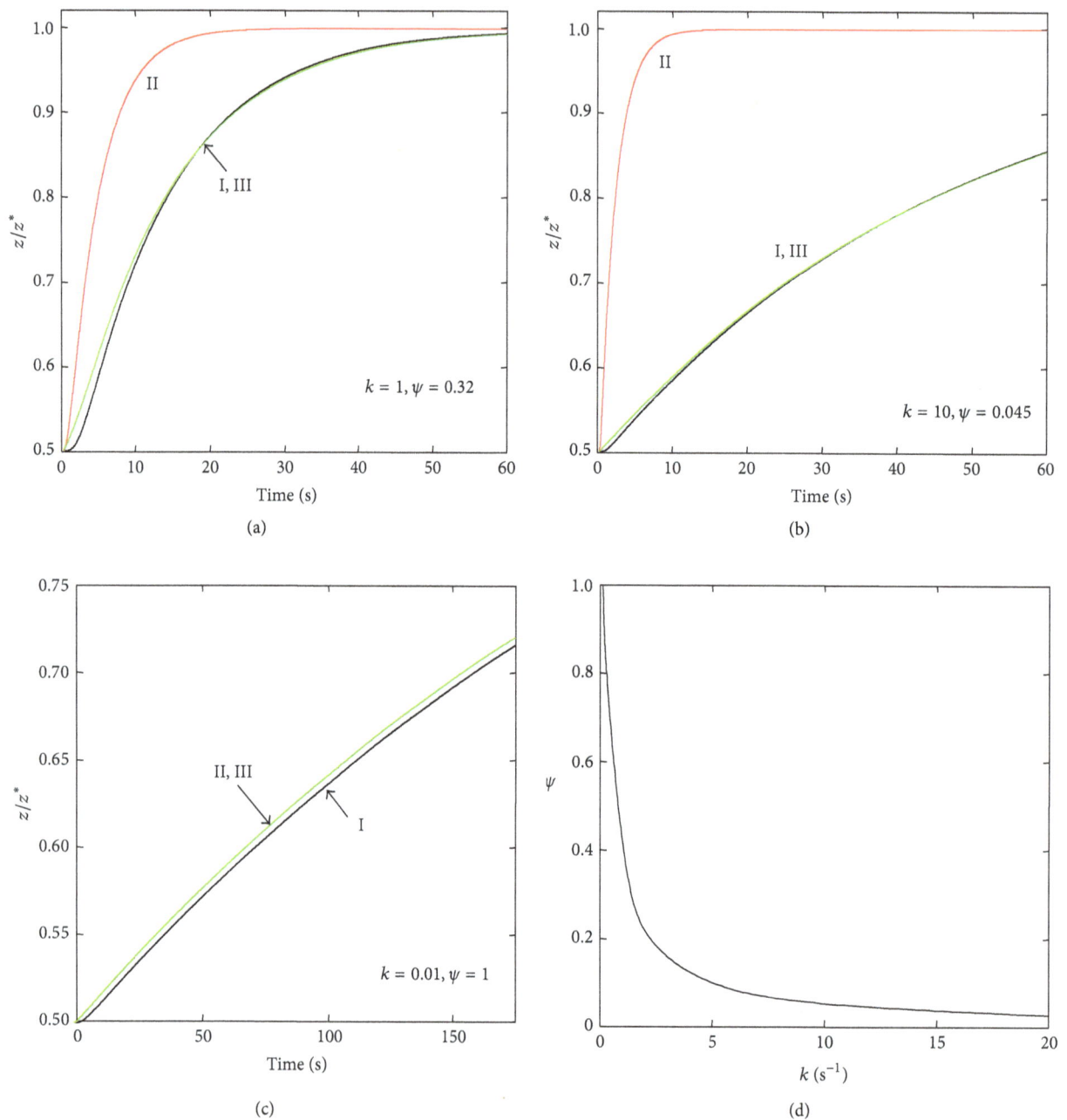

FIGURE 5: Similar to Figure 4, but for Model VI. In this case, Curve I corresponds to the numerical integration of the complete model shown in Figure 2(d) (explicit equations not shown in the text). In addition, (A.1) and (B.2) were employed to obtain Curve II.

be approximated by a more complex configuration, such as the ones represented by Models II to VI (Figure 2), or still more complex. As mentioned in Section 1, it must be emphasized that a discrete multicompartment description of the intracellular compartment can only be considered as an initial approach to represent this highly complex medium, an approach that may nevertheless be operative for the consideration of some specific issues, such as the one addressed in this study.

4. Conclusions

The results of this theoretical study permit to suggest that complex models of transepithelial transport of solutes may nevertheless exhibit dynamic properties undistinguishable from those of simple models. At least in the realm of linear models of transport, it was shown here that models incorporating several intermediate states of the solute in the intracellular compartment may, under the proper conditions, be reduced to simple two-state models that assume the existence of a unique concentration value for the intracellular solute. Even if those reduction conditions are not accomplished, the numerical studies also permitted to obtain approximate two-state dynamic models to mimic the original complex ones. Taken together, the results of this work permit to ascertain that, at least for the case of the elementary processes of epithelial transport of solutes, it may be possible to reconcile the finding of relatively simple transcellular transport dynamics with the current conception of the cell interior as a highly complex structural media.

This work has, therefore, focused on the case of the transport of solutes across epithelial cells to illustrate that complex models of transport can exhibit dynamic behaviors undistinguishable from those of simple ones. The occasional emergence of relatively simple dynamic properties may be a property encountered for the case of many other complex biological processes, such as transitions between macro-molecular states [19] and reactions in biochemical networks [23]. As illustrated in this work, at least in some cases, a possible means to understand the emergence of relatively simple behaviors in complex dynamical systems can be obtained in a rather straightforward fashion by employing standard techniques of reduction of dynamic models.

Appendices

A. Solution of the General Two-State Model of Transepithelial Solute Transport

As a reference to this study, this section resumes the basic properties of the general two-state model of transepithelial solute transport. This model is given by

$$\frac{dx}{dt} = \alpha - r_{12}x + r_{21}z,$$
$$\frac{dz}{dt} = r_{12}x - (r_{21} + \beta)z,$$
(A.1)

where α and β have the same meanings for all the models (Figures 1 and 2) and where the r's are, in general, reduced rate constants (see Appendix B). For the particular case of the simplest model ((1), Model I), $r_{12} = k_{1f}$ and $r_{21} = k_{1b}$.

The solution of the system given by (A.1) can be obtained employing standard procedures. For the case of z, the solution reads

$$z(t) = z^* + C_1 \exp(m_1 t) + C_2 \exp(m_2 t),$$
(A.2)

where z^* is the steady-state value of $z[z^* = \alpha/\beta]$ and where

$$m_{1,2} = \frac{\left[-\Phi \pm (\Delta)^{1/2}\right]}{2},$$
with $\Phi = (r_{12} + r_{21} + \beta)$,
$$\Delta = \Phi^2 - 4r_{12}\beta.$$
(A.3)

Since $\Phi > (\Delta)^{1/2}$, m_1 and m_2 are necessarily negative. Hence, from any initial value of z, the system given by A.1 converges asymptotically to the steady-state value z^*.

For the case that $z(0) = z^*/2$,

$$C_1 = -\left(\frac{z^*}{2}\right)(1+K), \qquad C_2 = \left(\frac{z^*}{2}\right)K,$$
(A.4)
With $K = \frac{[m_1(1+m_1) + r_{12}\beta]}{\{\Phi[m_2(1+m_2) - m_1(1+m_1)]\}}.$

In the numerical studies, an approximate solution to the dynamics of a complete (i.e., nonreduced) model can be obtained by introducing an adjust factor Ψ to the parameters m_1 and m_2 of the corresponding reduced model:

$$\mu_1 = \Psi m_1, \qquad \mu_2 = \Psi m_2.$$
(A.5)

The approximate solution reads

$$z(t) = z^* + C_1 \exp(\mu_1 t) + C_2 \exp(\mu_2 t),$$
(A.6)

where C_1 and C_2 are the same as in the corresponding reduced model (A.2).

B. Reduction of Linear Dynamic Models of Transepithelial Transport

This section summarizes the procedure to reduce dynamic models of transcellular transport, for the case that some states of the transported species are transient intermediates. The method described in this section is based upon the techniques originally developed by Hill [30] for the reduction of linear sequences of transitions in biochemical systems and further extended to more complex configurations [19]. Instead of deriving general expressions, the procedure is illustrated employing Models II and VI as examples (Figures 2(b) and 2(d)). The method described here can be adapted in a straightforward manner to handle more complicated schemes. For a more detailed exposition of the reduction technique, the reader may consult [19].

For the case of Model II, if y is a transient intermediate ((2)), we may assume that, at any time, $dy/dt = 0$. This requires that $k_{1b} \gg k_{1f}$ and $k_{2f} \gg k_{2b}$. From this condition, (2) can be transformed into a system of the form of (A.1), with the reduced constants r_{12} and r_{21} given in this case by

$$r_{12} = \frac{k_{1f}k_{2f}}{(k_{1b} + k_{2f})}, \qquad r_{21} = \frac{k_{1b}k_{2b}}{(k_{1b} + k_{2f})}.$$
(B.1)

Analogously, the more complex models (Models III to VI) can be reduced to the general two-state model given by (A.1) under the condition that all the states "y" are transient intermediates. For example, for the case of Model VI (Figure 2(d)) this is achieved if the rate constants k_{1f} and k_{2b} are significantly smaller than the other constants. The expressions for the reduced rate constants r_{12} and r_{21} that can be obtained for Model VI under this condition are the following (cf. Figure 2(d)):

$$r_{12} = \frac{D_{12}}{D}, \qquad r_{21} = \frac{D_{21}}{D}$$
(B.2)

with

$$D_{12} = k_{1f}\, k_{2f} \left[k_{11} l_{21} \left(k_{21} + l_{-11} \right) \right.$$
$$\left. + l_{11} k_{21} \left(l_{21} + k_{-11} \right) \right],$$

$$D_{21} = k_{1b}\, k_{2b} \left[k_{-11} l_{-21} \left(k_{21} + l_{-11} \right) \right.$$
$$\left. + l_{-11} k_{-21} \left(l_{21} + k_{-11} \right) \right],$$

$$D = \left(k_{1b}\, k_{2f} \right) \left(k_{21} + l_{-11} \right) \left(l_{21} + k_{-11} \right) \qquad \text{(B.3)}$$
$$+ k_{1b} \left[k_{-11} l_{-11} \left(k_{-21} + l_{-21} \right) + l_{-11} l_{21} k_{-21} \right.$$
$$\left. + k_{-11} k_{21} l_{-21} \right]$$
$$+ k_{2f} \left[k_{21} l_{21} \left(k_{11} + l_{11} \right) \right.$$
$$\left. + k_{11} l_{-11} l_{21} + k_{-11} k_{21} l_{11} \right].$$

Acknowledgments

This work was supported by Grants from the Programa para el Desarrollo de las Ciencias Básicas (PEDECIBA) and from the Comisión Sectorial de Investigación Científica (CSIC), Universidad de la República, Uruguay.

References

[1] C. Yeaman, K. K. Grindstaff, and W. J. Nelson, "New perspectives on mechanisms involved in generating epithelial cell polarity," *Physiological Reviews*, vol. 79, no. 1, pp. 73–98, 1999.

[2] W. J. Nelson, "Epithelial cell polarity from the outside looking in," *News in Physiological Sciences*, vol. 18, no. 4, pp. 143–146, 2003.

[3] G. Whittembury and L. Reuss, "Mechanisms of coupling of solute and solvent transport in epithelia," in *The Kidney: Physiology and Pathophysiology*, D. W. Seldin and G. Giebishch, Eds., pp. 317–360, Raven Press, New York, NY, USA, 2nd edition, 1992.

[4] S. G. Schultz, "A century of (epithelial) transport physiology: from vitalism to molecular cloning," *American Journal of Physiology*, vol. 274, no. 1, pp. C13–C23, 1998.

[5] L. G. Palmer and O. S. Andersen, "The two-membrane model of epithelial transport: Koefoed-Johnsen and Ussing (1958)," *Journal of General Physiology*, vol. 132, no. 6, pp. 607–612, 2008.

[6] D. S. Goodsell, "Inside a living cell," *Trends in Biochemical Sciences*, vol. 16, pp. 203–206, 1991.

[7] K. Luby-Phelps, "Cytoarchitecture and physical properties of cytoplasm: volume, viscosity, diffusion, intracellular surface area," *International Review of Cytology*, vol. 192, pp. 189–221, 2000.

[8] J. A. Dix and A. S. Verkman, "Crowding effects on diffusion in solutions and cells," *Annual Review of Biophysics*, vol. 37, pp. 247–263, 2008.

[9] H. X. Zhou, G. Rivas, and A. P. Minton, "Macromolecular crowding and confinement: biochemical, biophysical, and potential physiological consequences," *Annual Review of Biophysics*, vol. 37, pp. 375–397, 2008.

[10] D. L. Ermak and J. A. McCammon, "Brownian dynamics with hydrodynamic interactions," *The Journal of Chemical Physics*, vol. 69, no. 4, pp. 1352–1360, 1978.

[11] M. Dlugosz and J. Trylska, "Diffusion in crowded biological environments: applications of Brownian dynamics," *BMC Biophysics*, vol. 4, no. 1, article 3, 2011.

[12] C. C. W. Hsia, C. J. C. Chuong, and R. L. Johnson, "Red cell distortion and conceptual basis of diffusing capacity estimates: finite element analysis," *Journal of Applied Physiology*, vol. 83, no. 4, pp. 1397–1404, 1997.

[13] P. Bauler, G. A. Huber, and J. A. McCammon, "Hybrid finite element and Brownian dynamics method for diffusion-controlled reactions," *Journal of Chemical Physics*, vol. 136, no. 16, Article ID 164107, 2012.

[14] I. I. Moraru, J. C. Schaff, B. M. Slepchenko et al., "Virtual Cell modelling and simulation software environment," *IET Systems Biology*, vol. 2, no. 5, pp. 352–362, 2008.

[15] I. L. Novak, P. Kraikivski, and B. M. Slepchenko, "Diffusion in cytoplasm: effects of excluded volume due to internal membranes and cytoskeletal structures," *Biophysical Journal*, vol. 97, no. 3, pp. 758–767, 2009.

[16] B. Berkowitz and R. P. Ewing, "Percolation theory and network modeling applications in soil physics," *Surveys in Geophysics*, vol. 19, no. 1, pp. 23–72, 1998.

[17] C. S. Patlak, F. E. Hospod, S. D. Trowbridge, and G. C. Newman, "Diffusion of radiotracers in normal and ischemic brain slices," *Journal of Cerebral Blood Flow and Metabolism*, vol. 18, no. 7, pp. 776–802, 1998.

[18] S. M. Baylor and S. Hollingworth, "Model of sarcomeric Ca^{2+} movements, including ATP Ca^{2+} binding and diffusion, during activation of frog skeletal muscle," *Journal of General Physiology*, vol. 112, no. 3, pp. 297–316, 1998.

[19] J. A. Hernández and J. C. Valle Lisboa, "Reduced kinetic models of facilitative transport," *Biochimica Et Biophysica Acta*, vol. 1665, pp. 65–74, 2004.

[20] S. G. Schultz S. G, "Homocellular regulatory mechanisms in sodium-transporting epithelia: avoidance of extinction by "flush-through"," *American Journal of Physiology*, vol. 242, no. 6, pp. F579–F590, 1981.

[21] J. M. Diamond, "Transcellular cross-talk between epithelial cell membranes," *Nature*, vol. 300, no. 5894, pp. 683–685, 1982.

[22] L. Reuss and C. U. Cotton, "Volume regulation in epithelia: transcellular transport and cross-talk," in *Cellular and Molecular Physiology of Cell Volume Regulation*, K. Strange, Ed., pp. 31–47, CRC Press, Boca Raton, Fla, USA, 1994.

[23] M. R. Maurya, S. J. Bornheimer, V. Venkatasubramanian, and S. Subramaniam, "Reduced-order modelling of biochemical networks: application to the GTPase-cycle signalling module," *IEE Proceedings Systems Biology*, vol. 152, no. 4, pp. 229–242, 2005.

[24] T. J. Pedley and J. Fischbarg, "Unstirred layer effects on osmotic water flow across gallbladder epithelium," *Journal of Membrane Biology*, vol. 54, no. 2, pp. 89–102, 1980.

[25] K. R. Spring, "Routes and mechanism of fluid transport by epithelia," *Annual Review of Physiology*, vol. 60, pp. 105–119, 1998.

[26] A. M. Weinstein, "Dynamics of cellular homeostasis: recovery time for a perturbation from equilibrium," *Bulletin of Mathematical Biology*, vol. 59, no. 3, pp. 451–481, 1997.

[27] A. M. Weinstein, "Modeling epithelial cell homeostasis: assessing recovery and control mechanisms," *Bulletin of Mathematical Biology*, vol. 66, no. 5, pp. 1201–1240, 2004.

[28] J. A. Hernández, "Stability properties of elementary dynamic models of membrane transport," in *Bulletin of Mathematical Biology*, vol. 65, pp. 175–197, 2003.

[29] J. A. Hernández, "A general model for the dynamics of the cell volume," *Bulletin of Mathematical Biology*, vol. 69, no. 5, pp. 1631–1648, 2007.

[30] T. L. Hill, *Free Energy Transduction in Biology*, Academic Press, New York, NY, USA, 1977.

9

Transport Reversal during Heteroexchange: A Kinetic Study

V. Makarov,[1] L. Kucheryavykh,[2] Y. Kucheryavykh,[2] A. Rivera,[3] M. J. Eaton,[2]
S. N. Skatchkov,[2,3] and M. Inyushin[3]

[1] Department of Physics, UPR, San Juan, PR 00931, USA
[2] Department of Biochemistry, UCC, Bayamon, PR 00960, USA
[3] Department of Physiology, UCC, Bayamon, PR 00960, USA

Correspondence should be addressed to S. N. Skatchkov; sergueis50@yahoo.com and M. Inyushin; iniouchine@yahoo.com

Academic Editor: Kuo-Chen Chou

It is known that secondary transporters, which utilize transmembrane ionic gradients to drive their substrates up a concentration gradient, can reverse the uptake and instead release their substrates. Unfortunately, the *Michaelis-Menten* kinetic scheme, which is popular in transporter studies, does not include transporter reversal, and it completely neglects the possibility of equilibrium between the substrate concentrations on both sides of the membrane. We have developed a complex two-substrate kinetic model that includes transport reversal. This model allows us to construct analytical formulas allowing the calculation of a "heteroexchange" and "transacceleration" using standard Michaelis coefficients for respective substrates. This approach can help to understand how glial and other cells accumulate substrates without synthesis and are able to release such substrates and gliotransmitters.

1. Introduction

Unlike "primary" or ATP dependent transporters that create the major ionic gradients of K/Na/H and Cl/CO_2 ions across cellular membranes harnessing the energy reserved in ATP, the "secondary transporters" utilize the energy available from transmembrane ionic and/or pH gradients and membrane potential to drive their substrates up a steep concentration gradient. Transporters on neurons and astrocytes clearing neurotransmitters from the synaptic cleft and extracellular space mainly belong to different "secondary transporters" families. Recently, it has been shown that astrocytes and other glial cells accumulate monoamines [1] and polyamines [2, 3] while lacking the enzymes for their synthesis [1, 4–6]. One among many known representatives of the "secondary transporters" that utilize the transmembrane ionic gradients and membrane potential is the family of organic cation transporters (OCT). These transporters take up different mono- and polyamines [7], and cells expressing such transporters also release these substrates using possibly two pathways: (i) large pores and (ii) transport reversal. Here we analyze one of transport reversal mechanisms.

Energy Calculations. Experimentally, it has been shown that secondary transporters can reverse their uptake releasing their substrates instead [8–10]. Energy based calculations were introduced to analyze the conditions for substrate release or uptake for this kind of transporter [11, 12]. It was established that substrate transport depends on the energy balance of coupled transport of the substrate and simultaneously transported ions (see Appendix A). Most secondary transporters could be reversed by membrane potential and by changes in the principal ion gradients and substrate concentrations. Experimentally, the reversal was shown for the glutamate transporters (for the review see [13]), GABA (reviewed by [12]), and for glial organic cation transporters [14, 15]. Being reversed, electrogenic transporters usually change the direction of the net ion flow. We summarize the energy balance study, introduced by Rudnick [11] in Appendix A. This analysis only studies one substrate uptake/release by a secondary transporter.

Michaelis-Menten Scheme. The kinetic concept based on the Michaelis-Menten scheme proved very useful for transporter mediated substrate uptake and inhibition [16, 17]. This kinetic model predicts saturability and specificity of secondary transporters in many cases, and atypical transport kinetics can be explained by multiple binding sites [18]. We have summarized this classic concept in Appendix B. Unfortunately, as one can see, the Michaelis-Menten model does not include transporter reversal, and it completely neglects the reversal constant (see Appendix B). A more complex transporter kinetic model is needed to predict quantitatively at least the following well-established experimental observations.

(1) It has been shown that one transporter substrate can release another one already accumulated inside the cell. Sometimes this is called "heteroexchange." For example, dopamine, tyramine, and amphetamine, which are substrates for the neuronal dopamine transporter (DAT), can release the substrate named N-methyl-4-phenylpyridinium (MPP) through DAT [19], with releasing ability of these substances correlated with the elicited coupled transport current. Also, it was shown that L-glutamate and its transportable analogs (substrates for EAATs) specifically release L-aspartate (another EAAT substrate) through this transporter and can be blocked by nontransportable analogs [20].

(2) A special term was coined for the release of the (tracer) substrate by the same substrate, a process named "transacceleration." While the phenomenon is not kinetically different from the "heteroexchange" described in the previous paragraph, it is well established experimentally (see, e.g., [21]). As new transporter models arise (e.g., a channel-transporter model [22]), it might be important to get this phenomenon explained by a purely thermodynamic model, not by using kinematic assumptions.

Here we present kinetic algorithms that more accurately explain the behavior of a secondary transporter pumping two substrates simultaneously; it predicts transporter reversal by the application of an additional second substrate to the transporter already in equilibrium with the first substrate, "transacceleration" and other interactions.

2. Results

We modified the Michaelis-Menten kinetic model to include two different transportable substrates and also additional elementary steps, characterized by their kinetic coefficients, which are necessary for the transporter not only to uptake but also to release substrates. The model is presented in Appendix C by relations 1–8. This model can be considered as a system of kinetic equations describing the dynamics of the model (C.2)–(C.11). A general solution for this scheme is difficult to obtain analytically. But some particularly interesting cases can be resolved (see Appendices C.1, C.2, and C.3), and we are presenting them below.

2.1. Equilibrium Conditions for Both Substrates (See Appendix C.1). Practically, the initial concentrations of substrates S_1 and S_2 are considered known (i.e., S_{10} and S_{20}), and then substrate concentration can be measured in the outside solution (x_1 and x_2 in our notation for this section of

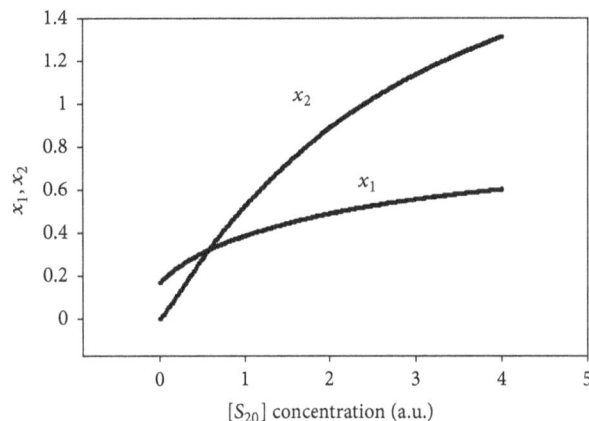

FIGURE 1: Dependence of x_1, x_2 on S_{20} at $S_{10} = 1.5$ a.u., $K_{11} = 1.5$, $K_{12} = 1.8$, $K_{21} = 1.3$, $K_{22} = 2$.

Appendix C). In this way, we tried to reduce all equations to measurable parameters.

It follows from relationships (C.30) (Appendix C) that at fixed concentration of S_{10} (initial concentration of first substrate S_1) and variable concentration of S_{20} (different initial concentrations of S_2), the equilibrium concentration of x_1 (the S_1 substrate outside) increases with increasing S_{20} (the effect of S_1 substrate *releasing* from the cell), and similarly, the equilibrium concentration of x_1 decreases with decreasing S_{20} (effect of S_1 substrate transport inside of the cell). The same behavior follows from (C.30) for the equilibrium concentration of x_2 at fixed concentration of S_{20} and variable concentration of S_{10}. These respective dependencies are shown in Figure 1.

Conclusions of Appendix C.1

(1) Effect of substrate being *released* in case of competition in the two-substrate system can be observed, if at equilibrium condition most of the transporter is coupled by both substrates of interest: $T_0 \gg y$.

(2) Efficiency of the substrate *releasing* process is dependent on equilibrium constant values describing processes of substrate-transporter intermediate complex formation.

(3) Correct sign of the square root term in relations (C.30) is defined by the conditions of

$$S_{10} \geq x_1 \geq 0,$$
$$S_{20} \geq x_2 \geq 0. \tag{1}$$

(4) Relationships (C.30) can be used for analysis of equilibrium substrate concentration dependence on initial substrate concentrations.

(5) The relationship of

$$x_2 = \frac{S_{10} + S_{20} - \alpha x_1}{\beta} \tag{2}$$

can be used to determine parameters α and β, if concentrations of x_1 and x_2 can be simultaneously measured as functions of S_{10} and S_{20}, the initial concentrations of the first and second substrates.

2.2. Two-Substrate System Dynamics at the Initial Time. Transporter velocities (transport rates) can be determined if (similar to Michaelis-Menten scheme) there is no equilibrium between substrate concentrations inside and outside and processes of the type

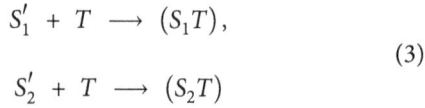

$$S_1' + T \longrightarrow (S_1 T),$$
$$S_2' + T \longrightarrow (S_2 T) \tag{3}$$

can be neglected. We also assume there are the initial conditions where S_1 and S_2 are added to the external solution, thus $x_1 = S_{10}$ and $x_2 = S_{20}$. In that case (see Appendix C.2),

(1)

$$v_{x1} = k_{12} K_1 \frac{S_{10} [T_0]}{1 + K_1 S_{10} + K_2 S_{20}}, \tag{4}$$

$$v_{x2} = k_{22} K_2 \frac{S_{20} [T_0]}{1 + K_1 S_{10} + K_2 S_{20}} \tag{5}$$

are analogous to the Michaelis-Menten formulation for a two-substrate system. If $S_{20} = 0$, we obtain the exact Michaelis-Menten formula for the first substrate velocity, and if $S_{10} = 0$, we obtain the exact Michaelis-Menten formula for the second substrate velocity. Note also that the term $k_{12}[T_0]$ can be interpreted as $V_{1\max}$, and $k_{22}[T_0]$ as $V_{2\max}$.

(2) The constants k_{12}, K_{M1}, k_{22}, and K_{M2} can be determined experimentally similar to the standard procedures used in the Michaelis formulation. There are some important equations:

(i) if $S_{10} \gg K_{M,2}$ and $S_{10} \gg \alpha S_{20}$:

$v_{1,s} = k_{12} [T_0]$, the velocity at maximum, $V_{1\max}$,

$$v_{x2} = k_{22} \frac{\alpha S_{20} [T_0]}{S_{10}}. \tag{6}$$

(ii) If $S_{10} \ll \alpha S_{20}$ and $\alpha S_{20} \gg K_{M,1}$:

$$v_{x1}' = \frac{dx_1}{dt} = k_{12} \frac{S_{10} [T_0]}{\alpha S_{20}},$$

$v_{2,s}' = k_{22} [T_0]$, the velocity at maximum, $V_{2\max}$

$$v_{1,s} v_{2,s} = k_{12} \frac{S_{10} [T_0]}{\alpha S_{20}} k_{22} \frac{\alpha S_{20} [T_0]}{S_{10}} \tag{7}$$

$$= k_{12} k_{22} [T_0]^2 \text{ (see Appendix C)}.$$

2.3. Effect of the Equilibrium Reverse Bias for a First Substrate When a Second One is Added to the System. If previously the equilibrium was established for a first substrate between outside concentration of the substrate and the inside concentration, the addition of a second substrate will produce a reverse bias (equilibrium shift). In the beginning, at initial time, some of the transporter molecules in the outside bind to the second substrate while inside there is still no second substrate. That means the availability of outside transporter for a first substrate becomes reduced. Thus equilibrium for a first substrate starts to break down; that is, the velocity of first substrate transport to outside (release) becomes bigger than its transport to the inside. At initial times during the start of the process and far from equilibrium for a second transporter, (8) allows the calculation of the velocity of the first substrate release due to transport reversal (See Appendix C.3):

$$v_{x1} = A \left[1 - \frac{K_{M,2}}{S_{20} + K_{M,2}} \right], \tag{8}$$

where $K_{M,2}$ is Michaelis constant for a second substrate and S_{20} is initial concentration of a second substrate,

$$A = \frac{k_{11}' S_{10}}{K_0 + 1} y_0,$$
$$y_0 = \frac{[T_0] (K_{11} K_{12} + 1)}{K_{11} S_{10} + K_{11} K_{12} + 1} = \frac{[T_0]}{K_S S_{10} + 1}, \tag{9}$$

where

$$K_S = \frac{K_{11}}{(K_{11} K_{12} + 1)}. \tag{10}$$

Thus, finally we have

$$A = \frac{k_{11}' S_{10}}{K_0 + 1} \frac{[T_0]}{K_S S_{10} + 1}. \tag{11}$$

Taking into consideration the relation (8) we have calculated the dependence of the velocity of first substrate release on the concentration of a second substrate, at initial times after it was added to the system. Functional dependence (8) is represented in Figure 2.

It can be seen from (8) and Figure 2 that the velocity of reversed transport (release) of a first substrate is 0 if $S_{20} = 0$, because there is equilibrium between the velocities of inward and outward flow of the first substrate through the transporter. Thus, the "net" velocity, is equal to zero. Also, from (8) and as seen in Figure 2, with increase of a second substrate concentration, when $S_{20} \gg K_{M,2}$, the velocity of a first substrate release becomes saturated and can be calculated as

$$v_{x1,S} = A = \frac{k_{11}' S_{10}}{K_0 + 1} \frac{[T_0]}{K_S S_{10} + 1}. \tag{12}$$

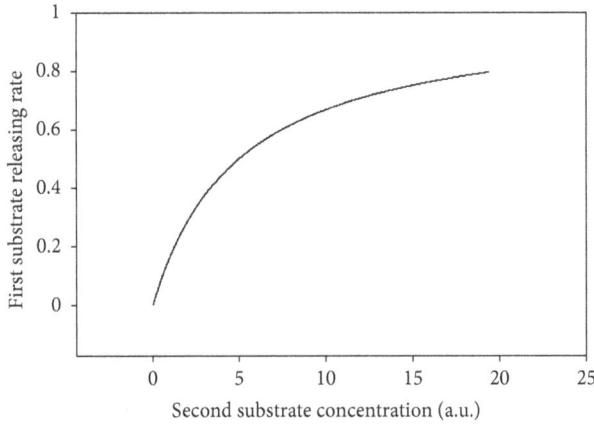

FIGURE 2

Release velocity depends on the first substrate concentration S_{10}, and at given value of S_{10} the value of A is a constant. Thus, (8) for the velocity at half maximal value at a certain concentration of second substrate S_{20} can be written as

$$v_{x1} = A \left[1 - \frac{K_{M,2}}{S_{20,1/2} + K_{M,2}} \right] = \frac{A}{2}, \tag{13}$$

and because of this equation it can be calculated as

$$K_{M,2} = S_{20,1/2}. \tag{14}$$

There is similarity between the formula of velocity of transporter reversal due to second substrate addition and Michaelis-like formulas for the velocity of substrate uptake.

The formula that predicts the velocity of substrate uptake (see Appendix C (C.54) or Appendix B (B.10)) can be written as

$$v_{x2} = \frac{k_{22} S_{20} [T_0]}{S_{20} + K_{M,2}}, \tag{15}$$

where the maximum velocity is represented by

$$v_{x2,max} = k_{22} [T_0]. \tag{16}$$

Thus, for the half maximal velocity,

$$\frac{v_{x2,max}}{2} = \frac{k_{22} S_{20,1/2} [T_0]}{S_{20,1/2} + K_{M,2}} = \frac{k_{22} [T_0]}{2}. \tag{17}$$

Thus we can write

$$K_{M,2} = S_{20,1/2}. \tag{18}$$

To say in plain words, the Michaelis constant for a second substrate can be determined in two ways: (i) from the standard Michaelis formulas at transport velocity measurements for the second substrate, or (ii) from the release velocity measurements of a first substrate, from our formula, where a second substrate produces release of the first one.

In the most important case, if $K_S S_{10} \gg 1$ (8)

$$A_S = \frac{k_{11}'}{K_0 + 1} \frac{[T_0]}{K_S} = \text{Const.},$$

$$v_{x1,S01} = A_S \left[1 - \frac{K_{M,2}}{S_{20} + K_{M,2}} \right], \tag{19}$$

then A_S can be interpreted as the release force for a first substrate after the addition of a second one. In the case of $K_S S_{10} \ll 1$, the release force can be written as

$$A (S_{10}) \approx A_S S_{10} = \frac{k_{11}'}{K_0 + 1} \frac{[T_0]}{K_S} S_{10}; \tag{20}$$

that is, in this case the release force for a first substrate after the addition of a second one has a linear dependence on the first substrate concentration.

3. Discussion and Conclusions

We have studied the extended kinetic model for a secondary transporter simultaneously dealing with two substrates, which includes direct (outside-in) and reverse transport (inside-out). The model was solved in different equilibrium conditions (See Appendices C.1, C.2, and C.3). We have shown that when both substrates are in equilibrium, addition of one of them leads to reequilibrium and release of the second substrate (Appendix C.1). This was emphasized in Appendix C.3, when the system was studied for conditions where a first substrate is in equilibrium (inside-outside concentrations) and a second one is just added and is far from equilibrium. This situation is of a special interest as it has been studied experimentally [19, 20]. Also, this is what probably happens when methamphetamine, ephedrine, or other similar substances induce dopamine (and other monoamine) release from monoamine neurons primarily via membrane transporters, reversing the dopamine transporter (DAT), norepinephrine transporter (NET), and/or serotonin transporter (SERT) [23–27] and also reversing VMAT vesicular transport [28]. In addition, it has been recently shown that astrocytes and other glial cells accumulate polyamines [2, 3] while lacking the enzymes for their synthesis [4–6], and OCT type of transporters (that are expressed in glia) take up different polyamines [7]. Polyamines are released in brain from glial cells, but the mechanisms of such release are unknown [29].

Actually, as we understand now from formula (8) it can be ANY transportable substrate. This formula allows us to classify experimental measurements of a "heteroexchange" related substrate release for substrate-transporter pairs, using standard Michaelis coefficients.

The special term for the release of the (tracer) substrate by the same substrate, a process named "transacceleration," can be explained by changes in equilibrium according to formula (8). There is no fundamental thermodynamic difference if the system has two chemically distinct substrates for the same transporter or there are radiolabelled and unlabelled chemically similar substrates. Thus, a new added substrate produces the release of a similar tracer substrate (labelled, e.g., with radioactive isotope) by equilibrium shift as shown in Appendix C.3.

We also have shown that if we assume both substrates are far away from equilibrium, and transporter reversal can be neglected (Appendix C.2), the formulas for the uptake velocity of both substrates become the same as in the Michaelis-Menten scheme (see Appendix C.2, (C.53) and (C.54)), with the respective inhibitory coefficients.

We suggest that formula (8) will be especially useful in the study of polyspecific transporters with known multiple substrates, such as the organic cation transporters (OCT) that participate in the transport of different monoamines [30], as well as polyamines [7].

Appendices

A. Classical Energy Based Calculations

Similar to the analysis of ion channels, substrate flux through transporters can be determined by the transmembrane electrochemical potential ($\Delta\tilde{\mu}$) which is the sum of the electrical potential ($\Delta\Psi$) and chemical potential (ΔG). For a single molecule X, the driving force is quantified as

$$\Delta\tilde{\mu}_x = \Delta\Psi_x + \Delta G_x = z_x \cdot F \cdot E_m + RT \cdot \ln\frac{[X]_{in}}{[X]_{out}}, \quad (A.1)$$

where z_x = the valence of X, F = Faraday's constant, E_m = membrane potential, R = universal gas constant, and T = temperature. It should be noted that when $\Delta\tilde{\mu} = 0$ (i.e., if X is at equilibrium), this equation reduces to the Nernst equation.

For the glial glutamate transporter GLT1, for example, all of the cotransported ions are coupled to each other as they cross the membrane (i.e., they are not independent). Therefore, the total electrochemical driving force for GLT1 is the sum of the linked contributions from each co-transported ion. Because one thermodynamic reaction cycle for GLT-1 involves coupled translocation of three sodium ions, one proton and one negative glutamate molecule and countertransport of one potassium ion are quantified as

$$\Delta\tilde{\mu}_{GLT-1} = 3 \times (\Delta\Psi_{Na} + \Delta G_{Na}) + 1 \times (\Delta\Psi_K + \Delta G_K)$$
$$+ 1 \times (\Delta\Psi_{Glu} + \Delta G_{Glu}) + 1 \sum (\Delta\Psi_H + \Delta G_H). \quad (A.2)$$

In equilibrium, when $\Delta\tilde{\mu}_{GLT1} = 0$ and knowing that the K^+ term must be negative as it is going in the opposite direction, this equation reduces to

$$E_m = -\frac{RT}{(3Z_{Na} + Z_H - Z_{Glu} - Z_K)F}$$
$$\times \left[\ln\frac{[Glu^-]_{in}}{[Glu^-]_{out}} + 3\ln\frac{[Na^+]_{in}}{[Na^+]_{out}} \right.$$
$$\left. - \ln\frac{[K^+]_{in}}{[K^+]_{out}} + \ln\frac{[H^+]_{in}}{[H^+]_{out}} \right]$$
$$\Longrightarrow E_m = -\frac{RT}{2F} \cdot \ln\left[\frac{[Glu^-]_{in}[Na^+]_{in}^3[H^+]_{in}[K^+]_{in}^{-1}}{[Glu^-]_{out}[Na^+]_{out}^3[H^+]_{out}[K^+]_{out}^{-1}} \right]$$
$$(A.3)$$

which defines the reversal potential for the transporter. It should be noted that this latter equation has a very similar form to the Goldman-Hodgkin-Katz equation for ion channels. The only difference is that ion fluxes are coupled unlike the fluxes for ion channels. The calculated driving force for a transporter can be viewed in the same way as the driving force for an ion channel; thus, there will be no net substrate flux when membrane potential is equal to the reversal potential [12, 13]. This last equation can be rearranged also like this, clearly showing the substrate gradient produced by the transporter:

$$\frac{[Glu^-]_{out}}{[Glu^-]_{in}} = \frac{[Na^+]_{in}^3[H^+]_{in}[K^+]_{in}^{-1}}{[Na^+]_{out}^3[H^+]_{out}[K^+]_{out}^{-1}}\exp^{-E_m(2F/RT)}. \quad (A.4)$$

B. Derivation of Michaelis-Menten Equation for the Transporter

Kinetic concept takes into account only steady-state velocities of transport that can be divided in adhesion, transport, and release of substrate on other side of the membrane. Let the transporter T and the substrate S first form the complex ST in the outer membrane, and then substrate is transported to the inner membrane and released. The reversal is not taken into consideration. This can be written as follows:

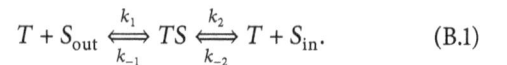

$$T + S_{out} \underset{k_{-1}}{\overset{k_1}{\rightleftharpoons}} TS \underset{k_{-2}}{\overset{k_2}{\rightleftharpoons}} T + S_{in}. \quad (B.1)$$

Formation of TS complex is proportional to the concentration of substrate and the free transporter:

$$\text{Formation} = k_1 \cdot T \cdot S, \quad (B.2)$$

when the complex transports the substrate and releases it inside the cell proportionally to the concentration of TS, and also some TS complex just releases substrate again without transporting it. Thus, the removal of TS from the system is

$$\text{Removal} = k_{-1} \cdot TS + k_2 \cdot TS. \quad (B.3)$$

After TS complex formation the quantity of free transporter left is

$$T = T_{\text{total}} - TS. \tag{B.4}$$

At steady state (at equilibrium), formation and removal of TS became the same:

$$k_1 \cdot T \cdot S_{\text{out}} = k_{-1} \cdot TS + k_2 \cdot TS \tag{B.5}$$

$$\implies \quad k_1 \cdot (T_{\text{total}} - TS) \cdot S_{\text{out}} = (k_{-1} + k_2) \cdot TS \tag{B.6}$$

$$\implies \quad k_1 \cdot T_{\text{total}} \cdot S_{\text{out}} = (k_{-1} + k_2 - k_1 \cdot S_{\text{out}}) \cdot TS \tag{B.7}$$

$$\implies TS = \frac{k_1 \cdot T_{\text{total}} \cdot S_{\text{out}}}{k_{-1} + k_2 - k_1 \cdot S_{\text{out}}} \implies \frac{T_{\text{total}} \cdot S_{\text{out}}}{((k_{-1} + k_2)/k_1) + S_{\text{out}}}. \tag{B.8}$$

As velocity of substrate transport is proportional to k_2 and $TS : v = k_2 \cdot TS_{\text{out}}$, we can write

$$v = \frac{k_2 \cdot T_{\text{total}} \cdot S_{\text{out}}}{(k_{-1} + k_2)/k_1 + S_{\text{out}}}. \tag{B.9}$$

As in abundance of the substrate all velocity depends only on transporter T_{total}, maximal velocity of the transport ($V_{\max}$) is $k_2 * T_{\text{total}}$ and the constant $(k_{-1} + k_2)/k_1 = K_m$, we can write

$$v = \frac{V_{\max} \cdot S_{\text{out}}}{K_m + S_{\text{out}}}. \tag{B.10}$$

The formula allows determination of K_m from the experimental curve: $K_m + S_{\text{out}} = ((V_{\max} \cdot S_{\text{out}})/v) \implies K_m = ((V_{\max} \cdot S_{\text{out}})/v) - S_{\text{out}} = S_{\text{out}}((V_{\max}/v) - 1)$, which means K_m becomes equal to S_{out} if $((V_{\max}/v) - 1) = 1$, and it happens when $v = (V_{\max}/2)$.

So, when the speed of transport is saturated and becomes $V_{\max}$ it is simple to find half of $V_{\max}$ that is equal to K_m. Michaelis constant $K_m = (k_{-1} + k_2)/k_1$ are used as a measure of substrate affinity to the transporter. Please, note that K_m does not include the K_{-2} constant, reflecting unidirectionality of transporters in the Michaelis-Menten approach.

Competitive Inhibition from the Michaelis-Menten Point of View. Transporters can have another substrate I. It can bind to the transporter, whether it is transported or not

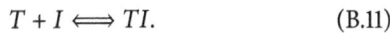

$$T + I \Longleftrightarrow TI. \tag{B.11}$$

The constant of dissociation of this complex can be written as follows:

$$K_{\text{diss}} = K_i = \frac{T \cdot I}{TI} \implies TI = \frac{T \cdot I}{K_i}. \tag{B.12}$$

On the other hand, from the equation

$$\frac{d\,[TS]}{dt} = T \cdot S_{\text{out}} \cdot k_1 - TS \cdot (k_{-1} + k_2) = 0$$

$$\implies \frac{T \cdot S_{\text{out}}}{TS} = \frac{k_{-1} + k_2}{k_1} = K_m \implies T = \frac{K_m}{S_{\text{out}}} \cdot TS, \tag{B.13}$$

if we replace T from (B.7) with the last formula, we will get

$$TI = \frac{T \cdot I}{K_i} \implies TI = \frac{K_m}{S_{\text{out}}} \cdot \frac{I}{K_i} TS. \tag{B.14}$$

The transporter can be in free form or can be occupied by inhibitor or by substrate; therefore, one can write

$$T_{\text{total}} = T + TI + TS,$$

$$T_{\text{total}} = \frac{K_m}{S_{\text{out}}} \cdot TS + \frac{K_m}{S_{\text{out}}} \cdot \frac{I}{K_i} TS + TS \tag{B.15}$$

$$\implies TS = \frac{T_{\text{total}} \cdot S_{\text{out}}}{K_m \left(1 + I/K_i\right) + S_{\text{out}}}.$$

This means the velocity of the transport of the main substrate S will be

$$v = k_2 \cdot TS = \frac{k_2 \cdot T_{\text{total}} \cdot S_{\text{out}}}{K_m \left(1 + I/K_i\right) + S_{\text{out}}} \tag{B.16}$$

$$= \frac{V_m \cdot S_{\text{out}}}{K_m \cdot \alpha + S_{\text{out}}}.$$

C. Kinetic Scheme with Two-Substrate Uptake-Release by the Same Transporter

In a general case, the simplest representation of the two-substrate scheme can be determined by set of elementary steps, characterized by their kinetic coefficients:

(1) $S_1 + T \rightarrow (S_1 T) : k_{11}$;—formation of the intermediate complex "first substrate-transporter" $(S_1 T)$ outside the cell;

(2) $(S_1 T) \rightarrow S_1 + T : k_{-11}$—$(S_1 T)$ dissociation outside the cell;

(3) $(S_1 T) \rightarrow S_1' + T : k_{12}$—dissociation of $(S_1 T)$ complex inside of cell, with substrate S_1' released inside;

(4) $S_1' + T \rightarrow (S_1 T) : k_{-12}$—formation of the first substrate complex with transporter inside the cell;

(5) $S_2 + T \rightarrow (S_2 T) : k_{21}$—formation of the intermediate complex "second substrate-transporter" $(S_2 T)$ outside the cell;

(6) $(S_2 T) \rightarrow S_2 + T : k_{-21}$—dissociation of $(S_2 T)$ complex outside of cell;

(7) $(S_2 T) \rightarrow S_2' + T : k_{22}$—dissociation of $(S_2 T)$ complex inside of cell;

(8) $S_2' + T \rightarrow (S_2 T) : k_{-22}$—formation of the second substrate complex with transporter inside the cell.

Let us introduce notations of

$$[S_1] = x_1, \qquad [T] = y,$$

$$[S_1 T] = z_1, \qquad [S_1'] = s_1,$$

$$[S_2] = x_2, \qquad [S_2 T] = z_2, \tag{C.1}$$

$$[S_2'] = s_2.$$

Series of kinetics equations describing dynamics of the system of interest can be represented as follows:

$$\frac{dx_1}{dt} = -k_{11}x_1 y + k_{-11}z_1, \tag{C.2}$$

$$\begin{aligned}\frac{dy}{dt} &= -k_{11}x_1 y + k_{-11}z_1 + k_{12}z_1 \\ &\quad - k_{-12}ys_1 - k_{21}x_2 y + k_{-21}z_2 \\ &\quad + k_{22}z_2 - k_{-22}ys_2,\end{aligned} \tag{C.3}$$

$$\frac{dz_1}{dt} = k_{11}x_1 y - k_{-11}z_1 - k_{12}z_1 + k_{-12}ys_1, \tag{C.4}$$

$$\frac{ds_1}{dt} = k_{12}z_1 - k_{-12}ys_1, \tag{C.5}$$

$$\frac{dx_2}{dt} = -k_{21}x_2 y + k_{-21}z_2, \tag{C.6}$$

$$\frac{dz_2}{dt} = k_{21}x_2 y - k_{-21}z_2 - k_{22}z_2 + k_{-22}ys_2, \tag{C.7}$$

$$\frac{ds_2}{dt} = k_{22}z_2 - k_{-22}ys_2. \tag{C.8}$$

Summarizing (C.3), (C.4), and (C.7), we will obtain

$$\begin{aligned}\frac{d(y + z_1 + z_2)}{dt} &= 0, \\ y + z_1 + z_2 &= \text{const.} = T_0, \\ y &= T_0 - z_1 - z_2.\end{aligned} \tag{C.9}$$

As well as we can write

$$\begin{aligned}S_{10} &= x_1 + z_1 + s_1, \\ s_1 &= S_{10} - x_1 - z_1, \\ S_{20} &= x_2 + z_2 + s_2, \\ s_2 &= S_{20} - x_2 - z_2.\end{aligned} \tag{C.10}$$

Therefore, the equation system represented above can be rewritten as follows:

$$\begin{aligned}\frac{dx_1}{dt} &= -k_{11}x_1 T_0 + (k_{11}x_1 + k_{-11})z_1 + k_{11}x_1 z_2, \\ \frac{dz_1}{dt} &= k_{11}x_1(T_0 - z_1 - z_2) - k_{-11}z_1 - k_{12}z_1 \\ &\quad + k_{-12}(T_0 - z_1 - z_2)(S_{10} - x_1 - z_1), \\ \frac{dx_2}{dt} &= -k_{21}x_2 T_0 + (k_{21}x_1 + k_{-21})z_2 + k_{21}x_2 z_1, \\ \frac{dz_2}{dt} &= k_{21}x_2(T_0 - z_1 - z_2) - k_{-21}z_2 - k_{22}z_2 \\ &\quad + k_{-22}(T_0 - z_1 - z_2)(S_{20} - x_2 - z_2).\end{aligned} \tag{C.11}$$

C.1. Equilibrium Conditions for Both Substrates. In equilibrium conditions, we can write

$$\begin{aligned}K_{11} &= \frac{z_1}{x_1 y} = \frac{z_1}{x_1(T_0 - z_1 - z_2)}, \\ K_{12} &= \frac{ys_1}{z_1} = \frac{(T_0 - z_1 - z_2)(S_{10} - x_1 - z_1)}{z_1}, \\ K_{21} &= \frac{z_2}{x_2 y} = \frac{z_2}{x_2(T_0 - z_1 - z_2)}, \\ K_{22} &= \frac{ys_2}{z_2} = \frac{(T_0 - z_1 - z_2)(S_{20} - x_2 - z_2)}{z_2},\end{aligned} \tag{C.12}$$

where

$$\begin{aligned}K_{11} &= \frac{k_{11}}{k_{-11}}, & K_{12} &= \frac{k_{12}}{k_{-12}}, \\ K_{21} &= \frac{k_{21}}{k_{-21}}, & K_{22} &= \frac{k_{22}}{k_{-22}}\end{aligned} \tag{C.13}$$

are equilibrium constants of processes of interest. Taking into account relations (C.12) for equilibrium constants, we can form combinations of such constants, which may be represented as follows:

$$\begin{aligned}K_{11}K_{12} &= \frac{S_{10} - x_1 - z_1}{x_1}, \\ K_{21}K_{22} &= \frac{S_{20} - x_2 - z_2}{x_2}.\end{aligned} \tag{C.14}$$

Let us consider the case, when $T_0 \gg y$,—all transporter is coupled to complexes of $(S_1 T)$ and $(S_2 T)$. In this case, we can write with a fair approximation the relations of

$$z_1 = [T_0] - z_2,$$

$$K_{11}K_{12} = \frac{S_{10} - x_1 - [T_0] + z_2}{x_1}, \tag{C.15}$$

$$K_{11}K_{12}x_1 = S_{10} - x_1 - [T_0] + z_2.$$

From the last relationship, the z_2 concentration can be represented as follows:

$$z_2 = K_{11}K_{12}x_1 - S_{10} + x_1 + [T_0]. \tag{C.16}$$

Applying the same approach to equations related to the second substrate and taking into account the last relationship for concentration of z_2 we can write

$$\begin{aligned}K_{21}K_{22} &= \frac{S_{20} - x_2 - (K_{11}K_{12}x_1 - S_{10} + x_1 + [T_0])}{x_2} \\ &= \frac{S_{20} - x_2 - K_{11}K_{12}x_1 + S_{10} - x_1 - [T_0]}{x_2}, \tag{C.17}\end{aligned}$$

$$x_2 = \frac{S_{20} - K_{11}K_{12}x_1 + S_{10} - x_1 - [T_0]}{1 + K_{21}K_{22}}.$$

If relationship of

$$S_{10} + S_{20} \gg [T_0] \qquad \text{(C.18)}$$

is satisfied, we can write

$$x_2 = \frac{S_{20} + S_{10}}{1 + K_{21}K_{22}} - \frac{1 + K_{11}K_{12}}{1 + K_{21}K_{22}}x_1, \qquad \text{(C.19)}$$

and finally we can write

$$(1 + K_{11}K_{12})\,x_1 + (1 + K_{21}K_{22})\,x_2 = S_{10} + S_{20},$$
$$\alpha x_1 + \beta x_2 = S_{10} + S_{20}, \qquad \text{(C.20)}$$

where

$$\alpha = (1 + K_{11}K_{12}),$$
$$\beta = (1 + K_{21}K_{22}). \qquad \text{(C.21)}$$

Let us consider an additional relationship, which can be formed by multiplication of all equilibrium constants, that is, by multiplication of relationships of

$$K_{11}K_{22} = \frac{(S_{20} - x_2 - z_2)\,z_1}{x_1 z_2},$$
$$K_{12}K_{21} = \frac{z_2\,(S_{10} - x_1 - z_1)}{x_2 z_1}. \qquad \text{(C.22)}$$

In this case, we will obtain relationship of

$$K_{11}K_{22}K_{12}K_{21} = \frac{(S_{20} - x_2 - z_2)(S_{10} - x_1 - z_1)}{x_1 x_2}$$
$$\approx \frac{(S_{20} - x_2)(S_{10} - x_1)}{x_1 x_2}, \qquad \text{(C.23)}$$

$$K_{11}K_{22}K_{12}K_{21}x_1 x_2 \approx (S_{20} - x_2)(S_{10} - x_1),$$

where with respect to relationship (C.18), we may assume that the relationships of

$$S_{10} \gg z_1,$$
$$S_{20} \gg z_2 \qquad \text{(C.24a)}$$

are satisfied, and we may finally write

$$(K_{11}K_{22}K_{12}K_{21} - 1)\,x_1 x_2 \approx S_{10}S_{20} - S_{20}x_1 - S_{10}x_2. \qquad \text{(C.24b)}$$

Taking into account relationship of

$$\alpha x_1 + \beta x_2 = S_{10} + S_{20}, \qquad \text{(C.25)}$$

we can write

$$x_2 = \frac{S_{10} + S_{20} - \alpha x_1}{\beta}, \qquad \text{(C.26)}$$

and if we will substitute the last value to relationship (C.24b), we will obtain

$$\gamma x_1 \frac{S_{10} + S_{20} - \alpha x_1}{\beta}$$
$$\approx S_{10}S_{20} - S_{20}x_1 - S_{10}\frac{S_{10} + S_{20} - \alpha x_1}{\beta}$$
$$- \frac{\alpha\gamma}{\beta}x_1^2 + \frac{\gamma}{\beta}(S_{10} + S_{20})\,x_1 + S_{20}x_1 - \frac{S_{10}\alpha}{\beta}x_1$$
$$- S_{10}S_{20} + \frac{S_{10}}{\beta}(S_{10} + S_{20}) = 0,$$
$$x_1^2 - \left[\frac{S_{10} + S_{20}}{\alpha} + \frac{\beta}{\alpha\gamma}S_{20} - \frac{S_{10}}{\gamma}\right]x_1$$
$$+ \frac{\beta}{\alpha\gamma}S_{10}S_{20} - \frac{S_{10}}{\alpha\gamma}(S_{10} + S_{20}) = 0, \qquad \text{(C.27)}$$

$$x_1^2 - a x_1 + b = 0, \qquad \text{(C.28)}$$

$$x_1 = \frac{a}{2} \pm \sqrt{\frac{a^2}{4} - b},$$

where

$$\gamma = (K_{11}K_{22}K_{12}K_{21} - 1),$$
$$a = \left[\frac{S_{10} + S_{20}}{\alpha} + \frac{\beta}{\alpha\gamma}S_{20} - \frac{S_{10}}{\gamma}\right], \qquad \text{(C.29)}$$
$$b = \frac{\beta}{\alpha\gamma}S_{10}S_{20} - \frac{S_{10}}{\alpha\gamma}(S_{10} + S_{20}).$$

Thus,

$$x_1 = \frac{1}{2}\left[\frac{S_{10} + S_{20}}{\alpha} + \frac{\beta}{\alpha\gamma}S_{20} - \frac{S_{10}}{\gamma}\right]$$
$$\pm \left(\frac{1}{4}\left[\frac{S_{10} + S_{20}}{\alpha} + \frac{\beta}{\alpha\gamma}S_{20} - \frac{S_{10}}{\gamma}\right]^2\right.$$
$$\left. - \left(\frac{\beta}{\alpha\gamma}S_{10}S_{20} - \frac{S_{10}}{\alpha\gamma}(S_{10} + S_{20})\right)\right)^{1/2}$$
$$= \frac{1}{2}\left[S_{10}\left(\frac{1}{\alpha} - \frac{1}{\gamma}\right) + \frac{S_{20}}{\alpha}\left(1 + \frac{\beta}{\gamma}\right)\right]$$
$$\pm \left(\frac{1}{4}\left[S_{10}\left(\frac{1}{\alpha} - \frac{1}{\gamma}\right) + \frac{S_{20}}{\alpha}\left(1 + \frac{\beta}{\gamma}\right)\right]^2\right.$$
$$\left. + \frac{1}{\alpha\gamma}(S_{10} - (\beta - 1)S_{20})^2 - \frac{S_{20}^2}{\alpha\gamma}(\beta - 1)^2\right)^{1/2},$$

$$x_1 = \frac{1}{2}\left[a_1 S_{10} + a_2 S_{20}\right]$$

$$\pm \left(\frac{1}{4}\left[a_1 S_{10} + a_2 S_{20}\right]^2\right.$$

$$\left. + a_3\left(S_{10} - a_4 S_{20}\right)^2 - a_3 a_4^2 S_{20}^2\right)^{1/2},$$

$$a_1 = \left(\frac{1}{\alpha} - \frac{1}{\gamma}\right),$$

$$a_2 = \left(1 + \frac{\beta}{\gamma}\right),$$

$$a_3 = \frac{1}{\alpha\gamma},$$

$$a_4 = (\beta - 1),$$

$$x_2 = \frac{S_{10} + S_{20}}{\beta} - \frac{\alpha}{\beta}$$

$$\times \left\{\frac{1}{2}\left[\frac{S_{10} + S_{20}}{\alpha} + \frac{\beta}{\alpha\gamma}S_{20} - \frac{S_{10}}{\gamma}\right]\right.$$

$$\pm \left(\frac{1}{4}\left[\frac{S_{10} + S_{20}}{\alpha} + \frac{\beta}{\alpha\gamma}S_{20} - \frac{S_{10}}{\gamma}\right]^2\right.$$

$$\left.\left. - \left(\frac{\beta}{\alpha\gamma}S_{10}S_{20} - \frac{S_{10}}{\alpha\gamma}(S_{10} + S_{20})\right)\right)^{1/2}\right\},$$

$$x_2 = \frac{S_{10} + S_{20}}{\beta} - \frac{\alpha}{\beta}$$

$$\times \left\{\frac{1}{2}\left[a_1 S_{10} + a_2 S_{20}\right]\right.$$

$$\left. \pm \sqrt{\frac{1}{4}\left[a_1 S_{10} + a_2 S_{20}\right]^2 + a_3\left(S_{10} - a_4 S_{20}\right)^2 - a_3 a_4^2 S_{20}^2}\right\}.$$

$$(C.30)$$

Solution is physically reasonable, if the relation of

$$\frac{1}{4}\left[\frac{S_{10} + S_{20}}{\alpha} + \frac{\beta}{\alpha\gamma}S_{20} - \frac{S_{10}}{\gamma}\right]^2$$
$$- \left(\frac{\beta}{\alpha\gamma}S_{10}S_{20} - \frac{S_{10}}{\alpha\gamma}(S_{10} + S_{20})\right) \geq 0 \qquad (C.31)$$

is satisfied. Since parameter $\gamma = (K_{11}K_{22}K_{12}K_{21} - 1)$, it can easy be shown that condition (C.31) is satisfied at all reasonable values of α and β parameters. It also follows from relationships (C.30) that at fixed concentration of S_{10} and variable concentration of S_{20}, equilibrium concentration of x_1 increases with increase of S_{20} (effect of S_1 substrate realizing from the cell) and decreasing of equilibrium concentration of x_1 with decreasing of S_{20} (effect of S_1 substrate transport inside of the cell). The same behavior follows from relationships of interest for equilibrium concentration of x_2 at fixed

concentration of S_2 and variable concentration of S_{10}. The respective dependences are shown in Figure 1.

Conclusions of Appendix C.1

(1) Effect of substrate realizing in case of transportation competition for two substrate system can be observable, if in equilibrium condition most of the transporter is coupled by both substrates of interest: $T_0 \gg y$.

(2) Efficiency of substrate realizing process is dependent on equilibrium constant values describing processes of substrate-transporter intermediate complex formation.

(3) Correct sign in front of square root in relations (C.30) is defined by the conditions of

$$S_{10} \geq x_1 \geq 0,$$
$$S_{20} \geq x_2 \geq 0. \qquad (C.32)$$

(4) Relationships (C.30) can be used for analysis of equilibrium substrate concentration dependence on initial substrate concentrations.

(5) The relationship of

$$x_2 = \frac{S_{10} + S_{20} - \alpha x_1}{\beta} \qquad (C.33)$$

can be used to determine parameters of α and β, if concentration of x_1 and x_2 can simultaneously be measured as function of S_{10} and S_{20} initial concentrations of first and second substrates.

Procedure of Experimental Measurements of $x_2 = (S_{10} + S_{20} - \alpha x_1)/\beta$ Relationship

(1) Measurements of equilibrium concentrations set of x_1, x_2 versus S_{10} and S_{20}.

(i) Let us choose a fixed set of concentrations $S_{10,1}$ from the interval $[S_{10}^{(0)}, S_{10}^{(1)}]$. For a fixed concentration $S_{10,1}$, let us measure the dependence of concentrations x_1 and x_2 versus $[S_{20}]$. The same procedure must be repeated for another fixed concentration $S_{10,2}$, the third fixed concentration $S_{10,3}$, and so forth.

(ii) Using the equation

$$x_2 = \frac{S_{10} + S_{20} - \alpha x_1}{\beta} = \frac{S_{10} + S_{20}}{\beta} - \frac{\alpha x_1}{\beta}, \qquad (C.34)$$

it can be rewritten as

$$S_{10} + S_{20} = \alpha x_1 + \beta x_2,$$
$$U = \alpha x_1 + \beta x_2. \qquad (C.35)$$

We can experimentally measure $\{U_i\}$, $\{x_{1i}\}$, and $\{x_{2i}\}$, with unknown parameters: α, β. Applying the least-squares method, we can write

$$\frac{\partial}{\partial p} \sum_i (U_i - \alpha x_{1i} - \beta x_{2i})^2 = 0 \quad p = \alpha, \beta. \tag{C.36}$$

Thus,

$$\sum_i (U_i - \alpha x_{1i} - \beta x_{2i}) x_{1i} = 0,$$

$$\sum_i (U_i - \alpha x_{1i} - \beta x_{2i}) x_{2i} = 0,$$

$$\sum_i (U_i - \alpha x_{1i} - \beta x_{2i}) x_{1i}$$

$$= \sum_i U_i x_{1i} - \alpha \sum_i x_{1i}^2 - \beta \sum_i x_{2i} x_{1i} = 0, \tag{C.37}$$

$$\sum_i (U_i - \alpha x_{1i} - \beta x_{2i}) x_{2i}$$

$$= \sum_i U_i x_{2i} - \alpha \sum_i x_{1i} x_{2i} - \beta \sum_i x_{2i}^2 = 0.$$

Main system determinants can be represented as follows:

$$\Delta = \begin{vmatrix} \sum_i x_{1i}^2 & \sum_i x_{2i} x_{1i} \\ \sum_i x_{2i} x_{1i} & \sum_i x_{2i}^2 \end{vmatrix}$$

$$= \left(\sum_i x_{1i}^2 \right) \left(\sum_i x_{2i}^2 \right) - \left(\sum_i x_{2i} x_{1i} \right)^2,$$

$$\Delta_\alpha = \begin{vmatrix} \sum_i U_i x_{1i} & \sum_i x_{2i} x_{1i} \\ \sum_i U_i x_{2i} & \sum_i x_{2i}^2 \end{vmatrix}$$

$$= \left(\sum_i U_i x_{1i} \right) \left(\sum_i x_{2i}^2 \right) \tag{C.38}$$

$$- \left(\sum_i U_i x_{2i} \right) \left(\sum_i x_{2i} x_{1i} \right),$$

$$\Delta_\beta = \begin{vmatrix} \sum_i x_{1i}^2 & \sum_i U_i x_{1i} \\ \sum_i x_{2i} x_{1i} & \sum_i U_i x_{2i} \end{vmatrix}$$

$$= \left(\sum_i x_{1i}^2 \right) \left(\sum_i U_i x_{2i} \right)$$

$$- \left(\sum_i U_i x_{1i} \right) \left(\sum_i x_{2i} x_{1i} \right).$$

Therefore,

$$\alpha = \frac{\Delta_\alpha}{\Delta}$$

$$= \frac{\left(\sum_i U_i x_{1i} \right) \left(\sum_i x_{2i}^2 \right) - \left(\sum_i U_i x_{2i} \right) \left(\sum_i x_{2i} x_{1i} \right)}{\left(\sum_i x_{1i}^2 \right) \left(\sum_i x_{2i}^2 \right) - \left(\sum_i x_{2i} x_{1i} \right)^2},$$

$$\beta = \frac{\Delta_\beta}{\Delta} \tag{C.39}$$

$$= \frac{\left(\sum_i x_{1i}^2 \right) \left(\sum_i U_i x_{2i} \right) - \left(\sum_i U_i x_{1i} \right) \left(\sum_i x_{2i} x_{1i} \right)}{\left(\sum_i x_{1i}^2 \right) \left(\sum_i x_{2i}^2 \right) - \left(\sum_i x_{2i} x_{1i} \right)^2}.$$

Earlier we had determined the parameters of interest as

$$\alpha = (1 + K_{11} K_{12}),$$
$$\beta = (1 + K_{21} K_{22}). \tag{C.40}$$

That is,

$$K_{11} K_{12} = \alpha - 1,$$
$$K_{21} K_{22} = \beta - 1. \tag{C.41}$$

As $K_{12}, K_{21}, K_{11}, K_{22}$:

$$K_{11} = \frac{k_{11}}{k_{-11}},$$

$$K_{12} = \frac{k_{12}}{k_{-12}},$$

$$K_{21} = \frac{k_{21}}{k_{-21}}, \tag{C.42}$$

$$K_{22} = \frac{k_{22}}{k_{-22}},$$

we can write

$$\frac{k_{11}}{k_{-11}} \cdot \frac{k_{12}}{k_{-12}} = \alpha - 1,$$

$$\frac{k_{21}}{k_{-21}} \cdot \frac{k_{22}}{k_{-22}} = \beta - 1. \tag{C.43}$$

C.2. Two-Substrate System Dynamics at the Initial Time. In this case we will not take into account processes (4) and (8):

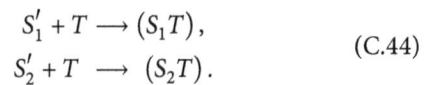

$$S_1' + T \longrightarrow (S_1 T),$$
$$S_2' + T \longrightarrow (S_2 T). \tag{C.44}$$

In this case, the series of kinetics equations can be represented as follows:

$$\frac{dx_1}{dt} = -k_{11} x_1 y + k_{-11} z_1,$$

$$\frac{dy}{dt} = -k_{11} x_1 y + k_{-11} z_1 + k_{12} z_1$$
$$- k_{21} x_2 y + k_{-21} z_2 + k_{22} z_2,$$

$$\frac{dz_1}{dt} = k_{11} x_1 y - k_{-11} z_1 - k_{12} z_1, \tag{C.45}$$

$$\frac{dx_2}{dt} = -k_{21} x_2 y + k_{-21} z_2,$$

$$\frac{dz_2}{dt} = k_{21} x_2 y - k_{-21} z_2 - k_{22} z_2.$$

For z_1, z_2, we may use quasi-stationary approximation:

$$\frac{dz_1}{dt} = k_{11}x_1 y - k_{-11}z_1 - k_{12}z_1 = 0,$$

$$\frac{dz_2}{dt} = k_{21}x_2 y - k_{-21}z_2 - k_{22}z_2 = 0. \tag{C.46}$$

Thus,

$$z_1 = \frac{k_{11}x_1 y}{k_{-11} + k_{12}},$$

$$z_2 = \frac{k_{21}x_2 y}{k_{-21} + k_{22}}. \tag{C.47}$$

Thus,

$$y = [T_0] - z_1 - z_2 = [T_0] - \frac{k_{11}x_1 y}{k_{-11} + k_{12}} - \frac{k_{21}x_2 y}{k_{-21} + k_{22}},$$

$$y = \frac{[T_0]}{1 + (k_{11}x_1)/(k_{-11}+k_{12}) + (k_{21}x_2)/(k_{-21}+k_{22})}. \tag{C.48}$$

For the initial time after the start of all processes, we can assume that $x_1 = S_{10}$ and $x_2 = S_{20}$; that is,

$$y = \frac{[T_0]}{1 + (k_{11}S_{10})/(k_{-11}+k_{12}) + (k_{21}S_{20})/(k_{-21}+k_{22})}. \tag{C.49}$$

Hence we can write

$$v_{x1} = \frac{dx_1}{dt} = -k_{11}x_1 y + k_{-11}z_1$$

$$= \left[-k_{11} + k_{-11}\frac{k_{11}}{k_{-11}+k_{12}}\right]$$

$$\times (S_{10}[T_0]) \times \left(1 + \frac{k_{11}S_{10}}{k_{-11}+k_{12}} + \frac{k_{21}S_{20}}{k_{-21}+k_{22}}\right)^{-1}$$

$$= \frac{k_{11}k_{12}}{k_{-11}+k_{12}}$$

$$\times (S_{10}[T_0]) \times \left(1 + \frac{k_{11}S_{10}}{k_{-11}+k_{12}} + \frac{k_{21}S_{20}}{k_{-21}+k_{22}}\right)^{-1}$$

$$= k_{12}K_{M,1}^{-1}\frac{S_{10}[T_0]}{1 + K_{M,1}^{-1}S_{10} + K_{M,2}^{-1}S_{20}}$$

$$= k_{12}\frac{S_{10}[T_0]}{K_{M,1} + S_{10} + K_{M,1}K_{M,2}^{-1}S_{20}}$$

$$= k_{12}\frac{S_{10}[T_0]}{K_{M,1} + S_{10} + \alpha S_{20}}, \tag{C.50}$$

$$v_{x2} = \frac{dx_2}{dt} = -k_{21}x_2 y + k_{-21}\frac{k_{21}x_2 y}{k_{-21}+k_{22}}$$

$$= \left[-k_{21} + k_{-21}\frac{k_{21}}{k_{-21}+k_{22}}\right]x_2 y$$

$$= \frac{k_{21}k_{22}}{k_{-21}+k_{22}}$$

$$\times (S_{20}[T_0]) \times \left(1 + \frac{k_{11}S_{10}}{k_{-11}+k_{12}} + \frac{k_{21}S_{20}}{k_{-21}+k_{22}}\right)^{-1}$$

$$= k_{22}K_{M,2}^{-1}\frac{S_{20}[T_0]}{1 + K_{M,1}^{-1}S_{10} + K_{M,2}^{-1}S_{20}}$$

$$= k_{22}\frac{S_{20}[T_0]}{K_{M,2} + K_{M,1}^{-1}K_{M,2}S_{10} + S_{20}}$$

$$= k_{22}\frac{S_{20}[T_0]}{K_{M,2} + \alpha^{-1}S_{10} + S_{20}}, \tag{C.51}$$

where

$$K_1^{-1} = \frac{k_{11}}{k_{-11}+k_{12}},$$

$$K_2^{-1} = \frac{k_{21}}{k_{-21}+k_{22}}, \tag{C.52}$$

$$\alpha = \frac{K_{M,1}}{K_{M,2}}$$

are Michaelis constants for the first and second substrate, respectively, and α—is a factor of transporter inhibition.

Conclusions for Substrate Dynamics Section

(1) Equations

$$v_{x1} = \frac{dx_1}{dt} = k_{12}\frac{S_{10}[T_0]}{K_{M,1} + S_{10} + \alpha S_{20}}, \tag{C.53}$$

$$v_{x2} = \frac{dx_2}{dt} = k_{22}\frac{S_{20}[T_0]}{K_{M,2} + \alpha^{-1}S_{10} + S_{20}} \tag{C.54}$$

are the analogs to the Michaelis formula for a two-substrate system. If $S_{20} = 0$, we get the Michaelis formula for the first substrate, and if $S_{10} = 0$, we get the formula for the second one.

(2) Constants k_{12}, $K_{M,1}$, k_{22}, $K_{M,2}$ can be determined experimentally:

(iii) $S_{10} \gg K_{M,1}$ and $K_{M,2}$: $S_{10} \gg \alpha S_{20}$:

$$v_{1,s} = k_{12}[T_0],$$

$$v_{x2} = k_{22}\frac{\alpha S_{20}[T_0]}{S_{10}}, \tag{C.55}$$

(iv) $S_{10} \ll \alpha S_{20}$: $\alpha S_{20} \gg K_{M,1}$:

$$v'_{x1} = \frac{dx_1}{dt} = k_{12}\frac{S_{10}[T_0]}{\alpha S_{20}},$$

$$v'_{2,s} = k_{22}[T_0], \tag{C.56}$$

$$v_{1,s}v_{2,s} = k_{12}\frac{S_{10}[T_0]}{\alpha S_{20}}k_{22}\frac{\alpha S_{20}[T_0]}{S_{10}} = k_{12}k_{22}[T_0]^2.$$

C.3. Effect of the Equilibrium Reverse Bias for a First Substrate When a Second One Is Added to the System. If previously the equilibrium was established for a first substrate between the outside concentration of the substrate and the inside concentration, the addition of a second substrate will produce a reverse bias (equilibrium shift). To study this, we first can write the equation representing the velocity of second substrate transport:

$$v_2 = \frac{dx_2}{dt} = -\frac{k_{21}S_{20}}{k_{21}S_{20} + k_{-21} + k_{22}}$$

$$\times \frac{[T_0](K_{11}K_{12} + 1)}{K_{11}S_{10} + K_{11}K_{12} + 1} \qquad (C.57)$$

$$= -\frac{S_{20}}{S_{20} + K_{M,2}} \times \frac{[T_0](K_{11}K_{12} + 1)}{K_{11}S_{10} + K_{11}K_{12} + 1},$$

where

$$K_{M,2} = \frac{k_{-21} + k_{22}}{k_{21}}. \qquad (C.58)$$

In the beginning, at initial time, some of the transporter molecules in the outside bind to the second substrate while inside there is still no second substrate. This means the availability of outside transporter for a first substrate becomes reduced. Thus, equilibrium for a first substrate starts to break down; that is, the velocity of first substrate transport to outside (release) becomes bigger than its transport to the inside. Our model allows evaluation of the release of a first substrate with simple approximation.

Unlike that in the Appendix C.1, where all the transport molecules were available for both substrates, let us approximate that the amount of transporter binding the first substrate outside is reduced by the second substrate also binding to it. Thus, using this rough approximation, the available outside transporter is reduced to the amount:

$$y = y_0 - z_2, \qquad (C.59)$$

where

$$y_0 = \frac{[T_0](K_{11}K_{12} + 1)}{K_{11}S_{10} + K_{11}K_{12} + 1},$$
$$z_2 = \frac{k_{21}S_{20}y_0}{k_{21}S_{20} + k_{-21} + k_{22}}, \qquad (C.60)$$

while inside (the cell) the amount of transporter available for the first substrate still remains y_0. Consequently, the velocity of the first substrate reverse flow can be represented as

$$v_{x1} = \frac{dx_1}{dt} = -k'_{11}x_{1,e}y + k'_{-11}s_{1,e}y_0, \qquad (C.61)$$

where k'_{11} and k'_{-12} are some effective parameters that can be only analytically solved in more complex model that will implicate intermediate transporter-substrate complexes and are out of the scope of this study. Here we can only get

formulas for the initial times during the start of the process and far from equilibrium for a second transporter. In (C.61) we had balanced the flow disparity between the first substrate in-flow (uptake) and out-flow (release), where $x_{1,e}$ and $s_{1,e}$ are the equilibrium concentrations of the first substrate outside and inside (the cell). Thus, these concentrations can be approximately found using the approximation that still the system is close to the equilibrium for a first substrate:

$$x_1 \Longleftrightarrow s_1,$$

$$K_0 = \frac{s_{1,e}}{x_{1,e}} = \frac{S_{10} - x_{1,e}}{x_{1,e}} = \frac{k'_{11}}{k'_{-11}},$$

$$x_{1,e} = \frac{S_{10}}{K_0 + 1}, \qquad (C.62)$$

$$s_{1,e} = \frac{K_0 S_{10}}{K_0 + 1}.$$

Thus,

$$v_{x1} = \frac{dx_1}{dt} = -k'_{11}x_{1,e}y + k'_{-11}s_{1,e}y_0$$

$$= -k'_{11}\frac{S_{10}}{K_0 + 1}y_0\frac{k_{-21} + k_{22}}{k_{21}S_{20} + k_{-21} + k_{22}}$$

$$+ k'_{-11}\frac{K_0 S_{10}}{K_0 + 1}y_0 \qquad (C.63)$$

$$= \frac{S_{10}}{K_0 + 1}y_0\left[k'_{-11}K_0 - k'_{11}\frac{k_{-21} + k_{22}}{k_{21}S_{20} + k_{-21} + k_{22}}\right]$$

$$= \frac{k'_{11}S_{10}}{K_0 + 1}y_0\left[1 - \frac{K_{M,2}}{S_{20} + K_{M,2}}\right].$$

For initial time conditions with relatively good precision, we can accept that

$$\frac{k'_{11}S_{10}}{K_0 + 1}y_0 \approx A = \text{Const.}, \qquad (C.64)$$

as

$$\frac{k'_{11}S_{10}}{K_0 + 1}y_0 = \frac{k'_{11}S_{10}}{K_0 + 1} \times \frac{[T_0](K_{11}K_{12} + 1)}{K_{11}S_{10} + K_{11}K_{12} + 1}$$

$$= \frac{k'_{11}S_{10}[T_0]}{K_0 + 1} \times \frac{1}{K_S S_{10} + 1} \qquad (C.65)$$

$$= \frac{k'_{11}S_{10}[T_0]}{K_0 + 1} \times K_Z \approx A = \text{Const.},$$

where

$$K_S = \frac{K_{11}}{K_{11}K_{12} + 1},$$

$$K_Z = \frac{1}{K_S S_{10} + 1},$$

$$v_{x1} = \frac{k'_{11} S_{10} [T_0]}{K_0 + 1} \times K_Z \left[1 - \frac{K_{M,2}}{S_{20} + K_{M,2}} \right] \qquad (C.66)$$

$$= A \left[1 - \frac{K_{M,2}}{S_{20} + K_{M,2}} \right].$$

Thus

$$v_{x1} = A \left[1 - \frac{K_{M,2}}{S_{20} + K_{M,2}} \right]. \qquad (72^*)$$

First substrate reverse flow (the release speed) velocity dependence on the external second substrate concentration is represented on Figure 2.

One can see that the reverse flow of the first substrate is absent if $S_{20} = 0$, because in that case the formula (72^*) gives us zero velocity. When the second substrate is added externally the second term inside square brackets becomes reduced and the velocity augments. Thus, the velocity of reversed transport grows with the second substrate concentration and until a limit:

$$v_{x1} = \frac{k'_{11} S_{10}}{K_0 + 1} y_0. \qquad (C.67)$$

Acknowledgments

This paper was possible by Grants from the National Center for Research Resources (5 G12 RR 003035-27) and the National Institute on Minority Health and Health Disparities (8G12 MD 007583-27) from the National Institutes of Health, as well as NIH-NINDS-R01-NS065201.

References

[1] M. Y. Inyushin, A. Huertas, Y. V. Kucheryavykh et al., "L-DOPA uptake in astrocytic endfeet enwrapping blood vessels in rat brain," *Parkinson's Disease*, vol. 2012, Article ID 321406, 8 pages, 2012.

[2] G. Laube and R. W. Veh, "Astrocytes, not neurons, show most prominent staining for spermine/spermidine-like immunoreactivity in adult rat brain," *Glia*, vol. 19, no. 2, pp. 171–179, 1997.

[3] S. N. Skatchkov, M. J. Eaton, J. Krušek et al., "Spatial distribution of spermine/spermidine content and K⁺- current rectification in frog retinal glial (Müller) cells," *Glia*, vol. 31, no. 1, pp. 84–90, 2000.

[4] V. I. Madai, W. C. Poller, D. Peters et al., "Synaptic localisation of agmatinase in rat cerebral cortex revealed by virtual pre-embedding," *Amino Acids*, vol. 43, no. 3, pp. 1399–1403, 2012.

[5] M. Krauss, K. Langnaese, K. Richter et al., "Spermidine synthase is prominently expressed in the striatal patch compartment and in putative interneurones of the matrix compartment," *Journal of Neurochemistry*, vol. 97, no. 1, pp. 174–189, 2006.

[6] M. Krauss, T. Weiss, K. Langnaese et al., "Cellular and subcellular rat brain spermidine synthase expression patterns suggest region-specific roles for polyamines, including cerebellar presynaptic function," *Journal of Neurochemistry*, vol. 103, no. 2, pp. 679–693, 2007.

[7] M. Sala-Rabanal, D. C. Li, G. R. Dake et al., "Polyamine transport by the polyspecific organic cation transporters OCT1, OCT2, and OCT3," *Molecular Pharmacology*, vol. 10, no. 4, pp. 1450–1458, 2013.

[8] M. Szatkowski, B. Barbour, and D. Attwell, "Non-vesicular release of glutamate from glial cells by reversed electrogenic glutamate uptake," *Nature*, vol. 348, no. 6300, pp. 443–446, 1990.

[9] D. Attwell, B. Barbour, and M. Szatkowski, "Nonvesicular release of neurotransmitter," *Neuron*, vol. 11, no. 3, pp. 401–407, 1993.

[10] G. Levi and M. Raiteri, "Carrier-mediated release of neurotransmitters," *Trends in Neurosciences*, vol. 16, no. 10, pp. 415–419, 1993.

[11] G. Rudnick, "Bioenergetics of neurotransmitter transport," *Journal of Bioenergetics and Biomembranes*, vol. 30, no. 2, pp. 173–185, 1998.

[12] G. B. Richerson and Y. Wu, "Dynamic equilibrium of neurotransmitter transporters: not just for reuptake anymore," *Journal of Neurophysiology*, vol. 90, no. 3, pp. 1363–1374, 2003.

[13] L. M. Levy, O. Warr, and D. Attwell, "Stoichiometry of the glial glutamate transporter GLT-1 expressed inducibly in a Chinese hamster ovary cell line selected for low endogenous Na⁺-dependent glutamate uptake," *Journal of Neuroscience*, vol. 18, no. 23, pp. 9620–9628, 1998.

[14] T. Budiman, E. Bamberg, H. Koepsell, and G. Nagel, "Mechanism of electrogenic cation transport by the cloned organic cation transporter 2 from rat," *Journal of Biological Chemistry*, vol. 275, no. 38, pp. 29413–29420, 2000.

[15] P. Arndt, C. Volk, V. Gorboulev et al., "Interaction of cations, anions, and weak base quinine with rat renal cation transporter rOCT2 compared with rOCT1," *American Journal of Physiology*, vol. 281, no. 3, pp. F454–F468, 2001.

[16] R. M. Wightman, C. Amatore, R. C. Engstrom et al., "Real-time characterization of dopamine overflow and uptake in the rat striatum," *Neuroscience*, vol. 25, no. 2, pp. 513–523, 1988.

[17] S. R. Jones, P. A. Garris, C. D. Kilts, and R. M. Wightman, "Comparison of dopamine uptake in the basolateral amygdaloid nucleus, caudate-putamen, and nucleus accumbens of the rat," *Journal of Neurochemistry*, vol. 64, no. 6, pp. 2581–2589, 1995.

[18] I. Tamai, T. Nozawa, M. Koshida, J.-I. Nezu, Y. Sai, and A. Tsuji, "Functional characterization of human organic anion transporting polypeptide b (OATP-B) in comparison with liver-specific OATP-C," *Pharmaceutical Research*, vol. 18, no. 9, pp. 1262–1269, 2001.

[19] H. H. Sitte, S. Huck, H. Reither, S. Boehm, E. A. Singer, and C. Pifl, "Carrier-mediated release, transport rates, and charge transfer induced by amphetamine, tyramine, and dopamine in mammalian cells transfected with the human dopamine transporter," *Journal of Neurochemistry*, vol. 71, no. 3, pp. 1289–1297, 1998.

[20] M. Funicello, P. Conti, M. De Amici, C. De Micheli, T. Mennini, and M. Gobbi, "Dissociation of [3H]L-glutamate uptake from L-glutamate-induced [3H]D-aspartate release by 3-hydroxy-4,5,6,6a-tetrahydro-3aH- pyrrolo[3,4-d]isoxazole-4-carboxylic acid and 3-hydroxy-4,5,6,6a-tetrahydro-3aH/ pyrrolo[3,4-d] isoxazole-6-carboxylic acid, two conformationally constrained

aspartate and glutamate analogs," *Molecular Pharmacology*, vol. 66, no. 3, pp. 522–529, 2004.

[21] S. L. Povlock and J. O. Schenk, "A multisubstrate kinetic mechanism of dopamine transport in the nucleus accumbens and its inhibition by cocaine," *Journal of Neurochemistry*, vol. 69, no. 3, pp. 1093–1105, 1997.

[22] A. Galli, C. I. Petersen, M. Deblaquiere, R. D. Blakely, and L. J. DeFelice, "Drosophila serotonin transporters have voltage-dependent uptake coupled to a serotonin-gated ion channel," *Journal of Neuroscience*, vol. 17, no. 10, pp. 3401–3411, 1997.

[23] C. L. Schonfeld and U. Trendelenburg, "The release of 3H-noradrenaline by p- and m-tyramines and -octopamines, and the effect of deuterium substitution in α-position," *Naunyn-Schmiedeberg's Archives of Pharmacology*, vol. 339, no. 4, pp. 433–440, 1989.

[24] M. Nakamura, A. Ishii, and D. Nakahara, "Characterization of β-phenylethylamine-induced monoamine release in rat nucleus accumbens: a microdialysis study," *European Journal of Pharmacology*, vol. 349, no. 2-3, pp. 163–169, 1998.

[25] R. B. Rothman, M. H. Baumann, C. M. Dersch et al., "Amphetamine-type central nervous system stimulants release norepinephrine more potently than they release dopamine and serotonin," *Synapse*, vol. 39, no. 1, pp. 32–41, 2001.

[26] R. B. Rothman, N. Vu, J. S. Partilla et al., "In vitro characterization of ephedrine-related stereoisomers at biogenic amine transporters and the receptorome reveals selective actions as norepinephrine transporter substrates," *Journal of Pharmacology and Experimental Therapeutics*, vol. 307, no. 1, pp. 138–145, 2003.

[27] D. Martinez, M. Slifstein, A. Broft et al., "Imaging human mesolimbic dopamine transmission with positron emission tomography. Part II: amphetamine-induced dopamine release in the functional subdivisions of the striatum," *Journal of Cerebral Blood Flow and Metabolism*, vol. 23, no. 3, pp. 285–300, 2003.

[28] D. Sulzer, T.-K. Chen, Y. Y. L. Yau Yi Lau, H. Kristensen, S. Rayport, and A. Ewing, "Amphetamine redistributes dopamine from synaptic vesicles to the cytosol and promotes reverse transport," *Journal of Neuroscience*, vol. 15, no. 5, pp. 4102–4108, 1995.

[29] G. M. Gilad and V. H. Gilad, "Polyamine uptake, binding and release in rat brain," *European Journal of Pharmacology*, vol. 193, no. 1, pp. 41–46, 1991.

[30] M. Inazu, H. Takeda, and T. Matsumiya, "The role of glial monoamine transporters in the central nervous system," *Japanese Journal of Neuropsychopharmacology*, vol. 23, no. 4, pp. 171–178, 2003.

10

The Principle of Stationary Action in Biophysics: Stability in Protein Folding

Walter Simmons[1] and Joel L. Weiner[2]

[1] Department of Physics and Astronomy, University of Hawaii at Manoa, Honolulu, HI 96822, USA
[2] Department of Mathematics, University of Hawaii at Manoa, Honolulu, HI 96822, USA

Correspondence should be addressed to Walter Simmons; was@phys.hawaii.edu

Academic Editor: Kuo-Chen Chou

We conceptualize protein folding as motion in a large dimensional dihedral angle space. We use Lagrangian mechanics and introduce an unspecified Lagrangian to study the motion. The fact that we have reliable folding leads us to conjecture the totality of paths forms caustics that can be recognized by the vanishing of the second variation of the action. There are two types of folding processes: stable against modest perturbations and unstable. We also conjecture that natural selection has picked out stable folds. More importantly, the presence of caustics leads naturally to the application of ideas from catastrophe theory and allows us to consider the question of stability for the folding process from that perspective. Powerful stability theorems from mathematics are then applicable to impose more order on the totality of motions. This leads to an immediate explanation for both the insensitivity of folding to solution perturbations and the fact that folding occurs using very little free energy. The theory of folding, based on the above conjectures, can also be used to explain the behavior of energy landscapes, the speed of folding similar to transition state theory, and the fact that random proteins do not fold.

1. Descriptive Introduction

Processes that proceed reliably from a variety of initial conditions to a unique final state, regardless of changing conditions, are of obvious importance in biophysics. Proteins in an appropriate solution fold to unique forms and serve as a flagship example of stable processes in biology.

In this paper, we suggest how the action principle in classical mechanics could be used to analyze the stability of the protein folding process, which is of obvious importance per se, but because the techniques described here follow from fundamental physics, this approach will also be useful in the study of the stability of other biophysical processes.

In this introduction, we present a number of technical issues in a descriptive style. Technical details are discussed in a later section.

The action principle is a traditional starting point for classical mechanics. The action is a path integral of the difference between kinetic and potential energy (the Lagrangian), between an initial and final time over a trajectory $S(t) = \int_{t_0}^{t} (T - V)dt$. (The trajectory is implicit here.) The action is a scalar. The energy terms are written in generalized coordinates which take into account some or all constraints on the motion. The use of generalized coordinates makes this formalism particularly suited to moving parts of a complicated mechanical system. Standard treatments of the action principle allow for time-dependent potentials, which is also convenient for complicated processes. Direct applications of the action principle (i.e., without necessarily using the equations of motion that arise when the first variation of the action is set equal to zero) usually entail successive approximations [1].

When applied to mechanics the vanishing of the first variation of the action immediately yields the (Newtonian) equations of motion. Thus the physical picture described is that of particles moving along trajectories according to the equations of motion.

The most important degrees of freedom in protein molecules are dihedral angles associated in pairs with amino acid residues. In a common protein there might be 500 or more such angles. In folding, the molecule starts in some random assortment of these angles and moves toward a specific native set of angles. We speak of this motion as taking place in the space of dihedral angles.

If one considers protein molecules moving along trajectories in dihedral angle space, then several things are clearly missing from the trajectory picture.

First, the trajectories evidently move toward the common end point or points along the way to the native state but there is nothing explicit in the trajectory itself to define such a convergence; an energy landscape [2–5] is usually invoked to funnel the trajectories toward the end state.

Second, trajectories coursing through a rough energy landscape would arrive at the end point over a range of times, that is, diffusion. In contrast, many molecules have narrow melting curves and fast folding times that seem more appropriate to gas phase chemistry (TST). This is currently approached by postulating that the energy landscape [3, 6] is sufficiently smooth.

Third, there is the stability of the folding process. Consider an unfolded molecule in a dilute solution of a suitable denaturing agent. Such an agent interferes with the stability of the native state of the molecule but, curiously, does not deflect the process into alternative folded forms. Similarly, many other perturbations have little or no impact upon the final folded form.

Fourth, there is the problem of initial conditions. In the current view, a folding-ensemble of denatured protein molecules begins at the top of a funnel shaped energy landscape and proceeds down the funnel to the unique native state. The various conformations at the top of the funnel are equivalent in the sense that setting various initial conditions or subjecting the molecules to various perturbations results in conformations that are still in the folding-ensemble. The trajectory picture, per se, does not address ensemble behavior; again [7, 8], the energy landscape is invoked to explain how all the molecular trajectories behave in the same way.

These issues can all be addressed from a fundamental physics starting point by considering the vanishing of the second variation of the action [9–12]. This is an approach which has had spectacular success in modern optics wherein light rays focus, especially to a caustic [13].

Before we proceed we need to define some terms. Recall that the action is a function on arcs. If the first variation with fixed endpoints is zero, then we call that arc a critical arc. If the second variation is positive for that arc, then that arc is a local minimum of the action. It is often convenient to regard the arc as part of a longer trajectory. If we fix one endpoint of the arc at a point of the trajectory and move the other along the trajectory away from the fixed endpoint, we may reach a point and thus determine an arc, for which the second variation is zero. If we move the movable endpoint even further away from the fixed endpoint, the second variation will become indefinite; that is, it can take on both positive and negative values. Typically when we have a family of trajectories starting from a fixed point or fixed initial curve or surface, they will form an envelope, that is, a curve or surface to which all the trajectories are tangent, and the points of tangency will be points along the trajectories at which the second variation vanishes. This envelope is referred to as "caustic." If the envelope happens to be a point, then we call that point a "focus."

When we have such an envelope, it dominates the motion in the sense that all of the trajectories meet it or pass through it. (In a later section of this paper we cover this subject again in more mathematical detail.)

Some excellent examples of caustics in classical mechanical systems can be found in [9].

The concept of convergence is not, as we have just said, contained explicitly in individual trajectories. Rather, the concept of convergence or focusing of mechanical trajectories is best described by considering families of trajectories. If the dynamics of particles entails a caustic, then it is possible in principle to understand how a family of trajectories can behave in a coherent manner.

We next proceed to explain how powerful theorems of R. Thom and V. Arnold can be used to understand this behavior quantitatively.

We shall not attempt to define the stability of a shape in this paper. (We refer the reader to Section 6) However, for this discussion of protein folding, stability means that the topology of a part of the native state (or of an intermediate state) is not altered by perturbations. See [14] for a similar concept.

We need two additional technical terms to be used in describing the action: state variable and control parameter. We do not require the mathematical definitions of these terms, but those definitions are readily available in the literature.

A simple way to look at state variable in a mechanical system is to think of space and time coordinates that are used to describe the motion. At a point in space and time, the description of the physical system will depend upon various control parameters. For example these may describe the interface with some apparatus. For our purposes, the control parameters in folding are not tightly defined. The shape of the caustic will be defined entirely in terms of control parameters. They are assumed to be constant after folding and may turn out to be measurable distances or angles in the native state. Excellent examples of state variables and control parameters can be found in the literature, for example, [15].

The mathematics tells us that under appropriate constraints there is a finite set of stable forms of the action near a critical point. Natural selection has evidently picked out these stable forms for biological molecules by choosing dynamics containing critical points. The stability arises because an ensemble of actions can change into one another as a result of a perturbation but the topology is not affected.

A simple example of this is the familiar cusp:

$$S(s) = \frac{1}{4}s^4 + \frac{1}{2}as^2 + bs. \qquad (1)$$

Consider the case where this is topologically stable and where s is a state variable and a and b are some variables (control parameters) that appear in the Lagrangian. Then the remarkable thing about this form is that for fixed values of a and b all possible perturbations have already been accounted for [11, 15]. So, if this is the action around the point $s = 0$, then perturbations that have the form of higher order polynomials in s do not change the shape of $S(s)$. This highly nonobvious result means that a trajectory or trajectories passing through a moderately rough energy landscape will not be topologically perturbed by small changes in terms of the state variables (other than the two included in (1)). The other remarkable fact is that this cusp is one of only seven polynomial forms which have this remarkable property. If we suppose that folding occurs only when the action takes a multitrajectory form (and natural selection has eliminated unstable folds), then there are only seven distinct types of critical points. The possible trajectories fall into families or ensembles having the property that perturbing one member of the ensemble changes it into another member of the same ensemble.

Finally, we note that the physics and mathematics show that, when critical points are present in a dynamic, the critical point dominates the motion.

We summarize what we have described so far.

(1) We start with the principle of stationary action applied to the dynamics of protein molecules.

(2) To account for folding we turn to a standard formalism for focusing.

(3) Two types of focus appear in the formalism: stable and unstable. We assume that natural selection has eliminated unstable foci.

(4) Thom's theorems now tell us that there are just seven possible functional forms for the action at a focus. Thom also tells us that these actions are stable against perturbations.

Said differently, we are shifting our attention away from individual trajectories in dihedral angle space, with particles propagating according to the equations of motion, and toward groups of trajectories that share a common multitrajectory action and which converge in dihedral angle space.

This completes a descriptive introduction to the idea of a critical point in the molecular dynamics. Before continuing this subject in more detail, we next discuss the folding process that is under examination in this paper.

2. Two-State Folder and Torsion Waves

In this section, we set up the folding problem that we wish to address in a subsequent section.

We shall focus our analysis upon two-state folders; in particular, we are interested in the nonequilibrium transitions between the denatured and folded or intermediate states [16, 17].

The molecules are not under overall tension, so transverse waves and resonances with wavelength comparable to the length of the chain are disfavored. Torsional motions, which might include some long range waves, are favored by the geometry. A plausible picture is that energy is released at various localized points resulting in waves of torsional contraction or expansion which propagate away from the production point, generally with attenuation.

The theory described here does not depend upon the torsional form of the waves.

The details will ultimately depend upon whether the waves scatter off one another. In an earlier work on a continuous backbone model, the present authors showed that solitons are a possibility. Solitons pass through one another without shape change [18].

As we have said, the action is a scalar which depends upon energy and upon the path taken by a particle. For the two state folders, the action will depend upon the path taken by a molecule from unfolded to folded states. This path may be thought of as occurring in dihedral angle space. The molecule starts with a set of dihedral angles. It changes conformation following a path through dihedral angle space for which the first variation of the action is zero.

3. Toy Model

At this juncture, we pause to introduce a toy model which is solely intended to illustrate our points (and not to address the hard realities of folding dynamics [19]).

Let us simplify the torsional wave motion to just one axial degree of rotational freedom, that is, an angle, θ, describing a torsional shear, which will serve as an overly simplified generalized coordinate, and it serves to allow us to construct a model action.

It is common in simulations of folding to introduce angular spring potentials $V(\theta) = k\theta^2$ for dihedral angles; these potentials depend upon the sequence. Critical points appear where we have a multitrajectory action, as we have emphasized above. To introduce that, we make the spring force asymmetrical. The force needed to turn a given dihedral angle depends upon the direction of turning and the angle at which it sits.

If we place the critical point $\theta = 0$ at the folded end of the dynamical path, then the kinetic energy at that end point is negligible.

Fortunately, for our purposes, a static version of this mechanical arrangement is well known in the catastrophe theory literature, where it is known as the Zeeman machine [15]. If natural selection rejects all unstable folding motions (and if this rotation is important in folding), then it turns out that the cusp in (1), with $\theta = s$, describes the potential energy, including any additional perturbations. Setting the kinetic energy aside for illustration we can now construct an action.

We get,

$$S(\theta) = \frac{1}{4}\theta^4 + \frac{1}{2}a\theta^2 + b. \qquad (2)$$

In this case, a and b will be sequence dependent. The exact relationship between those parameters and the spring constants and lever arms is found in [15]. Introducing mutations to a given molecule could either change a and/or b or disrupt the form (2) altogether.

Obviously, at $\theta = 0$, $\delta S = \delta^2 S = 0$ as per our hypothesis.

If this were a valid theory of the torsional response to a wave passing through, then that response would be independent of modest perturbations other than the last two terms.

We could also use this potential to construct probability distributions and to derive statistical moments such as $\langle \Delta\theta^2 \rangle$. The moments derived from (1) are generally of simple form and change with a and b. Note, of course, that we have not specified the action for the entire molecule here. The φ analysis of mutations will depend upon other parts of the molecule as well as this short segment. The details of moment analysis for various catastrophes are worked out in detail in [11].

We emphasize that this is a toy model which illustrates how a wave on a molecule can develop a critical point and be used in some calculations of measurable quantities and shows how the shape can be independent of perturbations, perhaps such as dissipative forces and energy rough spots.

The toy model has no detailed structure (i.e., no sequence). However, it has unsymmetrical forces that can give rise to critical points and thereby to stability. Note that the spring forces that are often used in simulations do not have these properties.

4. Addressing the Issues

With our descriptive introduction complete, we can now address the four issues listed above. We start with the assumptions that there is a critical point in the molecular dynamics and that natural selection has picked out stable folds.

The first point, that trajectories converge, is a direct implication of the presence of the critical point.

The explanation of the remaining three issues (the energy landscape is apparently smooth, the folding process is stable under modest perturbations, and the initial conditions in the denatured state do not matter very much) follows from the insensitivity of the action to perturbations. The energy landscape may have many rough spots but if they are not too extreme, then they do not change the multitrajectory action and hence do not change the time to reach the folded state.

The time to the folded state (or to an intermediate state) for a short segment of the protein is a result of two important factors: (i) a Boltzmann factor describing the escape from a potential well into the transition state and (ii) a microscopic local rate factor, γ, describing how long it takes atoms moving on a fixed trajectory to collide and bind. The rate in the unfolded to folded direction takes the form,

$$\text{Rate} \longrightarrow \gamma \exp\left(-\frac{\Delta G^\dagger}{k_B T}\right), \qquad (3)$$

where $\Delta G^\dagger$ is the transition state free energy.

This picture of motions, that occur along multiple similar trajectories until contact and bond formation take place, is compatible with the observations that the time to folding is roughly proportional to contact order [16, 20, 21]. The factor γ in (3) increases with distance along the chain and hence with contact order.

As it can be seen directly from (3), a linear dependence of the free energy upon the concentration of denaturant might look like $\Delta G^\dagger = (\Delta G_0^\dagger)C$ giving chevron plots. It is also apparent that some proteins can fold at very high rates since the atoms only have to move along specific trajectories allowed by the presence of the critical point. Said differently, diffusion takes place before the molecules cross the barrier but follow trajectories toward a specific point (in action space) thereafter. Of course, not all proteins fit this simple picture.

5. Analysis Continued

An observation that follows the semiquantitative description that we have presented so far is that some simplifications in folding result from the presence of a critical point in the molecular dynamics. For example, for two-state folders, the denatured and folded forms can both exist in equilibrium. The denaturing agent may impact the entropy but not the degrees of freedom associated with the folding. (This phenomenon is more general than protein folding. Catalysts that change the rate of a reaction by many orders of magnitude, by changing the heat flow to the thermal reservoir, without changing the reaction products, are well known [22].)

Torsion waves on molecules in solution are expected to dissipate energy. The reliability of folding in the presence of agents that change the entropy or viscosity suggests that the degrees of freedom that participate in folding in an essential way are not impacted by dissipation. The theory presented here explains using a combination of critical points and natural selection.

Another application of our theory is to address the question of why biological proteins fold to unique final forms while random polymers do not. Our theory suggests qualitatively that the former have critical points in the dynamics and fold along specific sets of trajectories while the latter do not have critical points and fold diffusively to various end shapes. Another way to look at this is that the energy landscape may be rough for random polymers and they do not share the immunity to perturbations of biological proteins.

The topomer-sampling theory of Debe and collaborators [23] considers folding in the restricted space of topomers (smooth transformations of the native conformation). That limits the number of degrees of freedom needed for diffusive search. (We remark that Thom's theory of generic stability uses functional forms that are interrelated through smooth coordinate changes (diffeomorphisms).) Our theory makes a stronger statement than topomer-sampling theory which is, that the action of the paths has critical points; allowable action must support multiple trajectories.

There are several alternative explanations for why trajectories that pass through a caustic continue to the native state. One is that the caustic is small and it sits in a steep part of the energy funnel not far from the minimum. A similar phenomenon is the formation of an alpha helix. There is an initial energy barrier, but once that is passed, the helix quickly falls into place.

Our theory has neither reached the point in development where the sequence dependence can be pinned down nor identified the dynamical relationship between critical points and specific folds. However, some comments are in order. For torsion waves, this theory clearly requires that the sequence influences the mechanical parameters of torsional motion. An important feature is multitrajectory action. A single trajectory theory, like TST for gas phase reactions, will not develop critical points; critical points are of essence due to multiple alternative paths.

A major difference between our theory and others is that here the vanishing of the second variation of the action is utilized to make connections to the existence of envelopes, that is, caustics and hence to catastrophe theory.

6. Physics, Mathematics, and the Literature

This section is a concise treatment of physics and mathematics. We document this with references, especially books, where appropriate.

General references are as follows:

For mathematics is [10].

For catastrophe theory is [11]. Other useful treatments of catastrophe theory are [15, 24] and for catastrophe theory in chemical kinetics is [25].

For physics are [9, 12, 13]. The physics and mathematics explained in the context of optics are found in [13].

For completeness we mention that the action principle, including the formation of caustics, can be derived as a short wavelength limit of quantum mechanics [11, 15]. This is explicit in the book by Schulman [26] wherein the Feynman path integral formulation of quantum mechanics is used (especially Chapter 15 on caustics in quantum mechanics). We do not use quantum mechanics in this paper.

For protein science we suggest [22, 27]. Additionally, various review papers, some of which are cited in the text, for example, [3, 5, 19, 28–30].

Early ideas underlying this work can be found in [31].

A particularly useful application of the calculus of variations to mechanics most commonly goes under the name of Hamilton's principle of least action or stationary action. Other comparable principles exist but will not be discussed here [9]. In simple situations, Hamilton's principle can be stated as follows: among all possible trajectories which take a single particle from a fixed initial position at a fixed initial time t_0 to another position at moment t_1, the realized motion is that for which the action integral

$$S = \int_{t_0}^{t_1} L(x, \dot{x}, t) \, dt \qquad (4)$$

is stationary. The integrand $L(x, \dot{x}, t) = 1/2 m\dot{x}^2 - V(x, t)$ is referred to as the Lagrangian and the variable x may be a generalized coordinate.

In order to apply Hamilton's principle we need to introduce variations of a particular arc that will be denoted by $x_0(t)$. We do this by introducing an arbitrary function $\xi(t)$, with $\xi(t_0) = \xi(t_1) = 0$, a real parameter ε which is usually viewed as being very small, and considering the family of curves:

$$x_\varepsilon(t) = x_0(t) + \varepsilon \xi(t). \qquad (5)$$

One thus obtains the action as a function of ε, $S(\varepsilon)$. The implication of the first variation being stationary, that is, $dS/d\varepsilon = 0$ for all possible families x_ε, is that the realized motions must satisfy what is called the Euler-Lagrange equation:

$$\frac{d}{dt} \frac{\partial L}{\partial \dot{x}} - \frac{\partial L}{\partial x} = 0. \qquad (6)$$

We compute the second variation for curves that satisfy the Euler-Lagrange equations; that is, we compute $(d^2S/d\varepsilon^2)(0)$. We get,

$$\frac{d^2 S}{d\varepsilon^2}(0) = \int_{t_0}^{t_1} \left[-\xi^2 \frac{\partial^2 V}{\partial x^2} + m\dot{\xi}^2 \right] dt. \qquad (7)$$

To deal with the integral one can expand $\xi(t)$ in terms of special functions appropriate to the behavior of x_0 and use their properties to simplify the above integral. We shall not explore that subject here.

When the second variation is positive, meaning that $(d^2S/d\varepsilon^2)(0) > 0$, for all x_ε, one can show that x_0 minimizes S compared to all arcs "close to" x_0 with the same endpoints.

To appreciate the sign of the second variation one wants to consider families of trajectories of realized motions. The simplest situation to consider is that for which the action S is independent of the parameterization. We then consider such a family, where each trajectory begins at a particular point, or along a particular curve, or on particular surface. Let us suppose all the motions are to begin along a curve which is parameterized using a parameter u. Then we can regard S as a function of u which identifies the initial point and the position x, which is thought of as the terminal point of a curve realizing the motion; thus we write $S(u, x)$. To obtain an arc that achieves a minimum value of the action from the initial curve to a particular x_1, we solve $(\partial S/\partial u)(u, x_1) = 0$ for u. If u_1 is a solution, then $(\partial S/\partial u)(u_1, x) = 0$ is the equation for an arc of a realized motion through x_1 that minimizes S if $(\partial^2 S/\partial u^2)(u_1, x_1) > 0$. If on the other hand $(\partial^2 S/\partial u^2)(u_1, x_1) = 0$, that is, the second variation is zero, this then identifies the presence of a caustic and x_1 is a member of the caustic.

The curves defined by $(\partial S/\partial u)(u, x) = 0$ form an envelope and x_1 is the point at which the curve $(\partial S/\partial u)(u_1, x) = 0$ through x_1 is tangent to the envelope. In higher dimensions, simple derivatives are replaced by partial derivatives and the second derivative is replaced by the Hessian, that is, the matrix of second derivatives. The vanishing of the ordinary second derivative is replaced by the vanishing of the determinant of the Hessian.

In typical applications of singularity theory, or catastrophe theory, one considers a given function of several variables, which are referred to as state variables and control variables. One focuses on the form of sets determined by setting to zero the first and second derivatives of the given function with respect to state variables and takes advantage of known generic solutions of those equations for certain numbers of control variables. The function $S(u, x)$ introduced above, where u is a state variable and x is a control variable, fits this situation since we are clearly interested in sets where $(\partial S/\partial u)(u, x) = 0$ and $(\partial^2 S/\partial u^2)(u, x) = 0$. This is how one can establish a connection between calculus of variations and catastrophe theory.

Noticing that geometric optics fails in this situation, Berry and Upstill comment that this failure is "catastrophic" because this is just the point at which catastrophe theory becomes applicable.

Where caustics are present we have a strong focusing of trajectories into a space that is very closely circumscribed by up to five control parameters [11]. A critical point, where the above matrix vanishes, dominates the dynamics in that neighborhood around that point.

The conditions on the partial derivatives just mentioned are the same as the conditions for catastrophe theory to obtain. There are many texts on catastrophe theory so we will just remark that stability of the catastrophe (caustic) against perturbations of the state variables (time and space) is the major result we have used.

Let us consider the limitations of this conjecture.

An important issue is the degree of sensitivity to perturbations. For example, the strength of the denaturing agents may be so great that our conjecture may not apply. There appears to be no general rule from catastrophe theory to quantify the limits of allowed perturbations; the answer is in the details.

Another limitation is the possible appearance or nonappearance of false minima. Further research will be required to understand this issue. In a special case, however, if there is a single, long-lived false minimum (with the molecules slowly leaking down to the thermodynamic minimum), then our conjecture may apply to the false minimum. We remark that prions may be such a case.

7. General Comments

We emphasize that in this paper we are not addressing the issue of protein structure [22, 30]; rather, we are addressing the issue of the stability of the folding process, especially the earliest stage from denatured ensemble forward. Prospectively, a full understanding of the early phases is potentially very useful in experiment design where it is necessary to evaluate the impact of various external factors on the shape of the native state (e.g., fluorescence resonance energy transfer (FRET), various denaturants, etc.), in the design or discovery of agents or environmental factors [32], that interfere with folding, and in engineering new proteins that fold in specific ways. Retrospectively, an understanding of the stability of the earliest phases of folding has potential value in the study of the last common ancestor and the origin of life.

8. Future Directions

The calculations to check these ideas in model molecules are not simpler than traditional folding simulations. However, the results are different.

If the putative caustic appears somewhere along the folding path, then a simulation of the action up to that point can reveal it. The quantitative test is the vanishing (or near vanishing) of the Hessian determinate, as described in the previous section. The number of computations is formidable but, for simple molecules, not beyond supercomputer capacity.

Should a caustic be indicated along the folding path? This can be confirmed by searching for a saddle point in the action as the integration is continued to a point just beyond the caustic. Again, the computations are formidable but not impossible.

Once a caustic is found in a model, it will become possible to tune model parameters to optimize folding.

9. Conclusions

The phenomenon studied is the motion of protein molecules in a variety of initial conditions, in the presence of various perturbations, terminating in a unique final state in spite of the relatively little free energy available.

The principle of stationary action leads immediately to equations of motion. However, equations of motion describe the propagation of the molecule along trajectories in dihedral angle space and tend to obscure the behavior of groups of trajectories. Moreover, as noted originally by Levinthal, the number of conformations in dihedral angle space is of cosmological proportions.

By treating the problem directly using the principle of stationary action and putting the equations of motion aside, it is possible to treat groups of trajectories that behave in a similar manner, in particular, trajectories that converge either to a focus or to a caustic.

The resulting treatments narrow the number of possible paths through dihedral angle space because all trajectories pass through very narrow caustics located somewhere along the folding path.

The result of direct analysis from the action is that two types of focusing emerge: stable and unstable. We assume that natural selection has eliminated the unstable focusing. This treatment leads immediately to the strong stability of the process of folding. Many features of the folding process emerge directly.

As in most biological processes, protein folding entails a large number of complicated forces and parameters that change with conformation (i.e., with time during folding) and, as just mentioned, it takes place in a space of very high dimension. Yet folding does indeed lead to unique final forms even in the presence of denaturing agents that are chosen to disrupt the final shape.

This complexity might be enough to send any theorist back to his coffee pot. However, when this is approached from the viewpoint of the direct application of principle of stationary action, an idea takes shape rather naturally. The idea is that the fundamental dynamics of molecular motion contains critical points that dominate the motion and make the motion less vulnerable to disruption by various changing forces and conditions. This dominance of dynamical behavior by critical points is well established in physics.

We have shown how this idea emerges from the action principle and have given semiquantitative explanations for many of the phenomena that have been documented in laboratories and in simulations over the past five or six decades.

For theories that start with the molecule in its native state, unfold it in the lab or in simulation and then refold it; the acid test is prediction of the native form. By starting with the denatured state and applying physics and mathematics to study the stability of the folding process, we have only a germ of a full theory of folding and we cannot predict structures, not even approximately.

No theories, which are consistent with classical mechanics, are in contradiction with the least action and the vanishing of the first variation of the action for the dynamics of the molecule during folding. The fundamental departure embodied in this work is the putative vanishing of the second variation of the action, implying that various trajectories can be treated as a unit, and the role of natural selection in eliminating unstable folds.

What we have accomplished is to show that these putative critical points provide a level of quantitative understanding of many observed features as follows: rapid rates (TST like behavior), smoothness of the energy landscape, nonfolding of random polymers, insensitivity to many perturbations, and some qualitative insights into other features of folding such as the importance of topology and contact order.

The obvious next steps in the development of this theory are to learn exactly how the critical points emerge in terms of sequence and to learn how the critical points relate to specific structures.

References

[1] G. L. Eyink, "Action principle in nonequilibrium statistical dynamics," *Physical Review E*, vol. 54, no. 4, pp. 3419–3435, 1996.

[2] J. N. Onuchic and P. G. Wolynes, "Theory of protein folding," *Current Opinion in Structural Biology*, vol. 14, no. 1, pp. 70–75, 2004.

[3] P. G. Wolynes, "Recent successes of the energy landscape theory of protein folding and function," *Quarterly Reviews of Biophysics*, vol. 38, no. 4, pp. 405–410, 2005.

[4] M. Oliveberg and P. G. Wolynes, "The experimental survey of protein-folding energy landscapes," *Quarterly Reviews of Biophysics*, vol. 38, no. 3, pp. 245–288, 2005.

[5] K. A. Dill, S. B. Ozkan, M. S. Shell, and T. R. Weikl, "The protein folding problem," *Annual Review of Biophysics*, vol. 37, pp. 289–316, 2008.

[6] K. W. Plaxco, K. T. Simons, I. Ruczinski, and D. Baker, "Topology, stability, sequence, and length: defining the determinants of two-state protein folding kinetics," *Biochemistry*, vol. 39, no. 37, pp. 11177–11183, 2000.

[7] R. A. Goldbeck, Y. G. Thomas, E. Chen, R. M. Esquerra, and D. S. Kliger, "Multiple pathways on a protein-folding energy landscape: kinetic evidence," *Proceedings of the National Academy of Sciences of the United States of America*, vol. 96, no. 6, pp. 2782–2787, 1999.

[8] D. T. Leeson, F. Gai, H. M. Rodriguez, L. M. Gregoret, and R. B. Dyer, "Protein folding and unfolding on a complex energy landscape," *Proceedings of the National Academy of Sciences of the United States of America*, vol. 97, no. 6, pp. 2527–2532, 2000.

[9] C. G. Gray and E. F. Taylor, "When action is not least," *American Journal of Physics*, vol. 75, no. 5, pp. 434–458, 2007.

[10] I. M. Gelfand and S. V. Fomin, *Calculus of Variations*, Dover, New York, NY, USA, 2000.

[11] R. Gilmore, *Catastrophe Theory for Scientists and Engineers*, John Wiley & Sons, New York, NY, USA, 1981.

[12] M. V. Berry, "Waves and Thom's theorem," *Advances in Physics*, vol. 25, no. 1, p. 1, 1975.

[13] M. V. Berry and C. Upstill, "IV catastrophe optics: morphologies of caustics and their diffraction patterns," *Progress in Optics C*, vol. 18, pp. 257–346, 1980.

[14] D. E. Makarov and K. W. Plaxco, "The topomer search model: a simple, quantitative theory of two-state protein folding kinetics," *Protein Science*, vol. 12, no. 1, pp. 17–26, 2003.

[15] T. Poston and I. Stewart, *Catastrophe Theory and Its Applications*, Pitman, London, UK, 1978.

[16] R. Zwanzig, "Two-state models of protein folding kinetics," *Proceedings of the National Academy of Sciences of the United States of America*, vol. 94, no. 1, pp. 148–150, 1997.

[17] H. Kaya and H. S. Chan, "Origins of Chevron Rollovers in non-two-state protein folding kinematics," *Physical Review Letters*, vol. 90, no. 25, Article ID 258104, 2003.

[18] W. Simmons and J. Weiner, "Protein folding: a new geometric analysis," http://arxiv.org/abs/0809.2079.

[19] S. C. Harrison, "Forty years after," *Nature Structural and Molecular Biology*, vol. 18, no. 12, pp. 1305–1306, 2011.

[20] K. W. Plaxco, K. T. Simons, and D. Baker, "Contact order, transition state placement and the refolding rates of single domain proteins," *Journal of Molecular Biology*, vol. 277, no. 4, pp. 985–994, 1998.

[21] K. Chou and C. Zhang, "Predicting protein folding types by distance functions that make allowances for amino acid interactions," *Journal of Biological Chemistry*, vol. 269, no. 35, pp. 22014–22020, 1994.

[22] A. Fersht, *Structure and Mechanism in Protein Science*, W.H. Freeman and Company, New York, NY, USA, 1998.

[23] D. A. Debe, M. J. Carlson, and W. A. Goddard III, "The topomer-sampling model of protein folding," *Proceedings of the National Academy of Sciences of the United States of America*, vol. 96, no. 6, pp. 2596–2601, 1999.

[24] P. T. Saunders, *An Introduction to Catastrophe Theory*, Cambridge University Press, Cambridge, UK, 1980.

[25] A. Okninski, "Catastrophe theory," in *Chemical Kinetics*, R. G. Compton, Ed., Elsevier, Amsterdam, The Netherlands, 1992.

[26] L. S. Schulman, *Techniques and Applications of Path Integration*, John Wiley & Sons, New York, NY, USA, 1981.

[27] A. V. Finkelstein and O. B. Ptitsyn, *Protein Physics*, Academic Press, New York, NY, USA, 2002.

[28] M. Karplus, "Aspects of protein reaction dynamics: deviations from simple behavior," *Journal of Physical Chemistry B*, vol. 104, no. 1, pp. 11–27, 2000.

[29] H. A. Scheraga, M. Khalili, and A. Liwo, "Protein-folding dynamics: overview of molecular simulation techniques," *Annual Review of Physical Chemistry*, vol. 58, pp. 57–83, 2007.

[30] K. C. Chou and C. T. Zhang, "Prediction of protein structural classes," *Critical Reviews in Biochemistry and Molecular Biology*, vol. 30, no. 4, pp. 275–349, 1995.

[31] W. Simmons and J. Weiner, "Toward a theory on the stability of protein folding: challenges for folding models," http://arxiv.org/abs/1112.6190.

[32] J. Tyedmers, M. L. Madariaga, and S. Lindquist, "Prion switching in response to environmental stress," *PLoS Biology*, vol. 6, no. 11, pp. 2605–2613, 2008.

11

Thermal Aggregation of Recombinant Protective Antigen: Aggregate Morphology and Growth Rate

Daniel J. Belton[1] and Aline F. Miller[2]

[1] Department of Chemical and Biological Sciences, University of Huddersfield, Queensgate, Huddersfield HD1 3DH, UK
[2] School of Chemical Engineering and Analytical Science and Manchester Interdisciplinary Biocentre, University of Manchester, 131 Princess Street, Manchester M1 7DN, UK

Correspondence should be addressed to Aline F. Miller; aline.miller@manchester.ac.uk

Academic Editor: P. Bryant Chase

The thermal aggregation of the biopharmaceutical protein recombinant protective antigen (rPA) has been explored, and the associated kinetics and thermodynamic parameters have been extracted using optical and environmental scanning electron microscopies (ESEMs) and ultraviolet light scattering spectroscopy (UV-LSS). Visual observations and turbidity measurements provided an overall picture of the aggregation process, suggesting a two-step mechanism. Microscopy was used to examine the structure of aggregates, revealing an open morphology formed by the clustering of the microscopic aggregate particles. UV-LSS was used and developed to elucidate the growth rate of these particles, which formed in the first stage of the aggregation process. Their growth rate is observed to be high initially, before falling to converge on a final size that correlates with the ESEM data. The results suggest that the particle growth rate is limited by rPA monomer concentration, and by obtaining data over a range of incubation temperatures, an approach was developed to model the aggregation kinetics and extract the rate constants and the temperature dependence of aggregation. In doing so, we quantified the susceptibility of rPA aggregation under different temperature and environmental conditions and moreover demonstrated a novel use of UV spectrometry to monitor the particle aggregation quantitatively, *in situ*, in a nondestructive and time-resolved manner.

1. Introduction

The study of protein aggregation is a burgeoning field of research driven by the urgent need to elucidate the mechanism of neurodegenerative diseases, the desire to understand and mimic natures' ability to create hierarchical complex nanostructures, and the necessity to understand and minimise product loss during the processing and formulation of biopharmaceuticals. Aggregation is of particular importance for therapeutic proteins as it can lead to a loss of product, reduce efficacy, alter biological activity and pharmacokinetics, and even raise safety concerns such as increased immunogenicity [1–3]. Aggregation can be induced by solution conditions such as protein concentration, pH, salinity, temperature, and the presence of additives [4, 5]. These variables are also known to affect the quantity and morphology of aggregate formed [5]. Stresses to the protein

such as over-expression, refolding, freeze-thaw cycles, agitation, or exposure to hydrophobic surfaces or air (including foaming) can also lead to the formation of aggregates [2, 4–6]. Each of these environmental factors is typically encountered during bioprocessing, downstream processing, storage, and also during and after *in vivo* delivery of biopharmaceutical actives. Hence, there is significant on going work channelled into exploring the onset of aggregation and the aggregation pathway. Understanding these factors subsequently allows the development of an informed strategy to minimise aggregation during biopharmaceutical production, for example, the inclusion of surfactants to influence protein monomer interactions.

The influence of solution pH on aggregation is one of the more studied and important parameters that controls the onset of aggregation and final aggregate morphology [5]. This is because pH alters the surface charge of the protein monomer and also the extent of any structural disruption prior to aggregation, and hence influences their propensity to self-assemble and the manner in which they go onto aggregate. For example, at pH values close to the isoelectric point, repulsive interactions between native monomers are reduced, making assembly more favourable. Under these conditions, particulates typically form [7]. At pHs far from the isoelectric point, increased charge repulsion within the protein destabilises the folded conformation, leading to the exposure of hydrophobic groups, which in turn can drive the self-assembly of these unfolded states, typically into β-sheet rich fibrillar structures [8, 9]. Temperature is another key factor, if not the most critical factor in commercial processes that induces aggregation, where increasing the temperature increases the vibrational motion and diffusion of proteins which is a necessary step for aggregation. Moreover, as the temperature nears the denaturation temperature of the protein, the protein partially unfolds, exposing hydrophobic regions, which induces aggregation [10].

The literature is awash with many studies that postulate different models for the self-assembly of proteins, and these can generally be divided into two main categories: empirical or mechanistic [11]. Mechanistic models are based on a reaction scheme and have parameters relating to the kinetics/thermodynamics of the process, whereas empirical models utilise functions that fit the data but have no physical meaning. A large array of different mechanistic models have been proposed, and these have been broadly categorised as monomer addition, reversible association, prion aggregation, minimalistic 2-step, or quantitative structure-activity relationships [11]. Alternatively, Roberts proposed a comprehensive generic mechanistic scheme for protein aggregation with associated generic equations, and he went onto show how these could be simplified for certain conditions and limiting cases [12]. Overall, most of these models are based on the idea of aggregation being mediated by a reactive intermediate which is in equilibrium with the native state; the intermediate is able to aggregate, initially via a nucleation step followed by a growth phase, that is, Native (N) $\leftrightarrow$ Intermediate (I) $\rightarrow$ Aggregate (A). Many groups have experimentally tested such models, and techniques have been developed and exploited to study aggregation both *ex situ* and *in situ* [13, 14]. In *ex situ*, the state of aggregation is measured in static samples, where long data acquisition times are needed, or samples need to be separated, dried, fixed or labelled, for example, using electron microscopy or mass spectrometry. Furthermore, the aggregation process has to be reliably halted and persevered for a series of samples representing different stages. In *in situ* measurements, aggregation events are monitored as they happen, but the technique used needs sufficient time resolution for the process being monitored and the data analysis can be complex, since there is typically a mixture of component sizes, that is, aggregates and monomers. Typical techniques currently used include circular dichroism, Fourier transform infra-red, and dynamic light scattering [13, 15]. For a technique to yield useful information about the aggregation process, it is crucial to be able to relate the measured parameter back to changes in the aggregation state such as aggregate size or monomer depletion.

Here, the thermal aggregation of an industrially relevant biopharmaceutical recombinant protective antigen (rPA) (active component in a second-generation anthrax vaccine [16]) has been examined visually and via turbidity measurements, before information on the size and the shape of the aggregates formed were obtained using a combination of optical and environmental scanning electron microscopies. Two environmental conditions were explored, where samples were prepared with and without the denaturant urea. Subsequently, the rate of aggregate growth at different isothermal temperatures was compared using ultraviolet light scattering spectroscopy (UV-LSS) [17–23], with particle size being extracted as a function of time. This was done by collecting spectra over time without the need to remove material for analysis, hence demonstrating a nondestructive, *in situ*, and time-resolved method for monitoring the aggregation process. Results for the aggregation of rPA will be discussed, and a model describing the kinetics and thermodynamics of aggregation is presented.

2. Materials and Methods

2.1. Materials. rPA was supplied by Avecia Biologics (UK) at $2\,mg\,mL^{-1}$ in a phosphate-buffered saline solution adjusted to pH 7.4. Doubly distilled water was obtained from an Elga PureLab Ultra (18.2 Ω). All other chemicals were purchased from either Sigma-Aldrich (UK) or Acros Organics (UK), where the reagent grade is at least 97% pure and used as recieved.

2.2. Sample Preparation. The rPA samples supplied were treated using the following procedure to provide a consistent starting material for analysis. Initially, rPA was precipitated from solution by heating (50°C for circa 5 min). The resulting gel was centrifuged at 6000 rpm for 1 minute, and the supernatant was discarded. The gel was resuspended by adding doubly distilled water and vortexing for 2 min. The sample was centrifuged as before, and the supernatant was discarded. The gel was resolubilized by adding urea and doubly distilled water to form an 8 M urea solution. rPA was refolded by dilution, using 1 part rPA in 8 M urea to 31 parts refold buffer. This was done in two stages, with a 1:7 dilution followed by 1:4 dilution 2 minutes later. The refold buffer contained 25 mM TRIS, 25 mM NaCl, 2 mM CaCl$_2$ and was adjusted to pH 7.4 with hydrochloric acid. The final buffered samples contained between circa 0.15 and 0.3 mg mL^{-1} rPA with 0.25 M urea and were analysed immediately after refolding. This concentration was chosen as it mimics the conditions of storage of some of the formulations of protective antigen vaccines. Samples without urea were prepared using the refolding method followed by dialysis using 3500 Dalton molecular weight cutoff Visking dialysis membrane (Medicell International Ltd) against 10 times excess of chilled refold buffer for 18 hours in the refrigerator (~4°C). After this, the external solution was replaced twice with fresh refold buffer and allowed to dialyse for a further 24 hours. The rPA sample was then recovered from the sealed membrane and stored in the refrigerator (circa 4°C) prior to use. This was done to reduce any degradation of the rPA during storage. Samples, however, were not frozen to avoid potential freeze-thaw damage.

2.3. Concentration Analysis. The concentration of the refolded rPA solution was determined from its UV absorbance at 280 nm (Shimadzu UV 2501-PC spectrophotometer), using a molar absorption coefficient of $72769 \, M^{-1} \, cm^{-1}$.

2.4. Visual Observations. The isothermal aggregation of rPA was monitored visually over time by incubating samples over a range of temperatures (43–49°C) close to the denaturation temperature of rPA (50.0°C) by using a recirculating water bath connected to a heating stage that contained the sample cell. The temperature was measured using a calibrated K-type thermocouple placed in the sample solution (accurate to ± 0.3°C).

2.5. Optical Microscopy. Images of rPA aggregates were obtained using a Zeiss Axioplan 2 in transmission mode with a 10x magnification objective and a digital camera. The aggregated samples were pipetted onto a microscope slide and a cover slip placed on top.

2.6. Environmental Scanning Electron Microscopy (ESEM). Aggregates were mounted for ESEM analysis simply by lifting them out of solution on a mica disc and placing them directly onto the sample stage. The samples were examined using a Philips FEI Quanta 200 ESEM with the electron gun accelerating voltage set to 30 kV, the sample stage at 5°C, and the chamber pressure at 6 torr.

2.7. Light Scattering. Light scattering spectra were recorded using a Shimadzu UV 2501-PC spectrophotometer set to record between 250 and 390 nm every 2 nm at a medium scan rate. The sample was heated *in-situ* as previously described.

2.8. Theory. The wavelength dependence of absorbance arising from light scattering by particles in solution follows the relationship [24]

$$A_\lambda = \alpha \lambda^{-\beta}, \tag{1}$$

where A_λ is the absorbance at wavelength λ, α is a constant and β is the scattering exponent. The scattering exponent can be related to particle size using Mie theory. Here, the Mie equations were solved over a range of particle sizes (nanometres to micrometres) for wavelengths between 320 and 390 nm using a FORTRAN programme adapted from Bohren and Huffman [24]. These calculations required the refractive index of the particles and the surrounding solution for all relevant wavelengths. The solution refractive index was taken to be that of water [25] and the particle refractive index was calculated by evaluating the Lorentz-Lorenz molar refraction [26] using the chemical formula of rPA and a density of $1.43 \, g \, cm^{-3}$. The density was based on the molecular weight of rPA (82667 Da) and its volume ($95.74 \, nm^3$) as calculated by VADAR [27] using the crystal

Table 1: Refractive index of rPA particles at different wavelengths calculated using the Lorentz-Lorenz molar refraction.

Wavelength/nm	Refractive index, n
434.0	1.684
486.1	1.677
589.3	1.665
656.3	1.661

structure of rPA [28], which is stored in the protein data bank (http://www.pdb.org/) [29] under PDB ID : 1ACC. This value of density was considered reasonable compared to the density of other proteins with a similar molecular weight [30]. The values of refractive index, n, calculated for the particles at different wavelengths, λ, within the visible region are given in Table 1. These values were subsequently extrapolated to the UV region using the Cauchy equation [26];

$$n = 1.642 + \frac{7980}{\lambda^2}. \tag{2}$$

The results obtained using Mie theory were analysed to give a theoretical scattering exponent versus diameter, providing a means of converting experimental scattering exponents to particle size:

$$d = -0.686\,\beta^6 + 6.869\,\beta^5 - 21.397\,\beta^4 + 23.795\,\beta^3$$
$$- 23.682\,\beta^2 - 111.896\,\beta + 642.207, \tag{3}$$

where d is the particle diameter (nm) and β is the scattering exponent between 320 and 390 nm.

3. Results and Analysis

3.1. Qualitative Aggregation. Solutions of rPA ($0.31 \, mg \, mL^{-1}$) were incubated over a range of temperatures (25–50°C), and their visual appearance was monitored over time. This temperature range was selected as the denaturation temperature of rPA is known to be ~50°C [31]. Sample appearance was observed to change when the temperature increased above ~43°C. A typical example of these changes is recorded in Figure 1, which shows the visual appearance of rPA in solution over time whilst held at a steady temperature of 47.6°C. Initially, the sample was clear, as can be seen from Figure 1(a), and remained clear over the first few minutes. After 8 minutes, the sample started to become cloudy, which is evident in Figure 1(b). The slightly cloudy homogeneous appearance of the sample indicates the presence of microscopic particles of sufficient size and concentration to noticeably scatter visible light. Such behaviour is indicative of the aggregation of rPA. Over the minutes that followed, the sample remained homogeneous but gradually became more turbid, suggesting further aggregation. This can be seen by comparing the images in Figures 1(b) and 1(c); the latter appears cloudier. After 32 minutes, the sample was no longer homogeneous, since particles large enough to be observed visually started to form (see Figure 1(e)). From this point forwards, the black background was removed to provide improved contrast for observing the macroscopic particles that were forming in the solution; the sample remained cloudy but appeared brighter, since light was able to enter from behind. The macroscopic particles were more easily observed and appeared as dark spots against the bright cloudy solution. Following their emergence, the macroscopic particles were observed initially to increase in size and number as can be seen from the images shown in Figures 1(e) and 1(f). Beyond this point, the number of individual particles appeared to fall whilst continuing to grow in size (Figure 1(g)). This observation suggests that the particles grew by clustering together. The final image of the sample, Figure 1(h), was taken with the black background replaced to enable comparison with the initial sample appearance. It shows that the sample had returned to a clear solution apart from the presence of large white particles, some of which had settled on the bottom and the sides of the cuvette. Such observations suggest that small microscopic particles form initially, causing the sample to appear turbid, and subsequently cluster to form large macroscopic particles that sediment due to gravity when they are above a critical size.

These visual results were complemented by turbidity measurements recorded under identical conditions (see Figure 2); the labels ("a" to "h") in Figure 2 are positioned to relate to the photographs taken of rPA aggregation shown in Figure 1. The turbidity was assessed by recording the optical density using light with a wavelength of 320 nm. Figure 2 shows a sharp rise in turbidity over the first 20 minutes of incubation. This is consistent with the appearance and growth of particles in the solution, as indicated by the visual observations. The subsequent fall in turbidity suggests that the particles were either decreasing in size (which might occur if the aggregation was reversible) or decreasing in concentration (which could occur if the particles clustered to form larger structures). The latter is consistent with our visual observation of particles growing in size and reducing in number over time, before settling to the bottom of the sample cuvette.

Similar visual and turbidity observations were noted for other samples incubated between 43° and 49°C, as all appeared to have a similar two-step mechanism. The time scales for each step varied widely; however, for the different temperatures, the time to reach peak turbidity ranged from circa 10 minutes at 49°C to 10 hours at 43°C. This dramatic increase in aggregation rate versus temperature could be driven by either increasing the translational kinetic energy of protein monomers, causing more frequent collisions between the particles, or increasing the internal kinetic energy of the protein monomers. The latter will drive monomers toward an unfolded state, with the associated exposure of hydrophobic patches increasing the likelihood that a collision results in self-association.

3.2. Aggregate Morphology. The morphology of the macroscopic particles formed after incubation was examined using optical and electron microscopy. Figure 3(a) shows a typical optical micrograph of the rPA aggregates formed after 16 hours at 47°C. The micrograph reveals the formation of

several aggregates with different shapes and sizes, but all appear to be composed of microscopic particles. This is consistent with the visual observation of a cloudy solution of microscopic particles which cluster to form large aggregates. The aggregates generally range in length from ~75 to 730 μm and have a width of ~5–150 μm. Three aggregates with fairly distinct structures have been labelled α, β, and γ in Figure 3(a). The smallest, labelled α, is 165 μm long and 30 μm at its widest point. This aggregate appears reasonably linear with growth predominantly in one direction with limited side branching. The other two aggregates, β and γ, are larger and have similar dimensions: ~380 μm by 150 μm. However, γ has a very dense structure, whilst β has a more open structure. The open structure of β consists of a branched system of linear components similar to aggregate α, suggesting that α is a precursor to the formation of β: either as one of many components that come together or as the starting point for further growth or a combination of both. The structure of γ is densely packed; however, there is still some evidence of a branched structure and these are similar to the structural features observed in β. Suggesting that β is possibly a precursor to the formation of γ. Such observations and inferences are also supported by previous work on fractal aggregates, reviewed by Meakin [32]. Such comparisons suggest that the rPA aggregates described here are fractal aggregates, confirming that they are formed by the clustering of microscopic particles. To obtain further information on the size and the shape of the microscopic particles, the aggregates were viewed under ESEM and a typical micrograph is given in Figure 3(b). The microscopic particles appeared reasonably spherical and uniform with a diameter of approximately 500 nm. This observation is consistent with the previously reported results for the thermal aggregation of proteins close to their isoelectric points, which showed the formation of monodisperse particulates [7]. Here, we are working at pH 7.4, which is close to the isoelectric point of rPA (pH 5.6) [16]; therefore, individual protein monomers will have reduced net charge. In addition, the salt present in the buffered media will screen any remaining charge on the proteins, thus any long-range charge-charge repulsions which could act as a barrier to aggregation will have been minimised. The protein monomer is likely to be partially unfolded as the temperature approaches the denaturation temperature (~50°C). The exposed hydrophobic regions will consequently drive nonspecific monomer aggregation under these conditions of reduced net charge. This in turn leads to the formation of the three-dimensional spherical aggregates, as observed in Figure 3(b). The approximately uniform size of the particles in each sample suggests that their concentration remained reasonably constant during their growth which is consistent with the features of protein aggregation summarised by Gosal and Ross-Murphy [10]. A slight increase in particle diameter from 360 to 500 nm was noted as the incubation temperature increased from 43 to 49°C.

3.3. Particle Growth Rate. The rate of aggregation at different isothermal temperatures was explored using UV light scattering spectroscopy (UV-LSS), where spectra were recorded

FIGURE 1: Photographs taken at different time points of rPA aggregates forming in solution at 47.6°C. Note that each of (a) to (c) and (h) has a black background placed behind the cuvette and (d) to (g) have no background allowing light to enter the sample from behind.

every 20 seconds initially and then every 20 seconds to 40 minutes depending on the aggregation rate. Such fast acquisition times provide a speed advantage over other light scattering techniques for analysing such aggregation kinetics. Typical results for incubation at 45°C are shown in Figure 4(a). It is clear that there was no absorbance initially (no chromophores in the protein absorb over this wavelength range), but the absorbance intensity increases over time, where the difference between successive spectra is large initially but reduces over longer times. Minimal difference was observed after ~200 minutes. This implies that the aggregates grew quickly initially when the rPA monomer concentration was at its highest, followed by slowing growth as the rPA monomers were consumed and falling in concentration. This suggests that the particle growth rate is limited by rPA monomer concentration.

As discussed previously, rPA solutions became turbid during incubation at temperatures ≥43°C; therefore, it can be assumed that multiple light scattering was occurring. It can be assumed, however, that the wavelength dependence of light scattering obtained by solving the Mie equations would hold

for the scattering occurring here. As such, the UV spectra in Figure 4(a) were analysed (see Section 2.8) to give aggregate diameter versus time; see Figure 4(b). It is clear that the particles grew quickly over the initial ~60 min before gradually slowing and reaching a final particle size after ~300 min at this temperature. The equilibrium sizes of particles formed increased slightly from 390 to 500 nm with increasing the incubation temperature, which correlates well with the ESEM observations (~360–500 nm). This confirms our assumption that the wavelength dependence of light scattering in a turbid solution can be adequately approximated by Mie theory.

To explore the effect of isothermal temperature on particle growth rate and consequently gain an insight into the aggregation kinetics, particle size was recorded as a function of time for a range of temperatures (43°–49°C) and the results are given in Figure 5. The aggregate diameter versus time profile was fitted with

$$d_{\mathrm{model}} = d_f - d_0 \cdot e^{-t/\tau}, \tag{4}$$

where d_{model} is the model fit to the particle diameter (nm), d_f is the final particle diameter that the function converges

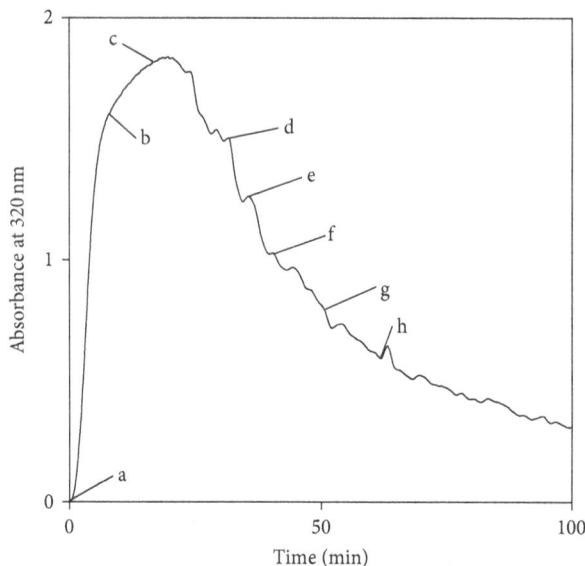

FIGURE 2: Absorbance at 320 nm versus time for rPA held at 47.7°C (×). Labels "a" to "h" relate to the photographs shown in Figure 1.

on (nm), d_0 is a fitting parameter (nm), τ is a time constant for the particle growth process (s), and t is time (s). The optimum fit was found using Newton's method to maximise the correlation coefficient, R^2, whilst allowing the value of d_f to vary. The values of d_0, τ, and correlation coefficient were calculated from a straight line fit (using least squares regression) to a plot of $\ln(d_f - d_{\text{model}})$ versus time. The model fit for each run is included in Figure 5.

All profiles show an increase in diameter over time; however, the time taken to reach the maximum diameter is markedly different for each temperature: the higher the temperature, the shorter the time scale to reach the maximum diameter. This is reflected in the results by a notable increase in the profile gradient when the temperature is increased, where the maximum gradient increases by 2.7 times on average for every 1°C increase in temperature. Similar results were obtained for a range of samples containing an additive: 0.25 M urea. In this case, similar trends were observed over time for a range of incubation temperatures, and the only difference arising was in the rate of particle growth being slightly faster for each sample in the presence of 0.25 M urea. All samples contained a homogeneous distribution of particle size (confirmed by ESEM) suggesting that the aggregate concentration remained constant over time. This means that the quantity of rPA in the aggregates can be determined by estimating the aggregate concentration and using the ratio of aggregate to monomer volume. From this, the difference between the quantity of rPA incorporated within aggregates and the quantity of monomer present initially gave monomer concentration over time, providing a possible means of assessing the aggregation kinetics. To this end, data over initial incubation times (where the scattering was increasing) were analysed using a generalised scheme of the protein aggregation pathway. This scheme is outlined in Figure 6, where N is the native state, I is the protein monomer

in an intermediate (unfolded or denatured) conformational state, A_j is an aggregate consisting of j protein molecules, $(A_m)_n$ is a cluster of particulate aggregates, and k_1 to k_5 are the rate constants for the different processes. This aggregation pathway can be simplified based on the aforementioned experimental observations; it was considered reasonable to exclude step d, since the spherical nature of the aggregates inferred that their growth was dominated by single monomer addition rather than clustering to form irregular structures, and step b was also considered not to have played a significant role, since the aggregates formed were reasonably uniform in size, for which the aggregate concentration would have had to be reasonably constant during their growth. It is reasonable to expect that if new aggregates had formed throughout the aggregation process, then a wide distribution of aggregate sizes would have been observed. The modelling of the aggregation kinetics was further simplified by considering two limiting cases: unfolding limited aggregation and association-limited aggregation. Both were tried and association-limited aggregation was found to be the most appropriate as it gave a better fit to the experimental data and more reasonable kinetic values in comparison to previous work [33, 34].

In the case of association-limited aggregation, step a is more rapid than step c, (k_1 and $k_2 \ggg k_4$). As such, N and I come to pseudoequilibrium, hence $k_1 C_N \approx k_2 C_I$. The total monomer concentration, C_M, is the sum of the concentrations of the monomers in the native state (N) and the structurally altered state (I), that is, $C_M = C_N + C_I$. Combining these two relationships gives

$$C_M = \frac{(1 + K) C_I}{K}, \tag{5}$$

where K is the equilibrium constant, $K = k_1/k_2$. Since step c is the rate limiting step, the resulting kinetic model is a second-order rate equation, which incorporates the equilibrium constant in order to be stated in terms of C_M,

$$(-r_M) = \frac{k_4 C_M C_A K}{(1 + K)}. \tag{6}$$

As stated previously, the aggregate concentration, C_A, is expected to have been reasonably constant; therefore, the model can be reduced to pseudo-first order (7), where the equilibrium constant is incorporated into the rate constant for the rate equation (8) as follows:

$$(-r_M) = k_{\text{pseudo}} C_M, \tag{7}$$

$$k_{\text{pseudo}} = \frac{k_4 C_A K}{(1 + K)}. \tag{8}$$

This model was subsequently used to estimate the aggregation kinetics of all samples using the experimental particle diameter, $d(t)$, (Figure 5) to calculate the number of monomers in each aggregate as a function of time ($A_N(t)$) from the ratio of the monomer volume to aggregate volume via

$$A_N(t) = \frac{\pi [d(t)]^3}{6 V_M}. \tag{9}$$

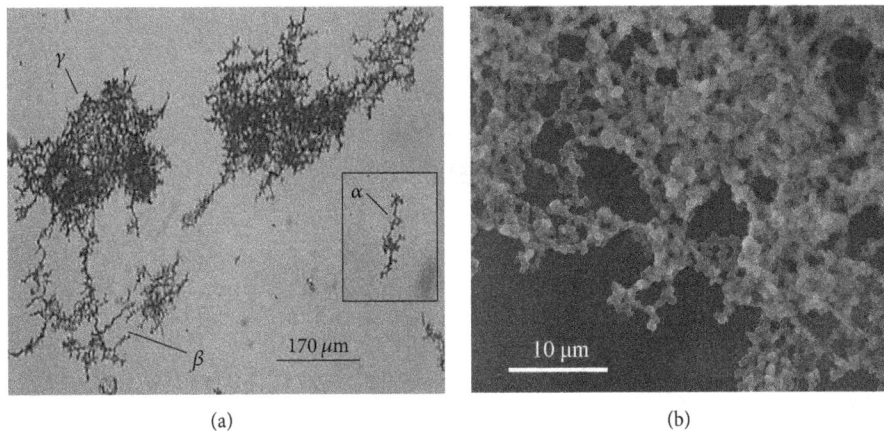

(a)   (b)

FIGURE 3: rPA aggregates formed after incubation at 47°C for 16 hours: (a) optical micrograph, 10x magnification (170 μm scale bar), and (b) ESEM micrograph taken with a sample temperature of 5°C, 6 torr chamber pressure, and 30 kV electron gun accelerating voltage (10 μm scale bar).

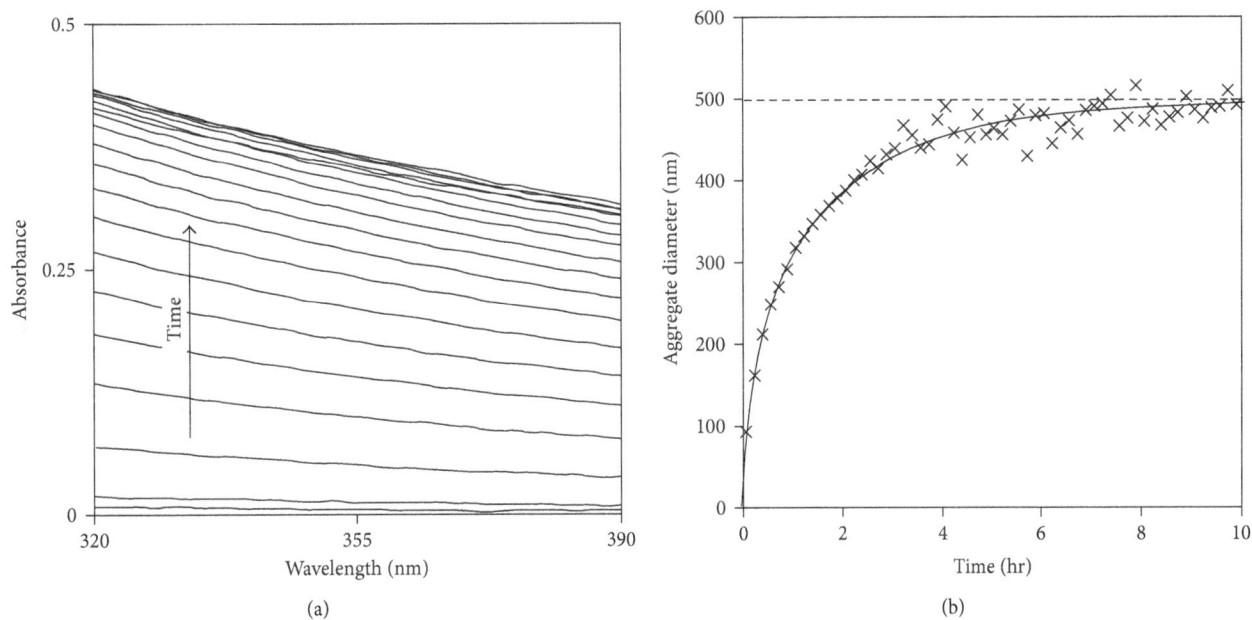

(a)   (b)

FIGURE 4: (a) Change in UV light scattering arising from the growth of rPA aggregates at 45°C, spectra collected in 10-minute intervals for 0–200 minutes, and (b) aggregate diameters calculated from light scattering spectra as a function of time.

The volume of a PA molecule (V_M) was taken to be 95.74 nm³, which was obtained using the crystal structure, as described previouslt. The free monomer concentration versus time ($C_M(t)$) was obtained by taking the difference between the initial monomer concentration (C_{M0}) and the amount of monomers in the aggregates, based on the previous assertion that the aggregate concentration remained constant,

$$C_M(t) = C_{M0} - [C_A \times A_N(t)]. \quad (10)$$

The final size of aggregate that the function converges on is assumed to be the size where aggregates would have stopped growing if the clustering process is not interfered with the particulate growth phase. As such, this is the point at which all the protein monomers would have been consumed and

all the proteins would have been present in the form of the particulate aggregates. The aggregate concentration is therefore given by;

$$C_A = \frac{C_{M0}\,6V_M}{\pi\,(d_f)^3}. \quad (11)$$

This procedure was used to generate concentration versus time profiles for the best fit pseudo first order kinetics over a range of temperatures, and the results for the 0.25 M urea samples are shown in Figure 7(a), and the values for the second-order k's are provided in Figure 7(b). It is evident that in each case the fitted data are in good agreement with those obtained experimentally and the magnitudes of the rate

TABLE 2: Changes in enthalpy (ΔH), entropy (ΔS), Gibbs free energy (ΔG) at 25°C, and denaturation temperature (T_m) for rPA unfolding ($N \leftrightarrow I$) with and without 0.25 M urea.

	ΔH/kJ mol^{-1}	ΔS/kJ mol^{-1} K^{-1}	ΔG at 25°C/kJ mol^{-1}	T_m ($\Delta G = 0$)/°C
rPA without additive	975.8 ± 40.5	3.02 ± 0.13	76.1	50.2
rPA with 0.25 M urea	949.7 ± 94.6	2.92 ± 0.30	77.7	51.6

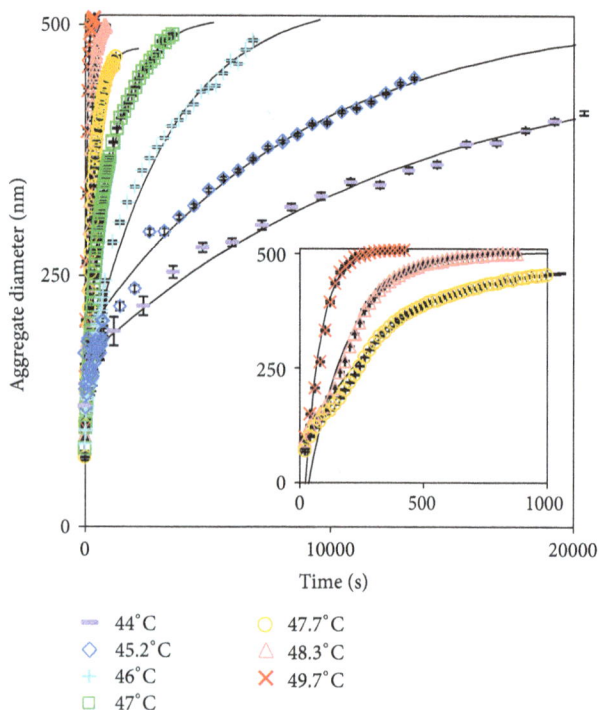

FIGURE 5: Comparison of aggregate diameter calculated from light scattering spectra versus time for rPA samples with 0.25 M urea held isothermally at various temperatures (the inset highlights higher temperature runs).

constants obtained from these data are reasonable, although a little high, compared to those reported elsewhere for the association-limited aggregation of protein (e.g., bovine granulocyte colony-stimulating factor) [33]; extracted values were in the region of 10^4 to 10^6 dm^3 mol^{-1} s^{-1} over a similar temperature range. It is clear from Figure 7(b) that the association-limited aggregation rate constants for samples with and without added urea increase exponentially with temperature. On average the rate constant increases three-fold for every 1°C increase. If the data are re-plotted as ln k versus $1/T$ (Figure 7(c)), it is clear that the data follows Arrhenius' Law, $k = A_f \exp(-E_{act}/RT)$, where A_f is the pre-exponential factor, T is the absolute temperature (K), E_{act} is the activation of the reaction (J mol^{-1}) and R is the ideal gas constant (J mol^{-1} K^{-1}). From the gradients of each slope the activation energies for the two sample types were calculated to be 942.0 ± 92.4 kJ mol^{-1} for rPA without additives and 928.8 ± 32.5 kJ mol^{-1} for rPA with 0.25 M urea. These values suggest that the addition of urea reduced the activation energy slightly, which correlates with the observation that the presence of urea increases the rate of aggregation. The

magnitude of these activation energies are reasonable compared to those reported elsewhere for the association limited aggregation of protein [12]; reported as being between ~420 and ~840 kJ mol^{-1}. These values are reported as "observed" activation energies and are accompanied by the suggestion that the temperature dependence of association limited rate constants does not follow true Arrhenius behaviour [12]. The frequency factors, A_f, for the two sample types were calculated to be 1.8×10^{161} dm^3 mol^{-1} s^{-1} for rPA without additives and 4.7×10^{158} dm^3 mol^{-1} s^{-1} for rPA with 0.25 M urea. These values are very high and unlikely to have any physical significance. Published values of the second-order frequency factors for various small molecule solution phase reactions are in the range of 10^2 to 10^{16} dm^3 mol^{-1} s^{-1} [35]. This suggests that the Arrhenius equation is no more than an empirical fit to the data here. This is not unexpected given the suggestion cited previously that the temperature dependence of association-limited rate constants does not follow true Arrhenius behaviour. However, the correlation coefficients between the data and the lines of the best fit show a reasonable fit: 0.963 for the samples without additives and 0.973 for the samples with 0.25 M urea.

An alternative and more appropriate approach to modelling the temperature dependence of the rate constant is to factor in the behaviour of the equilibrium constant, K. The analysis using the association-limited model has so far yielded an "observed" second-order rate constant for the aggregation process. The kinetic model for this analysis is represented by

$$(-r_M) = k_{obs} C_M C_A, \tag{12}$$

where k_{obs} is the observed rate constant. Comparing this with (6) reveals how k_{obs} is related to the actual rate constant (k_4) and the equilibrium constant K,

$$k_{obs} = \frac{k_4 K}{(1 + K)}. \tag{13}$$

The temperature dependence of K is given by [35]

$$\ln(K) = \frac{\Delta S}{R} - \frac{\Delta H}{RT}, \tag{14}$$

where ΔS is the entropy change (J mol^{-1} K^{-1}), ΔH is the enthalpy change for the process (J mol^{-1}), T is the absolute temperature (K), and R is the gas constant (J mol^{-1} K^{-1}). To fit the experimental data to the temperature dependence of the equilibrium constant, it was assumed that the rate constant, k_4, is independent of temperature. The physical significance of this is that the activation energy is assumed to be negligible for the process of perturbed monomers

$$N \underset{k_2}{\overset{k_1}{\rightleftarrows}} I \qquad \text{Structural perturbation—unfolding} \qquad\qquad nI \xrightarrow{k_3} A_n \qquad \text{Initial aggregate formation—nucleation}$$

(a)   (b)

$$I + A_j \xrightarrow{k_4} A_{j+1} \quad \text{Aggregate growth—monomer addition} \qquad nA_m \xrightarrow{k_5} (A_m)_n \quad \text{Aggregate clustering—gel formation}$$

(c)   (d)

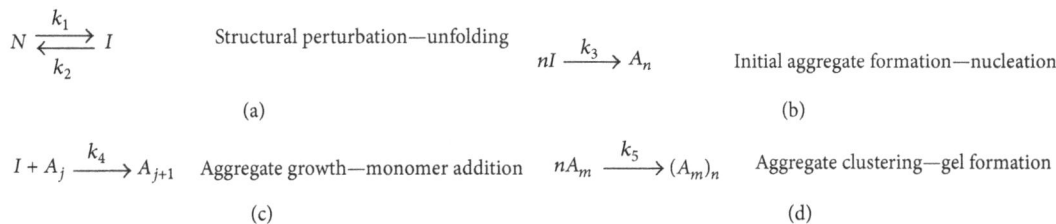

FIGURE 6: Schematic representing the 4 possible steps in the proposed kinetic model of rPA aggregation: (a) structural perturbation, (b) initial aggregate formation, (c) aggregate growth, and (d) aggregate clustering.

being added to the growing aggregate. This is a reasonable assumption, since the activation energies for the reaction of highly reactive free radicals can be close to zero [35, 36], and given that the perturbed monomers will have highly unfavourable hydrophobic patches exposed to water, they too are likely to be highly reactive and easily associated with an aggregate in order to reduce their free energy. Rearranging (13) and equating it to (14) yields

$$\ln\left(\frac{k_{\text{obs}}}{k_4 - k_{\text{obs}}}\right) = \frac{\Delta S}{R} - \frac{\Delta H}{RT}. \tag{15}$$

Therefore, plotting $\ln(k_{\text{obs}}/(k_4 - k_{\text{obs}}))$ versus $1/T$ will yield a straight line, the gradient of which will be $-\Delta H/R$ and the intercept $\Delta S/R$. The value of k_4 was found by searching for the best fit. The initial estimate for k_4 was picked by choosing a value greater than the largest value of k_{obs} so that the logarithmic term could be satisfied. When the 0.25 M urea data was plotted, it was found that the correlation coefficient for the fit between the line of the best fit and the data points improved as the value of k_4 was increased. Eventually, there was no change in the correlation coefficient as k_4 was increased. It was on this basis that an optimum value of 5.77×10^9 dm^3 mol^{-1} s^{-1} was selected for k_4. The value of k_4 effectively represents the value of the frequency factor, since the activation is assumed to be zero. As such, this optimum value of rate constant/frequency factor is much more likely to have a physical significance, since it is comparable to general values of the frequency factor found in the literature [35] discussed previously. The same value of k_4 was used for the analysis of both sets of data: rPA without additives and rPA with 0.25 M urea. This was done on the assumption that the collision rate would not be significantly altered in the presence or the absence of urea. A plot of the resulting data is shown in Figure 8. It is clear that the correlation coefficients between the data and the lines of the best fit are in reasonable agreement for both sets of samples. ΔH and ΔS were extracted from the graph for each sample and used to calculate the Gibbs free energy change, ΔG, at 25°C. The values obtained (Table 2) were compared reasonably well with those reported in the literature for the ΔG between folded and unfolded proteins, reported to typically be between 20 and 60 kJ mol^{-1} [37]. These values also indicate, as expected, that the presence of urea reduces the stability of rPA, most likely by helping to disrupt the native structure of the protein [4], causing it to become perturbed at temperatures lower than those in the absence of any urea.

To explore the effect of temperature on the extent of disruption of individual proteins, the fraction of protein in the perturbed intermediate state, X, was calculated using (16) and plotted as a function of temperature (Figure 9):

$$X = \frac{C_I}{C_M} = \frac{K}{(1 + K)}. \tag{16}$$

Extrapolation of this data to ambient temperatures shows that the fraction of protein in the perturbed state is negligible at low temperatures. For example, extrapolation to 25°C for the sample with 0.25 M urea shows that the fraction of protein in the perturbed state is 10^{-14}. This corroborates our experimental observations that rPA does not show observable aggregation over extended periods (circa 16 hours) at ambient temperature. The unfolded fraction has also been extrapolated to higher temperatures, using (11) and (12), and the results have been shown by the extrapolated line of the best fit in Figure 9. The inset of Figure 9 shows that the fraction of monomer in the perturbed state rises sigmoidally between 46°C and 55°C and that a higher proportion of protein is perturbed when urea is present. Both observations correlate with the susceptibility of the protein samples to aggregate, that is, the more protein perturbed, the higher the chances and the faster the kinetics of aggregation. It also links in well with other studies on the thermal unfolding of rPA. For example, 8-anilino-1-naphthalene sulfonate (ANS) and rPA reported elsewhere show that the hydrophobicity of rPA increases sigmoidally versus the temperature between 40° and 55°C [38], as was also the case when the thermal unfolding was examined by circular dichroism, which revealed a sigmoidal increase in unfolding versus the temperature between 45° and 55°C [31].

4. Conclusions

The susceptibility of a biopharmaceutical protein rPA to aggregate as a function of temperature and formulation conditions has been determined, and the kinetics and thermodynamics of the aggregation process have been modelled and quantified. Visual and turbidity experiments showed that the thermal aggregation of rPA occurs at incubation temperatures ≥43°C, which is close to its denaturation temperature. Under these conditions, the protein is likely to have increased translational kinetic energy, hence more collisions will take place, and also be at least partially unfolded, hence, it has some exposed hydrophobic regions which are known to

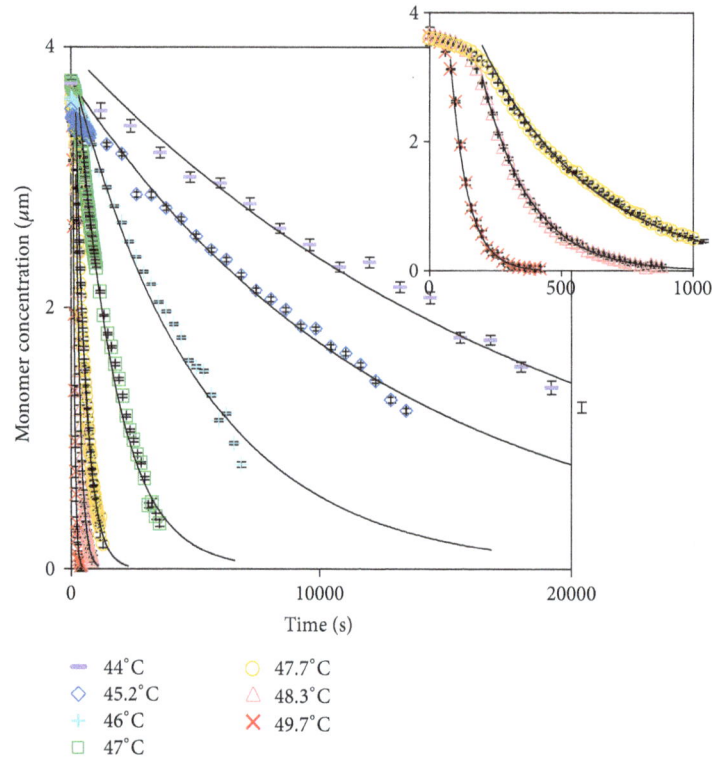

(a)

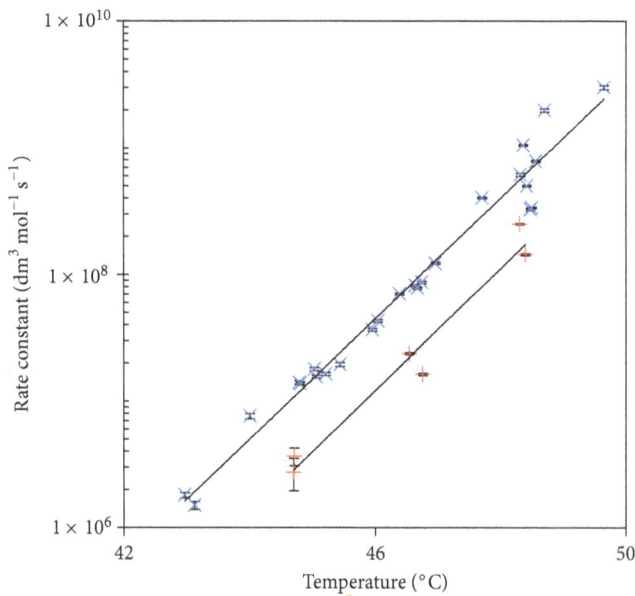

+ PA samples without additives
× With 0.25 M urea
— Exponential curves of the best fit

(b)

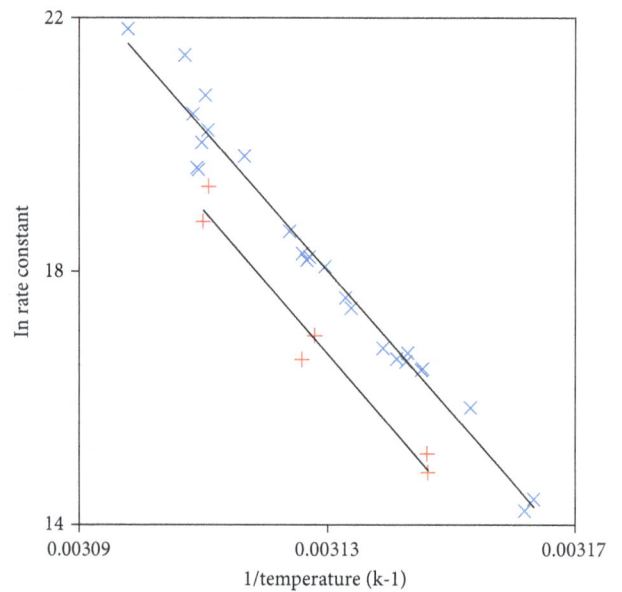

+ PA samples without additives
× With 0.25 M urea
— Exponential curves of the best fit

(c)

FIGURE 7: (a) Monomer concentration as a function of time for rPA samples with 0.25 M urea held isothermally at various temperatures (the inset highlights higher temperature runs). (b) Second-order association-limited aggregation rate constant versus temperature. (c) Arrhenius plot.

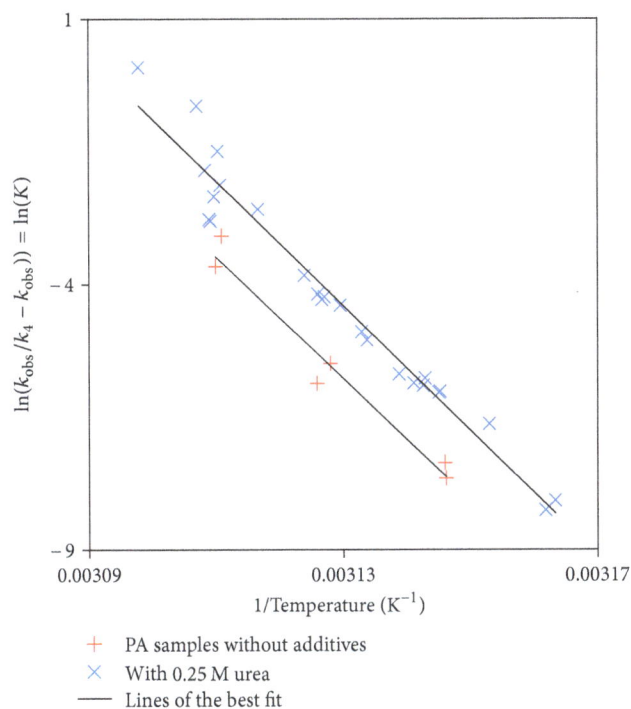

+ PA samples without additives
× With 0.25 M urea
— Lines of the best fit

FIGURE 8: Natural log of equilibrium constant versus the reciprocal of absolute temperature for the association-limited aggregation model.

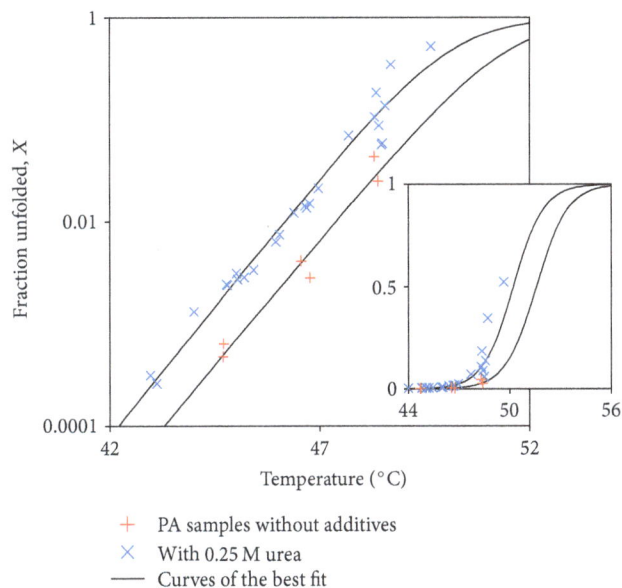

+ PA samples without additives
× With 0.25 M urea
— Curves of the best fit

FIGURE 9: Fraction of monomer in the perturbed state as a function of temperature for the association-limited aggregation model. Curves of the best fit follow (14) and (16) with ΔH and ΔS values from Table 2 (black curve) on logarithmic scale (the inset shows extrapolation to higher temperatures on normal scale).

induce rapid and nonspecific aggregation. Such aggregation was found to proceed in a stepwise manner, by first forming spherical microscopic particles followed by clustering to form fractal aggregates. Increasing the temperature more than 43°C increased the rate of aggregation dramatically and also the size of the diameter of the spherical microscopic particles formed from ~360 to 500 nm when increasing the temperature from 43° to 49°C. We went on to show that the growth of the microscopic particles can be monitored using UV-LSS. In particular, we used the increase in scattered light from the sample over time to elucidate aggregate size versus time, giving a quantitative measure of the aggregation. Moreover, the experiments were conducted over a range of temperatures, with and without 0.25 M urea, and as such, the results were analysed to determine the rate constant and the temperature dependence of the thermal aggregation process. Based on this analysis, we proposed that the aggregation process is association limited and that the temperature dependence relates to the equilibrium behaviour between native and perturbed states. We were also able to extract the thermodynamic parameters for aggregation in samples with and without urea, and these indicated that the presence of urea reduces the stability of rPA, hence increases its susceptibility to aggregation. The modelling tools developed here for analysis of data from the easily accessible UV-LSS technique provides a fast between *in situ* analysis method for comparing the stability of different formulations of protein when exposed to different environmental conditions. This method has an important speed advantage over other light scattering techniques when analysing such particle growth

kinetics. This work, therefore, provides a basis for quantitatively exploring the effect of additives and/or different processing conditions on the rate of aggregation of industrially relevant biopharmaceuticals with the aim of minimising any self-association during production, downstream processing, or storage.

Acknowledgments

The authors wish to thank EPSRC and Intertek for financial support and Avecia Biologics for supplying rPA samples. They would also like to thank Dr. Patrick Hill for his help with the ESEM.

References

[1] E. D. B. Clark, "Protein refolding for industrial processes," *Current Opinion in Biotechnology*, vol. 12, no. 2, pp. 202–207, 2001.

[2] J. L. Cleland, M. F. Powell, and S. J. Shire, "The development of stable protein formulations: a close look at protein aggregation, deamidation, and oxidation," *Critical Reviews in Therapeutic Drug Carrier Systems*, vol. 10, no. 4, pp. 307–377, 1993.

[3] A. Braun, L. Kwee, M. A. Labow, and J. Alsenz, "Protein aggregates seem to play a key role among the parameters influencing the antigenicity of interferon alpha (IFN-α) in normal and transgenic mice," *Pharmaceutical Research*, vol. 14, no. 10, pp. 1472–1478, 1997.

[4] W. Wang, "Instability, stabilization, and formulation of liquid protein pharmaceuticals," *International Journal of Pharmaceutics*, vol. 185, no. 2, pp. 129–188, 1999.

[5] W. Wang, S. Nema, and D. Teagarden, "Protein aggregation—pathways and influencing factors," *International Journal of Pharmaceutics*, vol. 390, no. 2, pp. 89–99, 2010.

[6] J. R. Clarkson, Z. F. Cui, and R. C. Darton, "Protein denaturation in foam: I. Mechanism study," *Journal of Colloid and Interface Science*, vol. 215, no. 2, pp. 323–332, 1999.

[7] M. R. H. Krebs, G. L. Devlin, and A. M. Donald, "Protein particulates: another generic form of protein aggregation?" *Biophysical Journal*, vol. 92, no. 4, pp. 1336–1342, 2007.

[8] C. M. Dobson, "Protein folding and misfolding," *Nature*, vol. 426, no. 6968, pp. 884–890, 2003.

[9] M. R. H. Krebs, K. R. Domike, and A. M. Donald, "Protein aggregation: more than just fibrils," *Biochemical Society Transactions*, vol. 37, no. 4, pp. 682–686, 2009.

[10] W. S. Gosal and S. B. Ross-Murphy, "Globular protein gelation," *Current Opinion in Colloid and Interface Science*, vol. 5, no. 3-4, pp. 188–194, 2000.

[11] A. M. Morris, M. A. Watzky, and R. G. Finke, "Protein aggregation kinetics, mechanism, and curve-fitting: a review of the literature," *Biochimica et Biophysica Acta*, vol. 1794, no. 3, pp. 375–397, 2009.

[12] C. J. Roberts, "Nonnative protein aggregation: pathways, kinetics, and stability prediction," in *Misbehaving Proteins*, R. M. Murphy and A. M. Tsai, Eds., pp. 17–46, Springer, New York, NY, USA, 2006.

[13] S. E. Bondos, "Methods for measuring protein aggregation," *Current Analytical Chemistry*, vol. 2, no. 2, pp. 157–170, 2006.

[14] M. E. M. Cromwell, C. Felten, H. Flores, J. Liu, and S. J. Shire, "Self-association of therapeutic proteins: implications for product development," in *Misbehaving Proteins*, R. M. Murphy and A. M. Tsai, Eds., pp. 313–330, Springer, New York, NY, USA, 2006.

[15] R. M. Murphy and C. C. Lee, "Laser light scattering as an indispensable tool for probing protein aggregation," in *Misbehaving Proteins*, R. M. Murphy and A. M. Tsai, Eds., pp. 147–165, Springer, New York, NY, USA, 2006.

[16] S. Jendrek, S. F. Little, S. Hem, G. Mitra, and S. Giardina, "Evaluation of the compatibility of a second generation recombinant anthrax vaccine with aluminum-containing adjuvants," *Vaccine*, vol. 21, no. 21-22, pp. 3011–3018, 2003.

[17] K. Chen, A. Kromin, M. P. Ulmer, B. W. Wessels, and V. Backman, "Nanoparticle sizing with a resolution beyond the diffraction limit using UV light scattering spectroscopy," *Optics Communications*, vol. 228, no. 1-3, pp. 1–7, 2003.

[18] A. J. Cox, A. J. DeWeerd, and J. Linden, "An experiment to measure Mie and Rayleigh total scattering cross sections," *American Journal of Physics*, vol. 70, no. 6, pp. 620–625, 2002.

[19] H. Fang, M. Ollero, E. Vitkin et al., "Noninvasive sizing of subcellular organelles with light scattering spectroscopy," *IEEE Journal on Selected Topics in Quantum Electronics*, vol. 9, no. 2, pp. 267–276, 2003.

[20] J. R. Mourant, T. Fuselier, J. Boyer, T. M. Johnson, and I. J. Bigio, "Predictions and measurements of scattering and absorption over broad wavelength ranges in tissue phantoms," *Applied Optics*, vol. 36, no. 4, pp. 949–957, 1997.

[21] A. M. K. Nilsson, C. Sturesson, D. L. Liu, and S. Andersson-Engels, "Changes in spectral shape of tissue optical properties in conjunction with laser-induced thermotherapy," *Applied Optics*, vol. 37, no. 7, pp. 1256–1267, 1998.

[22] L. B. Scaffardi, N. Pellegri, O. De Sanctis, and J. O. Tocho, "Sizing gold nanoparticles by optical extinction spectroscopy," *Nanotechnology*, vol. 16, no. 1, pp. 158–163, 2005.

[23] S. M. Scholz, R. Vacassy, J. Dutta, H. Hofmann, and M. Akinc, "Mie scattering effects from monodispersed ZnS nanospheres," *Journal of Applied Physics*, vol. 83, no. 12, pp. 7860–7866, 1998.

[24] C. F. Bohren and D. R. Huffman, *Absorption and Scattering of Light by Small Particles*, Wiley, New York, NY, USA, 1983.

[25] P. Schiebener, J. Straub, J. M. H. L. Sengers, and J. S. Gallagher, "Refractive index of water and steam as function of wavelength, temperature and density," *Journal of Physical and Chemical Reference Data*, vol. 19, pp. 677–717, 1990.

[26] A. Weissberger, "Physical methods of organic chemistry—part 2," in *Techniques of Organic Chemistry*, Wiley, New York, NY, USA, 1960.

[27] L. Willard, A. Ranjan, H. Zhang et al., "VADAR: a web server for quantitative evaluation of protein structure quality," *Nucleic Acids Research*, vol. 31, no. 13, pp. 3316–3319, 2003.

[28] C. Petosa, R. J. Collier, K. R. Klimpel, S. H. Leppla, and R. C. Liddington, "Crystal structure of the anthrax toxin protective antigen," *Nature*, vol. 385, no. 6619, pp. 833–838, 1997.

[29] H. M. Berman, J. Westbrook, Z. Feng et al., "The protein data bank," *Nucleic Acids Research*, vol. 28, no. 1, pp. 235–242, 2000.

[30] H. Fischer, I. Polikarpov, and A. F. Craievich, "Average protein density is a molecular-weight-dependent function," *Protein Science*, vol. 13, no. 10, pp. 2825–2828, 2004.

[31] D. A. Chalton, I. F. Kelly, A. McGregor et al., "Unfolding transitions of *Bacillus anthracis* protective antigen," *Archives of Biochemistry and Biophysics*, vol. 465, no. 1, pp. 1–10, 2007.

[32] P. Meakin, "Fractal aggregates," *Advances in Colloid and Interface Science*, vol. 28, pp. 249–331, 1987.

[33] C. J. Roberts, "Kinetics of irreversible protein aggregation: analysis of extended Lumry-Eyring models and implications for predicting protein shelf life," *Journal of Physical Chemistry B*, vol. 107, no. 5, pp. 1194–1207, 2003.

[34] B. S. Kendrick, J. L. Cleland, X. Lam et al., "Aggregation of recombinant human interferon gamma: kinetics and structural transitions," *Journal of Pharmaceutical Sciences*, vol. 87, no. 9, pp. 1069–1076, 1998.

[35] P. W. Atkins, *Physical Chemistry*, Oxford Unveristy Press, Oxford, UK, 2nd edition, 1982.

[36] R. L. Thommarson, "Alkyl radical disproportionation," *Journal of Physical Chemistry*, vol. 74, no. 4, pp. 938–941, 1970.

[37] R. Jaenicke, "Protein stability and protein folding," in *Ciba Foundation Symposium 161—Protein Conformation*, D. J. Chadwick and K. Widdows, Eds., pp. 206–221, Wiley, Chichester, UK, 2007.

[38] G. Jiang, S. B. Joshi, L. J. Peek et al., "Anthrax vaccine powder formulations for nasal mucosal delivery," *Journal of Pharmaceutical Sciences*, vol. 95, no. 1, pp. 80–96, 2006.

12

Potassium Current Is Not Affected by Long-Term Exposure to Ghrelin or GHRP-6 in Somatotropes GC Cells

Belisario Domínguez Mancera,[1,2] **Eduardo Monjaraz Guzman,**[1]
Jorge L. V. Flores-Hernández,[1] **Manuel Barrientos Morales,**[2] **José M. Martínez Hernandez,**[2]
Antonio Hernández Beltran,[2] **and Patricia Cervantes Acosta**[2]

[1] *Laboratorio de Neuroendocrinología, Instituto de Fisiología, Benemérita Universidad Autónoma de Puebla,*
  CP 7200, PUE, Mexico
[2] *Laboratorio de Biología Celular, Facultad de Medicina Veterinaria y Zootecnia, Universidad Veracruzana,*
  CP 91710, VER, Mexico

Correspondence should be addressed to Belisario Domínguez Mancera; beldominguez@uv.mx

Academic Editor: Eaton Edward Lattman

Ghrelin is a growth hormone (GH) secretagogue (GHS) and GHRP-6 is a synthetic peptide analogue; both act through the GHS receptor. GH secretion depends directly on the intracellular concentration of Ca^{2+}; this is determined from the intracellular reserves and by the entrance of Ca^{2+} through the voltage-dependent calcium channels, which are activated by the membrane depolarization. Membrane potential is mainly determined by K^+ channels. In the present work, we investigated the effect of ghrelin (10 nM) or GHRP-6 (100 nM) for 96 h on functional expression of voltage-dependent K^+ channels in rat somatotropes: GC cell line. Physiological patch-clamp whole-cell recording was used to register the K^+ currents. With Cd^{2+} (1 mM) and tetrodotoxin (1 μm) in the bath solution recording, three types of currents were characterized on the basis of their biophysical and pharmacological properties. GC cells showed a K^+ current with a transitory component (I_A) sensitive to 4-aminopyridine, which represents ~40% of the total outgoing current; a sustained component named delayed rectifier (I_K), sensitive to tetraethylammonium; and a third type of K^+ current was recorded at potentials more negative than −80 mV, permitting the entrance of K^+ named inward rectifier (K_{IR}). Chronic treatment with ghrelin or GHRP-6 did not modify the functional expression of K^+ channels, without significant changes ($P < 0.05$) in the amplitudes of the three currents observed; in addition, there were no modifications in their biophysical properties and kinetic activation or inactivation.

1. Introduction

The growth hormone is mainly under the control of two hypothalamic neuropeptides acting in opposition: one, the growth hormone releasing hormone (GHRH), as a stimulant, and the other, somatostatin, as an inhibitor [1, 2]. The GHRH specifically bind to its receptor on the plasmatic membrane of the somatotropes; this increments the activity of adenylate cyclase, which increases the generation of AMPc [3, 4]. This increase in the AMPc levels let to open the voltage-dependent Ca^{2+} channels [5, 6] and a rapid increase in the intracellular Ca^{2+} concentration $[Ca^{2+}]$, thus promoting the exocytosis of GH [7, 8]. The inhibitory effect of somatostatin involves the inhibition of adenylate cyclase activity and a reduction of $[Ca^{2+}]_i$ [9, 10].

In addition to the GHRH, a group of synthetic oligopeptides releasing GH (GHRPs) or GH secretagogues (GHS) are capable of stimulating the secretion of GH [2, 11, 12]. The GH-releasing peptide-6 (GHRP-6) is one of the most representative of those compounds [2, 11–13]. Research on the mechanism of GHS action upon the liberation of GH led to the discovery of the GHS receptor, the GHS-R, and later to the ghrelin, the endogenous ligand for the GHS-R [2, 11, 12, 14].

It has been suggested that diverse forms of signaling are activated by the action of GHS. After binding the ligand, the GHS-R receptor acts through the subunit of the G protein

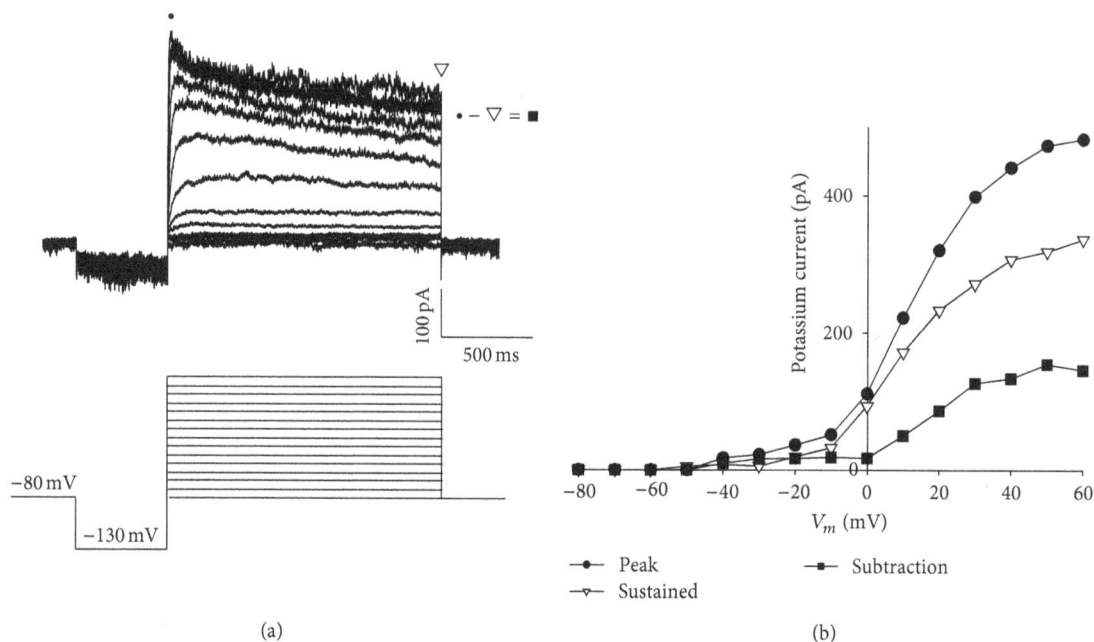

(a)   (b)

FIGURE 1: Voltage-dependent potassium current in the cellular line of rat GC somatotrope. (a) Family of representative traces of voltage-dependent K^+ current in GC cells, evoked by depolarizing voltage pulses with increases of 10 mV, starting from a holding potential in the prepulse of −130 mV with duration of 500 ms, followed by a voltage pulse lasting for 1.5 sec. The initial maintenance potential was fixed at −80 mV. In the lower part, the acquisition protocol is shown. (b) Current-voltage relationship for the transient component ($\bullet$; the first 50 ms), the sustained component ($\square$; 1.455–1.50 s), and the subtraction ($\bullet - \square = \blacksquare$).

to activate phospholipase C (PLC), which results in the hydrolysis of PIP$_2$ to produce inositol 1,4,5-triphosphate (IP$_3$) and diacylglycerol (DAG) [2]. As a consequence, there is an increase of $[Ca^{2+}]_i$ due to a transitory liberation of Ca^{2+} from the cytoplasmatic reserves sensitive to IP$_3$ and to the sustained influx of Ca^{2+} caused by the activation of the voltage-dependent Ca^{2+} channels this in addition to the blockage of the K^+ channels [16]; leads to a depolarization of the somatotrope membrane and the liberation of GH [2, 11]. The long-term effects of secretagogues like ghrelin and GHRPs on the cells secreting GH in association with the ionic channels explored have been few.

Cellular excitability depends on the opening and the closing of the ionic channels present in the plasmatic membrane as well as the level of membrane potential in repose. The conductance of K^+ is responsible for the membrane potential in repose [17]. It has been reported that somatostatin can increase different types of K^+ currents, including the voltage-gated K^+ currents in somatotropes from rats, ovines, and humans in order to hyperpolarize the membrane potential [18, 19]. On the other hand, it has been reported that synthetic secretagogues reduce the KIR current (potassium inward rectifier) [15] by means of a decrease in the protein that codifies for K^+; similarly, ghrelin has been reported to reduce the voltage-dependent K^+ in GH$_3$ cells via the GMP cycle [20]. Since most of the ionic currents passing through the plasmatic membrane in the membrane potential in repose are conducted by K^+, it is believed that the K^+ channels play an important role in the depolarization and repolarization

induced by ghrelin [21]. Therefore, the object of the following was to examine whether chronic treatment with ghrelin or GHRP-6 modifies the functional expression of voltage-dependent K^+ channels in the plasmatic membrane of GC cells during long-term treatments.

2. Material and Methods

2.1. Chemicals. Ghrelin (Cat. 55-0-03A) and GHRP-6 (Cat. 52-1-80B) were acquired from the American Peptide Company, Inc. (Sunnyvale, CA,USA). Tetrodotoxin (TTX; Cat. T550) was acquired from Almone Labs, Ltd. (Jerusalem, Israel). The chloride of tetraethylammonium (TEA; T2265) and the 4-aminopyridine (4-AP; A 0152) came from SIGMA (St. Louis, MO, USA); all other reagents were of a chemical grade.

2.2. Cell Culture. The cellular line of rat GC somatotropes was routinary maintained as a monolayer previously described [21] in a complete MegaCell DMEM culture medium (Sigma-Aldrich, St. Louis, MO, USA), supplemented with 3% of fetal bovine serum (Sigma-Aldrich) and 100 u.i. mL^{-1} of penicillin and 100 μg mL^{-1} of streptomycin (Sigma-Aldrich). The cultures were incubated at 37°C in a humidified atmosphere containing 5% CO$_2$. Once a week the cells were harvested by means of a Trypsin-EDTA treatment (Sigma-Aldrich) (0.05 w/v and 453 mM, resp.) and reseeded at densities of 2–2.5 $\times$ 10^5 cells per flask of 25 cm^2. For electrophysiological

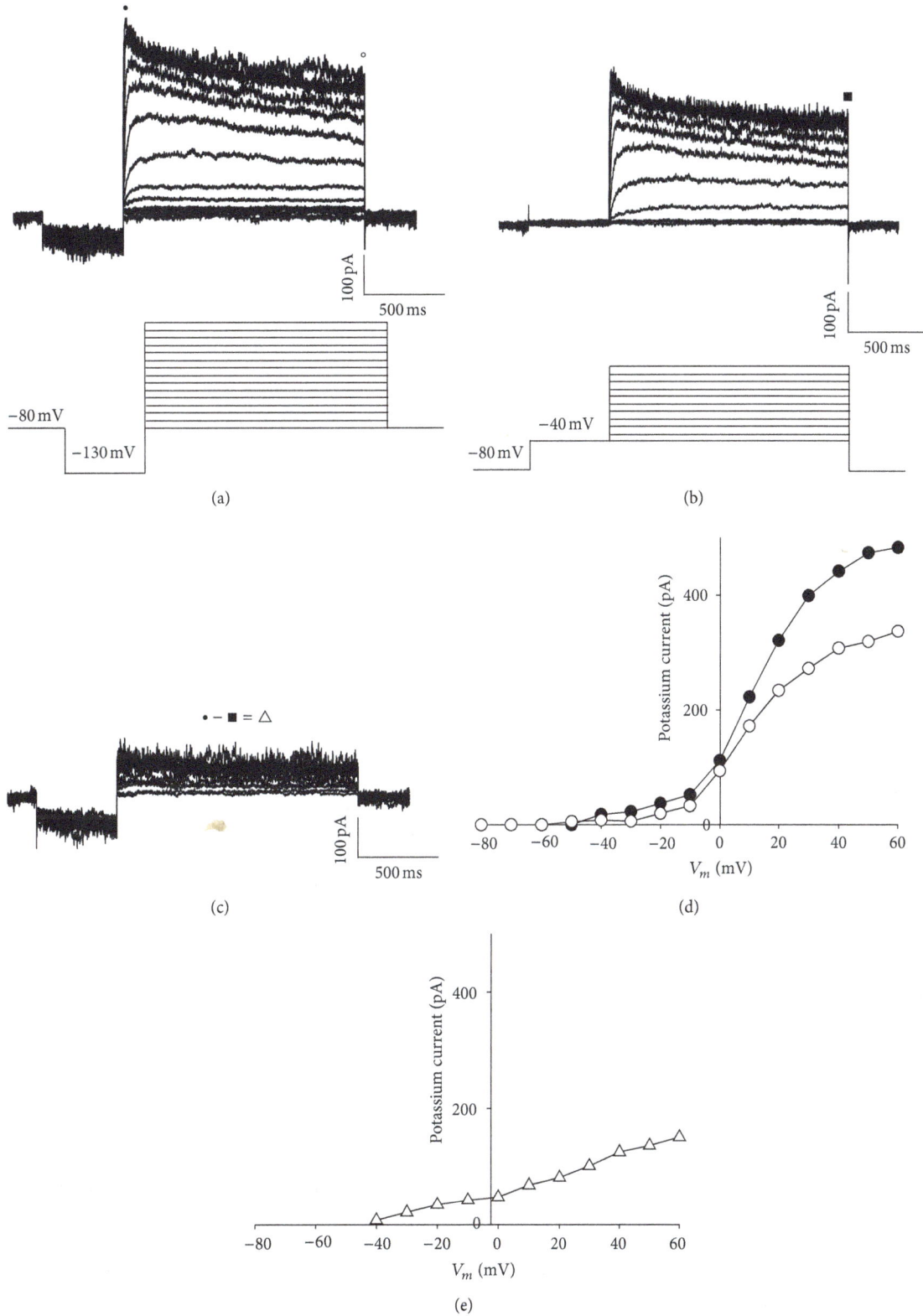

FIGURE 2: Components of the voltage-dependent potassium current in the cellular line of rat GC somatotropes. (a) Family of representative traces of the voltage-dependent K$^+$ current in the GC cells evoked by depolarizing voltage pulses lasting 1.5 sec, with increases of 10 mV starting from a prepulse of 500 ms and a holding potential of −130 mV, the initial holding potential have been fixed at −80 mV. The lower part shows the acquisition protocol. (b) Representative traces of the voltage-dependent K$^+$ current evoked by 1.5 s depolarizing pulses, starting from a prepulse lasting 500 ms and a holding potential of −40 mV, increasing by 10 mV at a time to 60 mV; the pulse protocol is shown in the lower part. (c) Transient K$^+$ current (I_A) isolated through a point-by-point subtraction of currents from traces (a) and (b): (• − ∎ = △). (d) Current-voltage relationship at peak of current; transient component (•; the first 50 ms), and sustained component (□; 1.455–1.5 s). (e) Current-voltage relationship of the point-by-point subtraction between (a) and (b): the measurement was taken in the first 50 ms (• − ∎ = □).

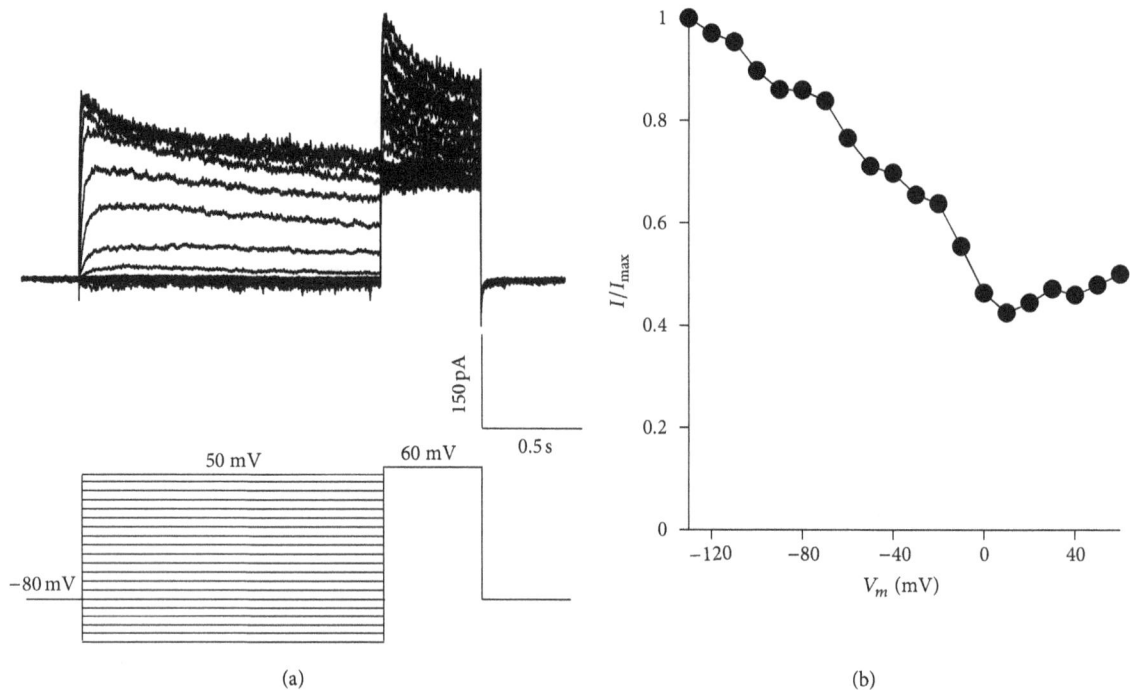

(a)   (b)

FIGURE 3: Inactivation in the steady state of the total I_K current in the cellular line of rat GC somatotropes. (a) Family of representative traces of the inactivation curve in the steady state for the peaks of voltage-dependent I_K current generated by a double-pulse protocol of −130 mV to +60 mV, with a duration of 1.5 seconds and increases of 10 mV before passing to a testing pulse at 70 mV with a duration of 500 ms; the initial holding potential was 80 mV. (b) The I_K peak evoked during the testing pulse at +60 mV was normalized in regard to the maximum (I/I_{max}) and graphed in relation to the conditioning pulse voltage.

recordings, the cells were seeded in culture dishes with poly-L-lysine for better cellular adhesion. The culture medium (with or without the secretagogues ghrelin or GHRP-6) was replaced every day.

2.3. Electrophysiology.
The K$^+$ currents were recorded in the cellular line of rat GC somatotropes under control conditions or after a chronic treatment (96 h) with ghrelin (10 nM) or GHRP-6 (100 nM) in the whole-cell recording configuration (WCR) of the patch-clamp technique [22]. For this, a patch-clamp amplifier Axopath 200B (Molecular Devices, Foster City, CA, USA) was used, as well as an interface Digidata 1322A with pClamp9 software (Molecular Devices) for the acquisition of on-line data. After the whole cell configuration was established, the capacitive transients were cancelled with the amplifier. The leakage current and the residual capacitive were subtracted online with a P/4 protocol. The current signals were filtered at 5 kHz (internal 4-pole Bessel filter) and digitalized at 10–100 kHz. The bath solution for recording contained (in mM): 145 NaCl, 5 KCl, 5 CaCl$_2$, 10 Hepes, 1 CdCl$_2$, 0.001 TTX, and 5 glucose; (pH 7.30, adjusted with NaOH). The internal solution for recording consisted of (in mM) 100 KCl, 30 NaCl, 2 MgCl$_2$, 1 CaCl$_2$, 10 EGTA, 10 Hepes, 2 ATP, 0.05 GTP, and 5 glucose (pH 7.30 adjusted with KOH). The experiments were performed at room temperature (~22°C). Both the control cells and the treated ones were rinsed in a culture medium free of peptides and were kept for ~60 min before the membrane currents were recorded

in order to avoid acute effects. The membrane capacitance (Cm) was determined as previously described [23] and was used to obtain the current density. Briefly, the Cm values were determined by applying a series of consecutive pulses (20 ms; −20 mV) having a maintenance potential (−80 mV), and by integrating the trace of the current obtained by subtracting the capacitive transient traces associated with the pipette patch (cell-attached conditions) from the total of the capacitive transients obtained immediately after the interior of the cell was broken (whole-cell conditions).

2.4. Hormonal Quantification.
The release of GH from GC cells treated with ghrelin (10 nM) or GHRP-6 (100 nM) for 96 h was quantified using a commercial enzyme-linked immunosorbent assay (ELISA) kit (SPI Bio, Massy Cedex, France) as previously described [21]. The color intensity of the reaction product (proportional to the concentration of GH) was measured by spectrophotometry in a plate reader using a 450 nm filter (Stat Fax 2100; Awareness Technology, Palm City, FL, USA). Intensity values of the samples were compared with values in a standard curve using SigmaPlot 11.0 software (Systat Software, Chicago, IL, USA).

2.5. Data Analysis.
The data were analyzed and graphed by combining pClamp software (previously mentioned) and SigmaPlot software v.11 (SPSS, Chicago, IL, USA); they are shown as the mean ± standard error. The statistical differences

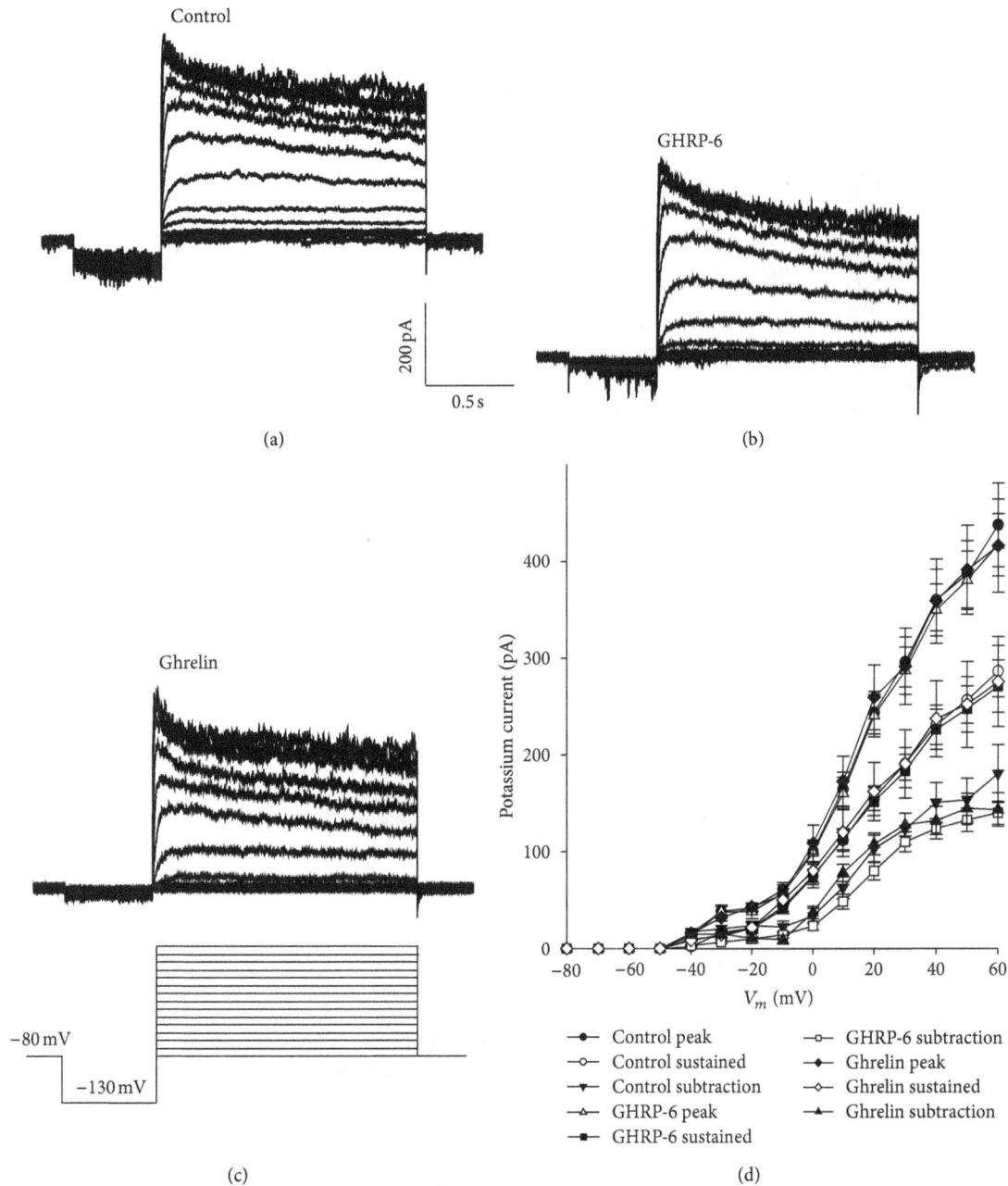

FIGURE 4: Effect of secretagogues (ghrelin and GHRP-6) on the K$^+$ current. (a), (b), and (c): family of representative traces of the K$^+$ current on GC cells in control conditions, treated with ghrelin (10 nM), or GHRP-6 (100 nM), for 96 hours. (d) Current-voltage relationship for each one of the experimental conditions (control; $n = 21$, ghrelina 10 nM; $n = 15$, and GHRP-6 100 nM; $n = 21$); the measurements were taken at the peak (the first 50 ms), the sustained component (the last 5 ms; 1.455–1.5 s), and the subtraction of peak minus sustained component.

between the means were determined with the Student t-test ($P < 0.05$). The adjustments to the curves were made by using nonlinear procedures by least squared included in the SigmaPlot program. The conductance of each testing potential was estimated by measuring the amplitude of the current at its peak. The conductance-voltage curves (G-V) for analyzing the activation were adjusted by means of a Boltzmann-type equation in the form of $G = G_{max}/(1 + \exp[(V_m - V_{1/2})/k]^{-1})$, where G_{max} is the maximum of conductance, V_m is the testing potential, $V_{1/2}$ is the potential

for the half of G_{max} (midpoint), and k is a slope factor. To construct the inactivation curve a protocol of two pulses was used; the first of these was denominated as conditioning, with duration of 1.5 s and variable amplitude (−130 to 60 mV), and the second pulse as testing, the latter having a membrane potential at 70 mV with a duration of 500 ms; the amplitude was graphed at the peak of the type I_A current induced by the testing pulse, dependent on the voltage of the conditioning pulse. Subsequently, the data obtained from each cell were adjusted with a Boltzmann-type equation in

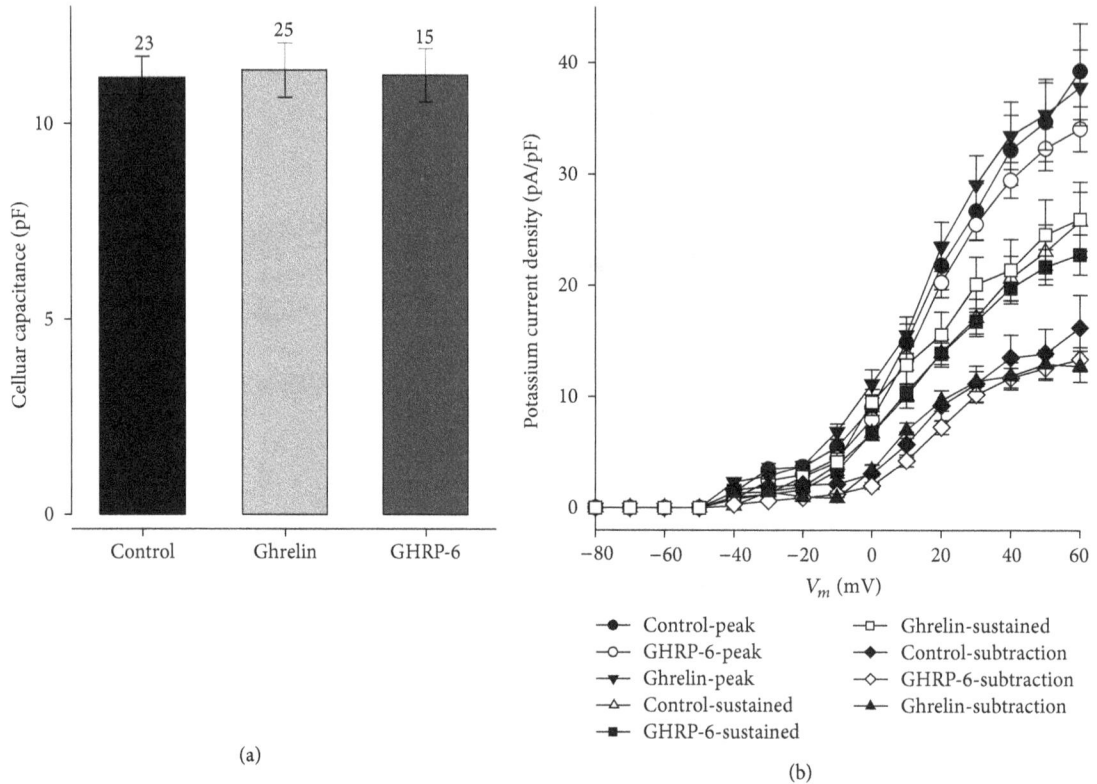

(a)

(b)

FIGURE 5: Density of K$^+$ current in the cellular line of rat GC somatotropes in control conditions and treated chronically with secretagogues (ghrelin or GHRP-6). (a) Cellular capacitance in each one of the experimental conditions. The numbers beside the error bar represent the number of cells analyzed. (b) Density of current-voltage relationship in the GC cells in control conditions and chronically treated (96 hrs) with ghrelin (10 nM) or GHRP-6 (100 nM).

the form of $I = I_{max}/(1 + \exp{[(V_m - V_{1/2})/k]^{-1}})$, where I_{max} is the maximum current, V_m is the testing potential, $V_{1/2}$ is the potential for the half of I_{max}, and k is a slope factor.

3. Results

3.1. General Properties of the K$^+$ Current in the Cellular Line of Rat GC Somatotropes under Control Conditions and Treated with Secretagogues (Ghrelin or GHRP-6). The total current of voltage-dependent K$^+$ in the GC cells was examined through the application of a pulse protocol, starting with a prepulse (duration of 500 ms) that fixes the holding potential at −130 mV; there followed a series of voltage pulses with a duration of 1.5 seconds, starting from −80 mV to 60 mV with increases of 10 mV (Figure 1(a), lower panel). In the GC cells, two components were observed in the potassium current, one transient component in the first 50 ms, and a sustained component that is maintained during the 1.5 seconds of the depolarizing pulse (Figure 1(a)); both the peak and the sustained part were activated at membrane potentials more depolarizing than −30 mV (Figure 1(b)). Once activated, the total current of voltage-dependent potassium inactivates slowly during the 1.5-second passing of depolarizing voltage (Figure 1(a)).

In order to isolate the slow-inactivation potassium current (I_K) component, the transient potassium current (I_A) was eliminated through a maneuver to change the holding potential from −130 mV to −40 mV for 500 ms in the prepulse, prior to the depolarizing pulses of 1.5 seconds ranging from −40 to 60 mV (Figure 2(b)). Subsequently, the point-by-point subtraction in the two families of current traces was performed for each potential, thereby obtaining the current of rapid activation and rapid inactivation known as I_A.

To determine the proportion of the total K$^+$ current in the GC cells evoked at different membrane potentials, the inactivation properties in a steady state of the K$^+$ current were examined by using a double-pulse protocol. The protocol consisted of a series of conditioning pulses of varying amplitude, with a duration of 1.5 seconds, ranging from −130 to 60 mV in 10 mV increases and followed by a conditioning pulse at 70 mV with a duration of 500 ms (Figure 3(a)).

A simple Boltzmann relationship could not be adjusted to the data because the presence of rapid and slow inactivation components was indicated. The current at the peak of the testing pulse between the maximum current (I/I_{max}) was normalized and was graphed in relation to the conditioning pulse voltage to obtain the participation percentages of the I_A and I_K currents, Figure 3(b) showing that the participation of the I_A current is below 0.4 (40%).

(a)

Control (peak); $n = 9$
GHRP-6 (peak); $n = 6$

Control (sustained); $n = 9$
GHRP-6 (sustained), $n = 6$

(b)

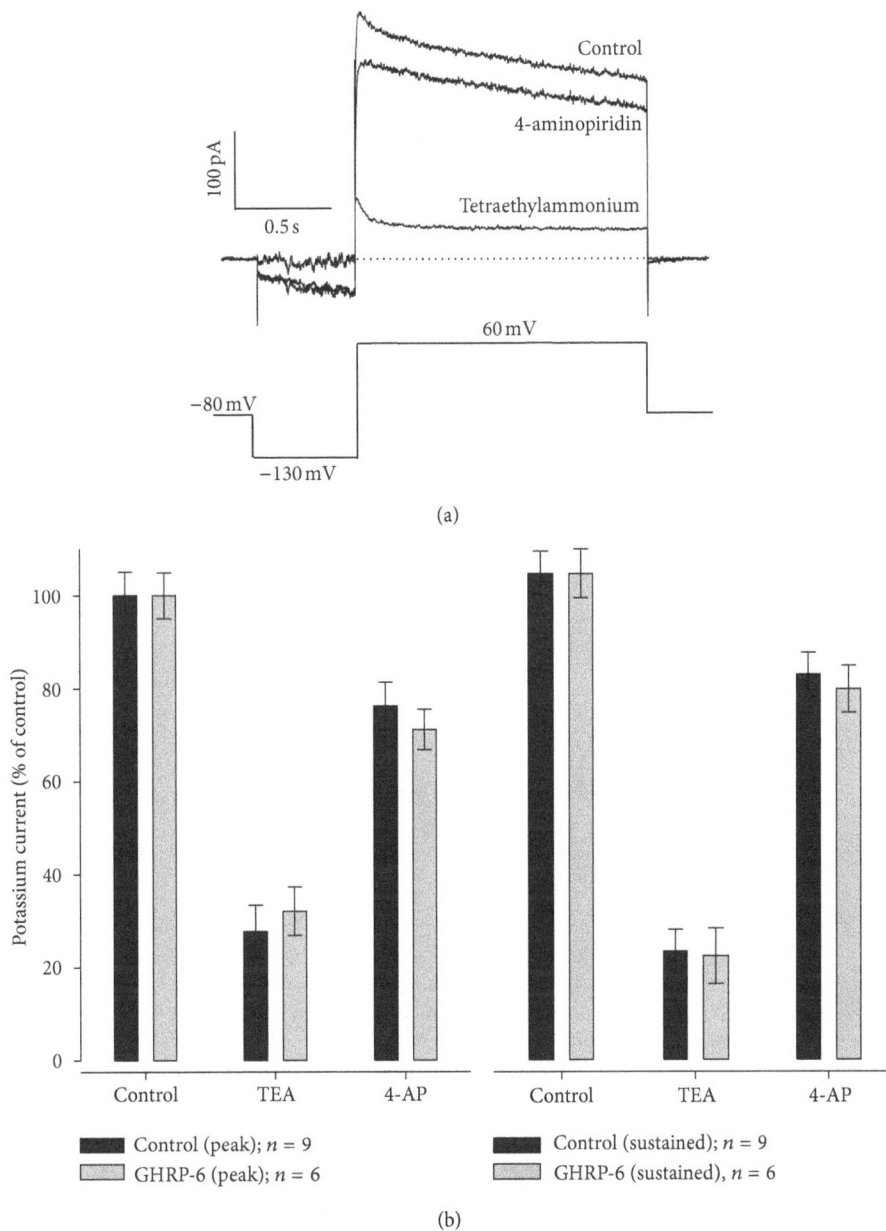

FIGURE 6: Effect of chronic treatment with secretagogues (ghrelin or GHRP-6) on the percentage of the components of the voltage-dependent potassium current on the cellular line of rat GC somatotropes. (a) Representative traces of a control cell. The current was evoked upon passing from a holding potential from −80 mV to −130 mV for 500 ms, from there to depolarize the cell at 60 mV for 1.5 seconds in order to obtain the totality of K$^+$ current. Subsequently, an extracellular solution supplemented with specific blockers for the I_K (tetraethylammonium; TEA 30 mM) and I_A currents (4-aminopyridine; 4-AP 5 mM) was perfused with the aim of desiccating the components of the K$^+$ current. It should be noted that, before a blocker was used, the control solution was employed for washing the effect of the blocker. (b) Summary of the results obtained by using blockers in the control cells ($n = 9$) and in the cells treated chronically with GHRP-6 100 mM ($n = 6$). The measurements were made at the peak in the first 50 ms, and in the sustained component of the trace in the last 5 ms. The peak and sustained values of the current in the control cells were 314 ± 47 and 200 ± 27 pA, respectively, and for the cells treated chronically with GHRP-6 (100 mM), they were 297 ± 35 and 198 ± 21 pA, respectively.

3.2. Effect of Chronic Treatment with Secretagogues (Ghrelin or GHRP-6) on the Voltage-Dependent Potassium Current in the Cellular Line of Rat GC Somatotropes. The chronic effect promoted by ghrelin or GHRP-6 on the voltage-dependent potassium current was evaluated in the GC cells,

which were treated for 96 h with ghrelin (10 nM) or GHRP-6 (100 nM). Figure 4 shows that the chronic treatment with ghrelin or GHRP-6 has no effect on the voltage-dependent potassium current in GC cells. The measurements of the current were performed at

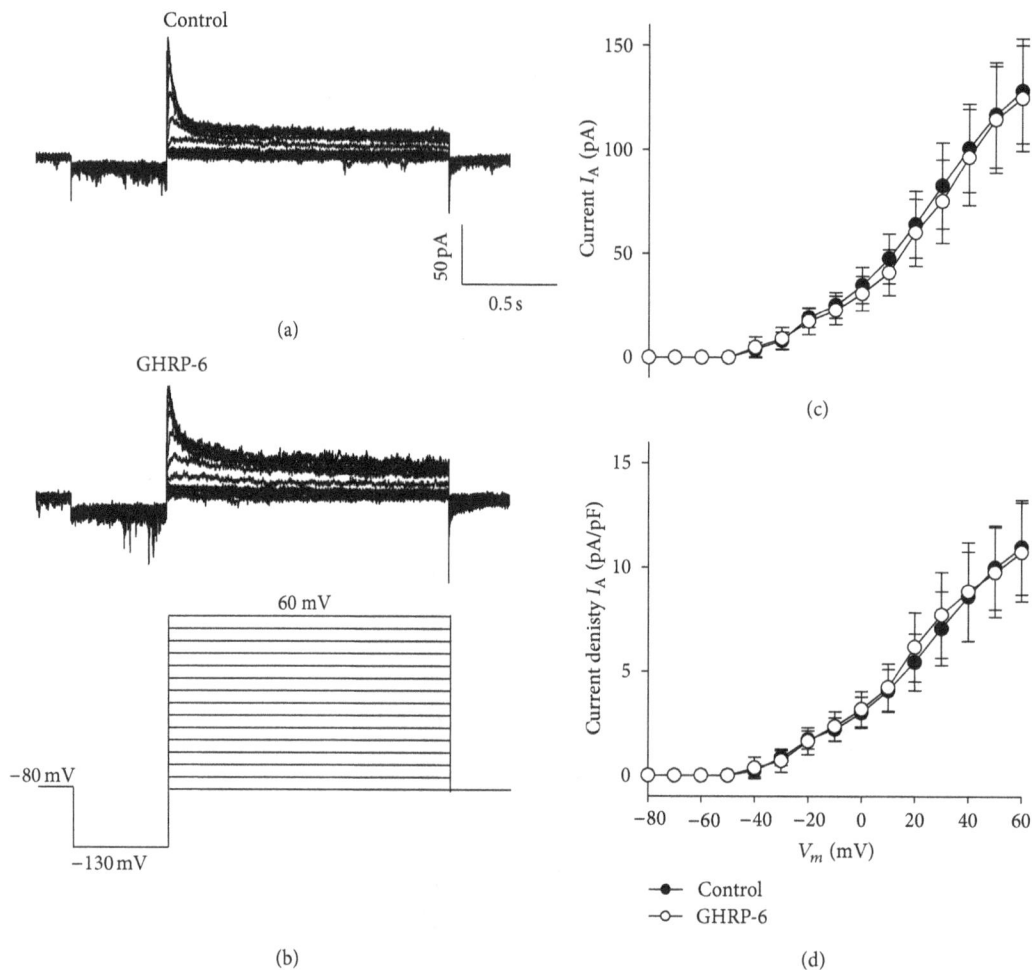

FIGURE 7: The transient potassium current (I_A) in the cellular line of rat GC somatotropes under control conditions and treated with GHRP-6. (a) and (b): family of representative traces of the I_A current of a control cell and a cell treated with GHRP-6 (100 nM) for 96 h. The current was evoked by means of a protocol of depolarizing pulses, starting from a holding potential of −80 mV. Thereafter, the cell was hyperpolarized at −130 mV for 500 ms, and then depolarized from −80 to 60 mV with increases of 10 mV lasting 1.5 s; the recording protocol is shown in the lower portion of panel (b). The external recording solution for potassium currents contained Tetraethylammonium 30 mM. (c) and (d): Current-voltage relationship and density of current-voltage, respectively, for each one of the experimental conditions; the peak current was measured in the first 50 ms of the trace. The number of cells analyzed in control condition was 16, with a capacitance value of 11.46 ± 2.86 ee; 14 cells that had been treated with GHRP-6 100 nM for 96 h were also analyzed, their capacitance value being 9.93 ± 2.65 ee.

the peak (the first 50 ms), the sustained component (the last 5 ms; 1.455–1.5 s), and the subtraction of both measurements.

Together with the results from the voltage-dependent K^+ current, the density value of the current was obtained in order to eliminate the cell size as a source of variation. The results obtained from this measurement show that the current density is not modified in regard to the control value through chronic treatment with ghrelin or GHRP-6 (Figure 5).

In order to examine in more detail whether ghrelin or GHRP-6 modified the transient (I_A) or delayed-rectifier (I_K) proportion of the potassium current, specific blockers were employed for each of the currents that comprise the total current recorded. A depolarizing pulse of 60 mV was applied, starting from a prepulse of 500 ms with a membrane potential fixed at −130 mV, the holding membrane potential was −80 mV.

When the total potassium current had been obtained, it was prefused with an external solution supplemented with 4-aminopyridine at a concentration of 5 mM to block the I_A current, then an extra cellular solution supplemented with tetraethylammonium (TEA) at a concentration of 30 mM was applied to block the I_K current. The results are shown in Figure 6.

In Figure 6, it can be observed that the chronic treatment with GHRP-6 does not significantly affect the proportion of the voltage-dependent K^+ current in its transient (I_A) component, sensitive to 4-aminopyridine, and its sustained (I_K) component, sensitive to TEA.

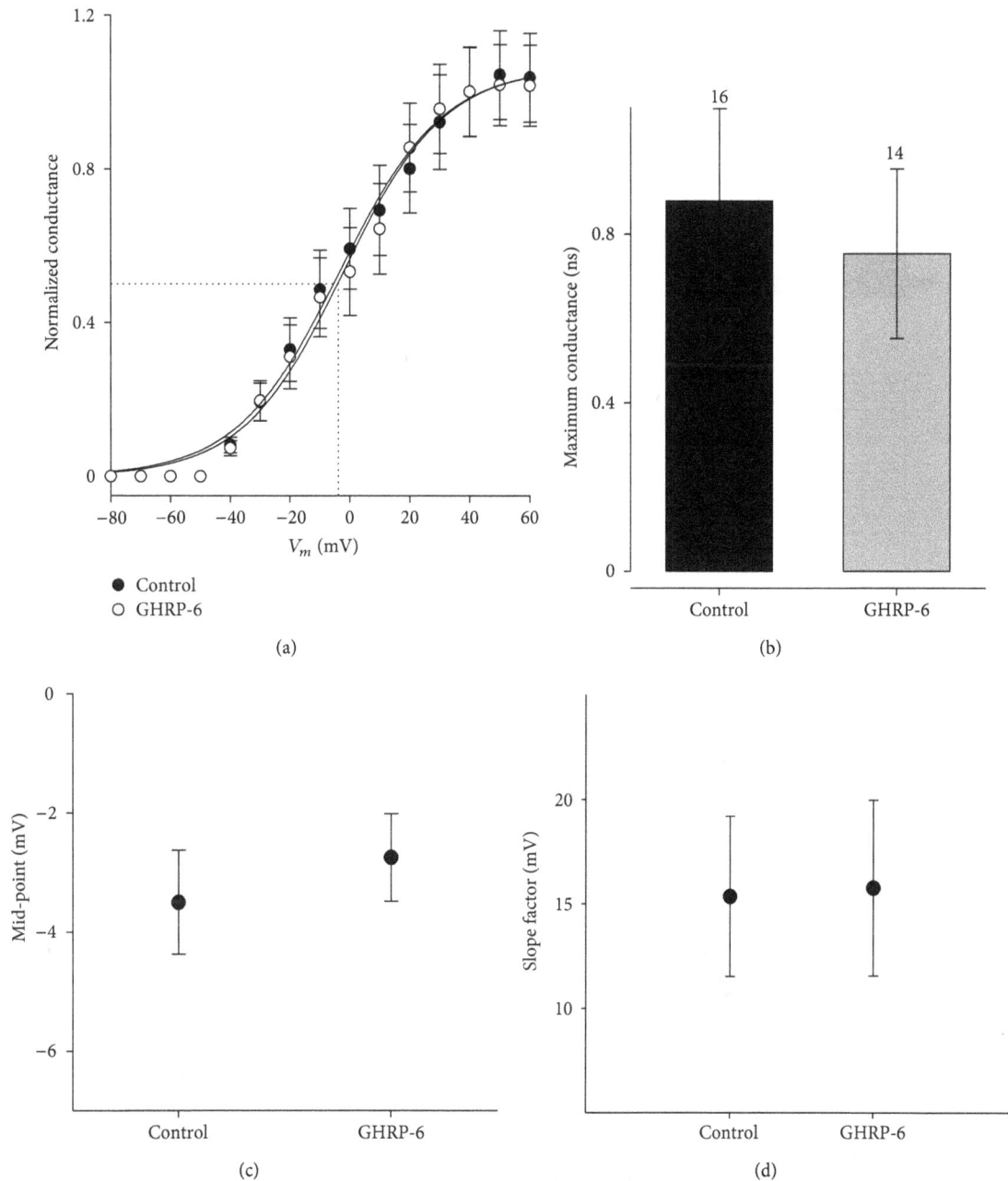

FIGURE 8: Activation curves of the voltage-dependent potassium current I_A in the cellular line of rat GC somatotropes. (a) Activation curves of the I_A current for control conditions and treated with GHRP-6 100 nM for 96 h. The conductances were obtained and these were normalized in regard to the maximum value. The continuous lines represent the adjustment of data with a Boltzmann-type equation. (b) Maximum conductance value for each one of the experimental conditions; the numbers beside the error bar represent the number of cells analyzed. (c) and (d): Boltzmann-type adjustment values, midpoint ($V_{1/2}$) and slope factor (k), respectively. The adjustments were obtained from the cell-current data shown in Figure 7.

Subsequently, it was decided to analyze the effect of GHRP-6 on the transient (I_A) current in more detail by using Tetraethylammonium at a concentration of 30 mM in the external recording solution for the purpose of blocking the sustained (I_K) component, which would make it possible to determine whether or not GHRP-6 affects the potassium current in its transient component. The results obtained from these experiments are shown in Figures 7(a) and 7(b).

Chronic treatment (96 h) with the synthetic analogue of ghrelin, GHRP-6 100 nM, does not modify the transient K$^+$ current (I_A). The peak value (Figures 7(c) and 7(d)) was obtained by measuring the current during the first 50 ms of the trace.

From this same series of experiments, it was possible to obtain the activation curves of I_A current by adjusting the data to a Boltzmann-type equation (Figure 8). A value of

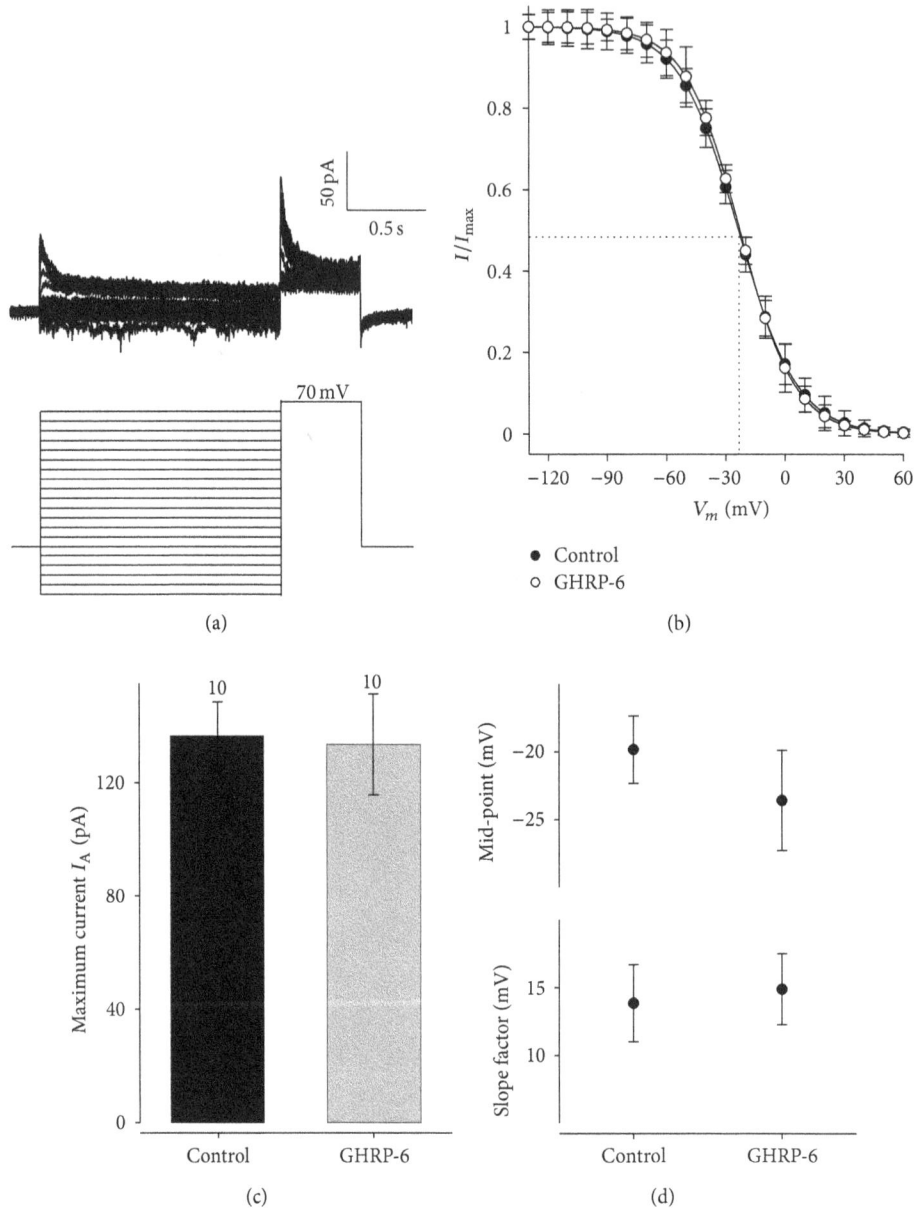

FIGURE 9: Inactivation in the steady state of I_A current in the cellular line of rat GC somatotropes. (a) Family of representative traces of the inactivation in the steady state of I_A current. The current was evoked by a double-pulse protocol; the first of these, named the conditioning pulse, starts from −130 mV and has a duration of 1.5 seconds and a varied amplitude (10 mV); whereas the second one, named testing, carries the membrane potential at 70 mV and lasts 500 ms. The holding potential was fixed at −80 mV. (b) Adjustment of the current at the peak of the testing pulse in accordance with the conditioning pulse voltage and normalization of the latter as regards to the maximum current; the continuous lines represent the adjustment of the data with a Boltzmann-type function. (c) Maximum current evoked by the testing pulse in experimental conditions; the numbers beside the error bar represent the number of cells analyzed. (d) Boltzmann-type adjustment of values: midpoint ($V_{1/2}$) and slope factor (k).

−74.89 mV was obtained for the equilibrium potential of the potassium ion (E_K) with Nerst's equation from our recording solutions. As one can observe, the parameters of the activation curve, $V_{1/2}$ (midpoint) and k (slope factor), which determine the position and the form of the adjusted curve in the voltage axis, were not modified; likewise, the maximum conductance suffered no significant change ($P > 0.05$) in the chronic treatment with GHRP-6 100 nM.

In the same way, the inactivation of the I_A current in control conditions and treated with GHRP-6 100 nM for 96 h was evaluated in order to discard any change in the kinetics of the potassium channels in charge of type I_A current (Figure 9). A double-pulse protocol was used to construct the inactivation curve, the first pulse being designated as conditioning and the second one as testing (see data analysis). The peak amplitude of the I_A-type current induced by

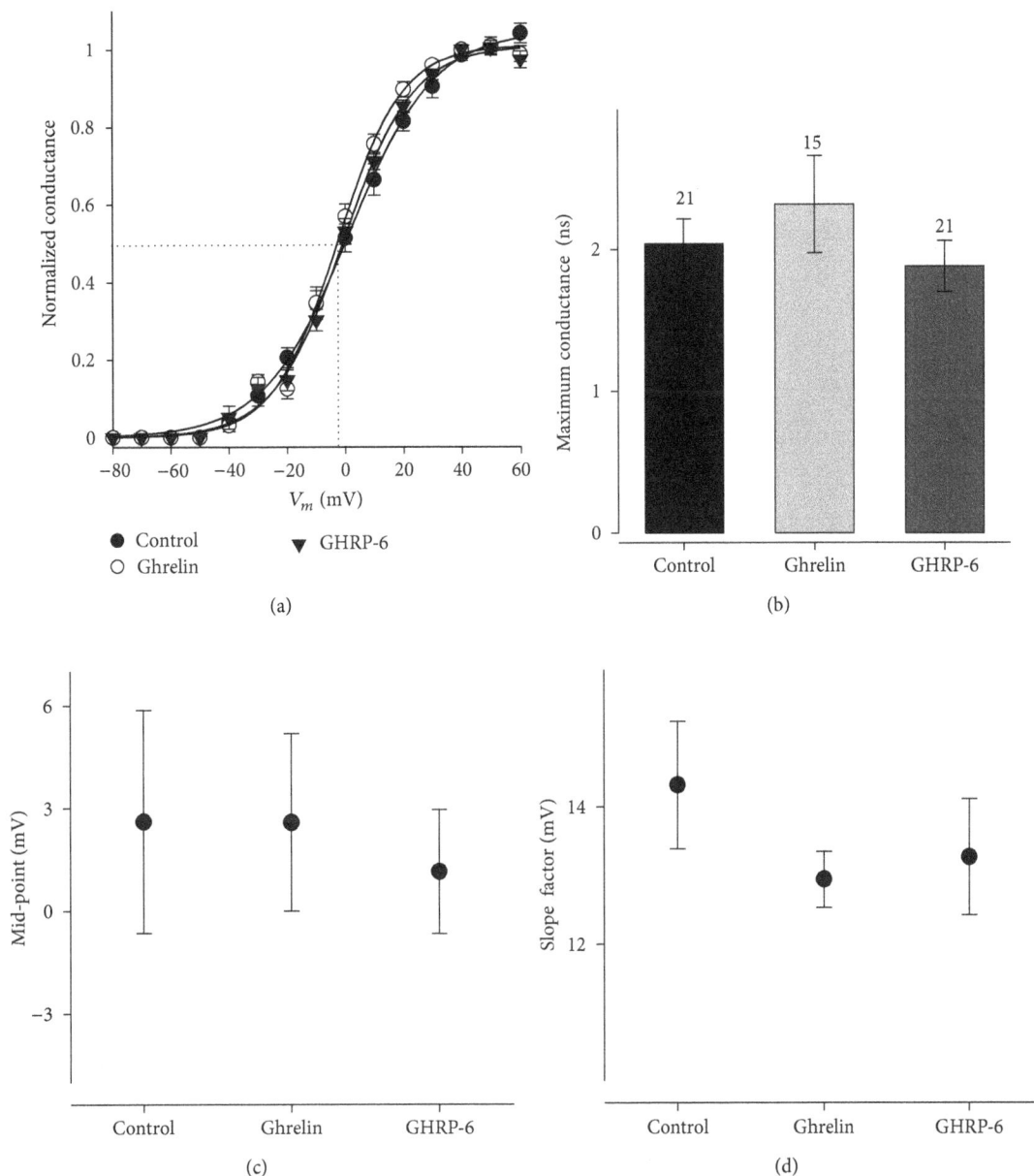

FIGURE 10: Activation curve of the sustained component of the voltage-dependent K^+ current in the cellular line of rat GC somatotropes. (a) Adjustments to the normalized conductance of the GC cells under control conditions, treated with ghrelin 10 nM and GHRP-6 100 nM for 96 h. The data were obtained from the current-voltage curves in Figure 4, the continuous lines representing the adjustment of data with a Boltzmann-type function. (b) Maximum conductance value for each of the experimental conditions. (c) and (d): values of the Boltzmann-type adjustment parameters for each of the experimental conditions, midpoint ($V_{1/2}$) and slope factor (k), respectively.

the testing pulse was graphed according to the voltage of the conditioning pulse and the data obtained from each cell were adjusted with a Boltzmann-type equation; then the data and the adjusted function were normalized in regard to the I_{max} value. The result of the inactivation curves with the I_A current in a steady state is shown in Figure 9. Chronic treatment with GHRP-6 100 nM does not modify the kinetic macroscopic properties of type I_A K^+ current, since the parameters of $V_{1/2}$ and of k are not affected (Figures 9(c) and 9(d)).

3.3. Chronic Effect of Secretagogues (Ghrelin or GHRP-6) on the Delayed-Rectifier K^+ Current (I_K) in GC Cells. We evaluated the chronic effect exercised by ghrelin and by its synthetic analogue GHRP-6 on the sustained component of the voltage-dependent potassium current named delayed rectifier (I_K). The values obtained from the current-voltage curves shown in Figure 4 were used to calculate the activation curve. The data obtained were normalized in regard to the maximum conductance value and were later adjusted to a Boltzmann-type function, the results of which are shown in Figure 10.

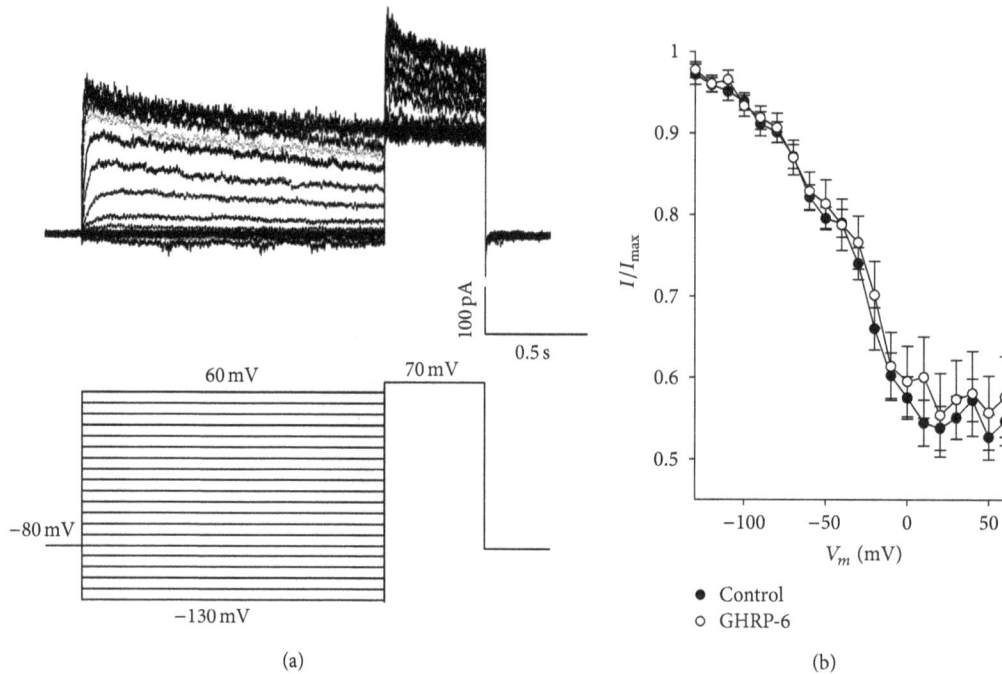

(a)

(b)

FIGURE 11: Inactivation in steady state of the total voltage-dependent K$^+$ current in the cellular line of rat GC somatotropes. (a) Family of representative traces of the steady-state inactivation for the total voltage-dependent K$^+$ current. The current was evoked by a double-pulse protocol. The first pulse, named conditioning, starts from a membrane potential of −130 mV, having duration of 1.5 sec and varying amplitude; the second one, testing, carries a membrane potential of 70 mV with duration of 500 ms. (b) The current at the peak of the testing pulse was graphed according to the voltage of the conditioning pulse, this being normalized in regard to the maximum current.

Figure 10 shows the results of the activation curves of the I_K current. The chronic treatment with ghrelin or GHRP-6 does not modify the kinetic properties of I_K current, since the voltage value at which 50% of the channels are activated is not significantly different ($P > 0.05$), the same as the value of k, which determines the form of the activation curve.

Otherwise, it was decided to evaluate the kinetics of inactivation in a steady state of the total potassium current under control conditions and treated with GHRP-6 (100 nM) for 96 h. A double-pulse protocol was used for this purpose, and the current was evoked by a double-pulse protocol (see data analysis) (Figure 11(a)). The current was graphed at the peak of the testing pulse in accordance with the voltage of the conditioning pulse, this being normalized in regard to the maximum current (Figure 11(b)).

As one can observe in Figure 11, chronic treatment with GHRP-6 100 nM does not modify the inactivation kinetics of the total K$^+$ current, the conclusion being that neither ghrelin nor GHRP-6 modifies the potassium current.

3.4. The Effect of Chronic Treatment with GHRP-6 on the Potassium-Inward-Rectifier (KIR) Current in the Cellular Line of Rat GC Somatotropes. Figure 12(a) shows the recording protocol [15] for evoking the rectifying current entering the GC cells. The protocol consists of hyperpolarizing pulses from −160 to −40, starting from a holding potential of −50 m and a duration of 250 ms in steps of 10 mV, with the same recording solution used for the K$^+$ current (I_K) Figure 12(b)

shows a family of currents presenting two components of the KIR current, an initial (transient) component of rapid activation and inactivation in the first 2-3 ms, followed by a (sustained) component of slow activation that remains during the 250 ms of the pulse. The transient component was measured 1 ms (empty and full circles) after the end of the capacitive component in order to avoid contamination from the latter, which had a duration of ~0.3 ms; the sustained component was measured 5 ms (empty and full inverted triangles) before the end of the current trace (242–250 ms). Figure 12(c) shows an insert of the same trace shown in part Figure 12(b), which was expanded in time in order to observe the transient component in more detail. The current-voltage curves of the two KIR components exhibit kinetics of inward rectification, especially for the sustained part of the KIR current (Figure 12(d)).

The chronic treatment with GHRP-6 (100 nM) does not significantly modify ($P < 0.05$) the current or the density of the potassium KIR current in its two components, transient and sustained, measured in the step of voltage at −160 mV (Figures 12(e) and 12(f)).

Finally, in order to investigate the possible contribution of the voltage-dependent potassium channels to the secretion mediated by the GH secretagogues in the GC cells, ELISA experiments using drugs affecting the activity of the channels were carried out. To investigate the participation of the potassium channels, the GC cells were incubated for 96 hours under control conditions or treated by ghrelin (10 nM) or GHRP-6 (100 nM), which tend to increase the secretion of

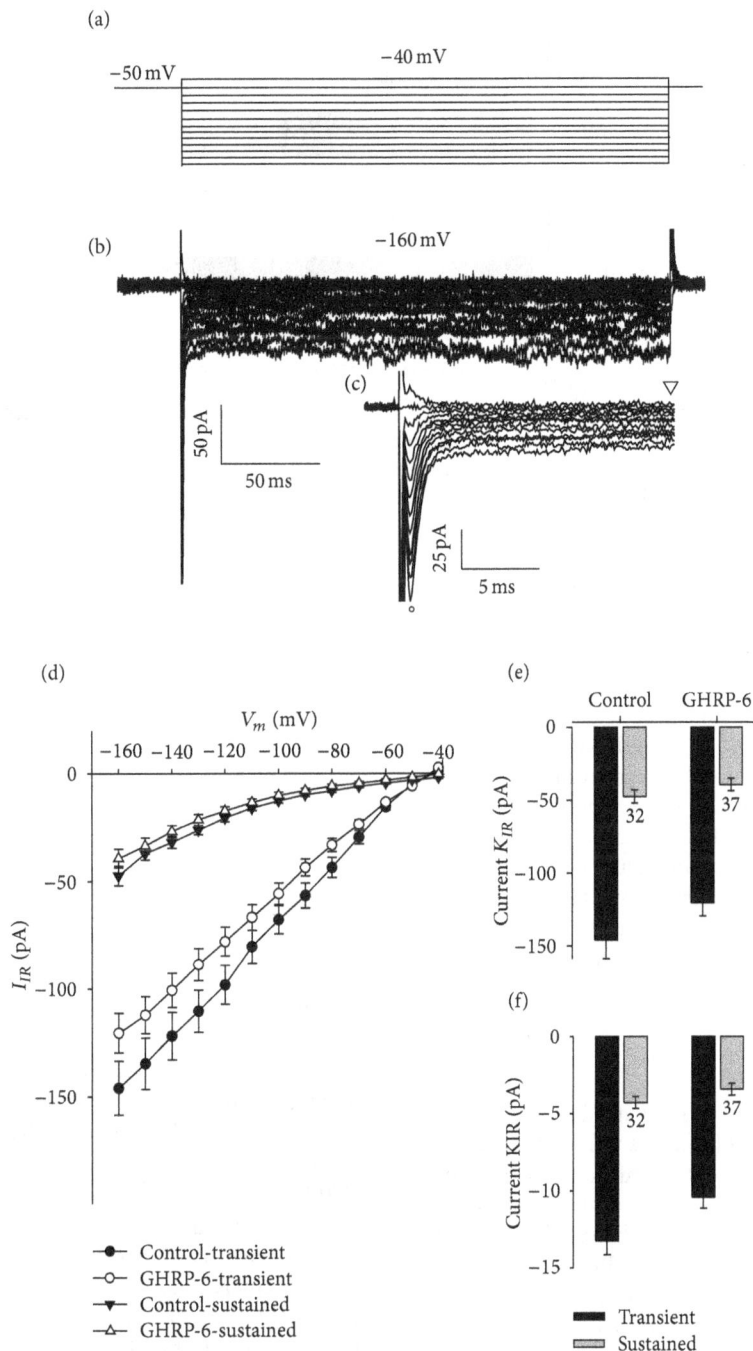

FIGURE 12: Effect of GHRP-6 on the K$^+$ inward-rectifier current (KIR) in the cellular line of rat GC somatotropes. (a) Recording protocol to evoke the inward-rectifier current in GC cells, taken from Xu et al. [15], The protocol consists of hyperpolarizing pulses from −160 to −40, starting from a holding potential of −50 mV and a duration of 250 ms in steps of 10 mV, with the same recording solution for the K$^+$ I_K current. (b) Family of KIR current traces showing two components, an initial (transient) one of rapid activation and inactivation in the first 2-3 ms, followed by a (sustained) component of slow activation that remains for the 250 ms duration of the pulse. The transient component was measured 1 ms after the end of the capacitive component to avoid contamination from the component, the latter having a duration of ~0.3 ms; the sustained component was measured 5 ms before the end of the current trace (245–250 ms). (c) Insert of the same trace shown in (b), which was expanded in time so as to observe the transient component of the KIR current. (d) Current-voltage curves of the two KIR components. (e) and (f): current and density of current measured in the trace at −160 mV. The capacitance value for the control cells was 11.02 ± 0.50 and for the cells treated with GHRP-6 it was 11.50 ± 0.58; the numbers beside the error bars show the number of cells analyzed.

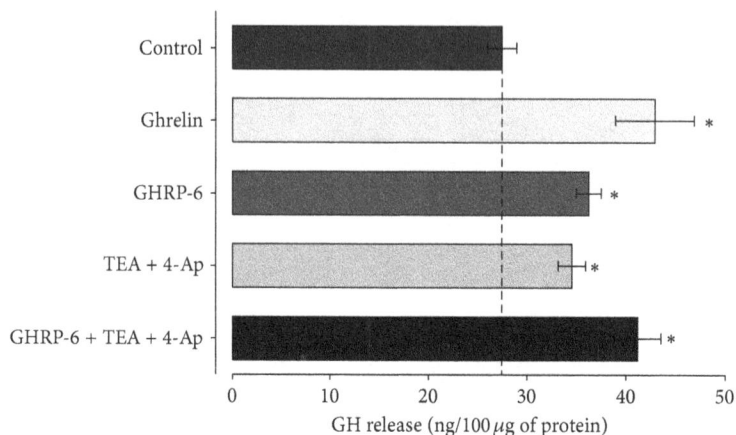

FIGURE 13: The treatment with secretagogues (ghrelin or GHRP-6) does not affect the functional expression of voltage-dependent K^+ channels in the cellular line of rat GC somatotropes. The bars in the graph illustrate the regulation of GH liberation by ghrelin (10 nM) or GHRP-6 (100 nM), applied alone or in the presence of a selective blocker of potassium channels (TEA; 30 mM and 4 Ap; 5 mM). Each value represents the mean ± EE of the determinations carried out in triplicate of three independent experiments. The asterisks denote significant differences ($P < 0.05$) compared with the untreated (control) cells.

GH (~35%) with regard to the control value (Figure 13), were used alone and in combination with a selective blocker of potassium channels (TEA, 30 mM, and 4 Ap, 5 mM). The secretion of GH in cells treated with GHRP-6 combined with the potassium blockers was not different from the secretion provoked by GHRP-6 or ghrelin alone. Upon analyzing the GH secretion in cells treated with TEA and 4 Ap, an increase is observed in regard to control value which, however, does not differ from that occasioned by GH secretagogues. Therefore, we may conclude that the secretagogues do not affect the functional expression of the voltage-dependent potassium channels.

4. Discussion

GH is an anabolic hormone that regulates both growth and development in many species [24]. It is well established that the secretion of GH occurs under the neuroendocrine control of GHRH and of somatostatin at the pituitary level, with an additional regulation proportioned by ghrelin [2, 11, 12]. Ghrelin, an endogenous ligand for GHS-R, is apparently involved in an additional neuroendocrine pathway to control the secretion of GH [12, 14]. For this reason, ghrelin strongly stimulates the liberation of GH, in vitro and in vivo, in a wide range of species, including humans [2, 11, 12, 14].

The present study was undertaken to examine the effect of the long-term incubation with secretagogues of GH (ghrelin or GHRP-6) on the functional expression of potassium channels in the cellular line of rat GC somatotropes, which is a subclone of GH$_3$ cell line. The results indicate that the voltage-dependent K^+ currents (I_A, I_K and KIR) do not suffer any significant changes, neither in their kinetic properties of activation nor in their inactivation through treatment with secretagogues. Previously, Chen [11] demonstrated that the K^+ current increases through long-term treatments (48 hrs) with GHRP-2 by means of an increase in protein synthesis for

the K^+ channel in ovine somatotropes, this effect is mediated by the pathway PKC system.

The effect of ghrelin on GH secretion is thought to be connected with various systems of signaling. In pigs, the liberation of GH in response to ghrelin depends on signaling systems that involve cAMP/PKA as well as PLC/PKC pathways, and on the influx of extracellular Ca^{2+} [4]. In rats, the liberation of ghrelin-dependent GH is provoked by both intracellular Ca^{2+} liberation and extracellular Ca^{2+} influx [25]; the influx of Ca^{2+} is via voltage-dependent Ca^{2+} channels that are activated through depolarization [25, 26]. In tumorous GC cells, it has been reported that long-term exposure (96 h) with ghrelin stimulates the functional expression of the voltage-dependent Na^+ and Ca^{2+} channels involved in the influx of extracellular calcium to promote the secretion of GH [26, 27]; in contrast, other researchers have reported that ghrelin reduces the voltage-dependent Ca^{2+} current via GMP in GH$_3$ cells [28].

The electrical activity of the somatotropes depends on the properties as well as on the functional expression of the ionic channels present in the plasmatic membrane and on the potential of the membrane in repose. The K^+ current flowing through the plasmatic membrane is responsible for the potential of the membrane in repose, although other ions such as Na^+ and Ca^{2+} may be involved [17]. Based on their properties through differences in time, voltage dependency, and pharmacological sensitivity, various types of K^+ current have been identified in the somatotropes [29, 30]; this currents include inward rectifiers (KIR), transient current (I_A), delayed rectifier current (I_K), sometimes called IRD, K^+ currents activated by $[Ca^{2+}]_i$ [30, 31]. Both currents, I_A and I_K, are involved in the electrical activity of somatotropes [29–31]. The I_A is partly responsible for maintaining the membrane potential in repose and participating in the repolarization process of the potential in action [29–32]. The role of the I_K current carried by voltage-dependent K^+ channels in electrical activity has been examined in various cellular types; in GH_3 cells, the inhibition of this channel by TEA has been seen to increment the duration of the action potential [33], as well as the amplitude of the spontaneous transients of $[Ca^{2+}]_i$ [34]; in rat lactotropes TEA does not modify the firing pattern [33]. Somatostatin increases both the I_K and the KIR in rat, ovine and human somatotropes [18, 19, 35, 36].

On the other hand, GHRH reduces the K^+ current in human adenoma cells as well as in GH_4C_1 cells [37, 38]; the synthetic analogue of ghrelin, GHRP-6, diminishes both the transient and the delayed rectifier currents in rat somatotropes [39]. However, an increase in the voltage-dependent K^+ current (both I_K and I_A) has been reported, occasioned by an increase in the synthesis of protein that codifies for the K^+ channel provoked by another GHS, GHRP-2 [40]. Up to now, there are few reports regarding the effect of ghrelin or GH secretagogues on the functional expression of ionic channels and the routes they may be occupying [26, 27, 40, 41]. It has been reported that ghrelin inhibits the inward rectifier of the K^+ channel coupled with protein G in neurons of the tuberous mammillary nucleus (KIR_3) [42]. In the present work, we demonstrate that ghrelin does not affect the K^+ currents (I_K and I_A) during a long-term exposure (96 h) in the cellular line of rat GC somatotropes. In work realized on GH_3 cells, ghrelin acutely (in the bath) reduces the voltage-dependent K^+ current; this effect of ghrelin, mediated by means of the cGMP/PKG system [20], occurs through the activation of a GHS receptor, since GHRH-R is not present in these cells [43].

In primary cultures of pituitary cells and in GH_3 cells, it has been shown that the inhibition of the KIRs can generate a higher firing rate of action potentials and subsequently an increase in the secretion of the prolactin hormone and GH [44]. Treatment with ghrelin or GHRP-6 does not significantly modify the KIR current in these cells at physiological voltages; however, a reduction of KIR current has been reported in ovine somatotropic cells with GHRP-2 through the PKA-AMPc pathway [15]. To summarize, chronic treatment with ghrelin or GHRP-6 does not modify the functional expression of the K^+ channels underlying the I_K, I_A, and KIR currents in the cellular line of rat GC somatotropes, subclone of GH_3.

References

[1] M. T. Bluet-Pajot, V. Tolle, P. Zizzari et al., "Growth hormone secretagogues and hypothalamic networks," *Endocrine*, vol. 14, no. 1, pp. 1–8, 2001.

[2] A. W. Root and M. J. Root, "Clinical pharmacology of human growth hormone and its secretagogues," *Current Drug Targets*, vol. 2, no. 1, pp. 27–52, 2002.

[3] A. O. L. Wong, B. C. Moor, C. E. Hawkins, N. Narayanan, and J. Kraicer, "Cytosolic protein kinase A mediates the growth hormone (GH)-releasing action of GH-releasing factor in purified rat somatotrophs," *Neuroendocrinology*, vol. 61, no. 5, pp. 590–600, 1995.

[4] M. M. Malagón, R. M. Luque, E. Ruiz-Guerrero et al., "Intracellular Signaling Mechanisms Mediating Ghrelin-Stimulated Growth Hormone Release in Somatotropes," *Endocrinology*, vol. 144, no. 12, pp. 5372–5380, 2003.

[5] M. Kato and M. Suzuki, "Inhibition by nimodipine of growth hormone (GH) releasing factor-induced GH secretion from rat anterior pituitary cells," *Japanese Journal of Physiology*, vol. 41, no. 1, pp. 63–74, 1991.

[6] A. P. Naumov, J. Herrington, and B. Hille, "Actions of growth-hormone-releasing hormone on rat pituitary cells: intracellular calcium and ionic currents," *Pflugers Archiv European Journal of Physiology*, vol. 427, no. 5-6, pp. 414–421, 1994.

[7] R. W. Holl, M. O. Thorner, G. L. Mandell, J. A. Sullivan, Y. N. Sinha, and D. A. Leong, "Spontaneous oscillations of intracellular calcium and growth hormone secretion," *Journal of Biological Chemistry*, vol. 263, no. 20, pp. 9682–9685, 1988.

[8] M. O. Thorner, R. W. Holl, and D. A. Leong, "The somatotrope: a endocrine cell with functional calcium transients," *Journal of Experimental Biology*, vol. 139, pp. 169–179, 1988.

[9] M. J. Cronin, S. T. Summers, M. A. Sortino, and E. L. Hewlett, "Protein kinase C enhances growth hormone releasing factor (1-40)-stimulated cyclic AMP levels in anterior pituitary. Actions of somatostatin and pertussis toxin," *Journal of Biological Chemistry*, vol. 261, no. 30, pp. 13932–13935, 1986.

[10] D. L. Lewis, F. F. Weight, and A. Luini, "A guanine nucleotide-binding protein mediates the inhibition of voltage-dependent calcium current by somatostatin in a pituitary cell line," *Proceedings of the National Academy of Sciences of the United States of America*, vol. 83, no. 23, pp. 9035–9039, 1986.

[11] C. Chen, "Growth hormone secretagogue actions on the pituitary gland: multiple receptors for multiple ligands?" *Clinical and Experimental Pharmacology and Physiology*, vol. 27, no. 5-6, pp. 323–329, 2000.

[12] R. G. Smith, "Development of growth hormone secretagogues," *Endocrinology Review*, vol. 26, no. 3, pp. 346–360, 2005.

[13] C. Y. Bowers, F. A. Momany, G. A. Reynolds, and A. Hong, "On the in vitro and in vivo activity of a new synthetic hexapeptide that acts on the pituitary to specifically release growth hormone," *Endocrinology*, vol. 114, no. 5, pp. 1537–1545, 1984.

[14] M. Kojima, H. Hosoda, Y. Date, M. Nakazato, H. Matsuo, and K. Kangawa, "Ghrelin is a growth-hormone-releasing acylated peptide from stomach," *Nature*, vol. 402, no. 6762, pp. 656–660, 1999.

[15] R. Xu, Y. Zhao, and C. Chen, "Growth hormone-releasing peptide-2 reduces inward rectifying K^+ currents via a PKA-cAMP-mediated signalling pathway in ovine somatoropes," *Journal of Physiology*, vol. 545, no. 2, pp. 421–433, 2002.

[16] J. Herrington and B. Hille, "Growth hormone-releasing hexapeptide elevates intracellular calcium in rat somatotropes by two mechanisms," *Endocrinology*, vol. 135, no. 3, pp. 1100–1108, 1994.

[17] S. M. Simasko, "A background sodium conductance is necessary for spontaneous depolarizations in rat pituitary cell line GH3," *American Journal of Physiology-Cell Physiology*, vol. 266, no. 3, pp. C709–C719, 1994.

[18] S. M. Sims, B. T. Lussier, and J. Kraicer, "Somatostatin activates an inwardly rectifying K^+ conductance in freshly dispersed rat somatotrophs," *Journal of Physiology*, vol. 441, pp. 615–637, 1991.

[19] C. Chen, "Gi-3 protein mediates the increase in voltage-gated K^+ currents by somatostatin on cultured ovine somatotropes," *American Journal of Physiology*, vol. 275, part 1, no. 2, pp. E278–E284, 1998.

[20] F. H. Xue, L. Z. Yun, M. Hernandez, D. J. Keating, and C. Chen, "Ghrelin reduces voltage-gated potassium currents in GH3 cells via cyclic GMP pathways," *Endocrine*, vol. 28, no. 2, pp. 217–224, 2005.

[21] B. Dominguez, R. Felix, and E. Monjaraz, "Ghrelin and GHRP-6 enhance electrical and secretory activity in GC somatotropes," *Biochemical and Biophysical Research Communications*, vol. 358, no. 1, pp. 59–65, 2007.

[22] A. Marty and E. Neher, "Tight-seal whole-cell recording," in *Single-Channel Recording*, B. Sakmann and E. Neher, Eds., pp. 31–52, Plenum Press, New York, NY, USA, 1995.

[23] G. Avila, A. Sandoval, and R. Felix, "Intramembrane charge movement associated with endogenous K^+ channel activity in HEK-293 cells," *Cellular and Molecular Neurobiology*, vol. 24, no. 3, pp. 317–330, 2004.

[24] L. A. Frohman, T. R. Downs, and P. Chomczynski, "Regulation of growth hormone secretion," *Frontiers in Neuroendocrinology*, vol. 13, no. 4, pp. 344–405, 1992.

[25] M. Yamazaki, H. Kobayashi, T. Tanaka, K. Kangawa, K. Inoue, and T. Sakai, "Ghrelin-induced growth hormone release from isolated rat anterior pituitary cells depends on intracellular and extracellular Ca^{2+} sources," *Journal of Neuroendocrinology*, vol. 16, no. 10, pp. 825–831, 2004.

[26] B. Dominguez, T. Avila, J. Flores-Hernandez et al., "Up-regulation of high voltage-activated Ca^{2+} channels in GC somatotropes after long-term exposure to ghrelin and growth hormone releasing peptide-6," *Cellular and Molecular Neurobiology*, vol. 28, no. 6, pp. 819–831, 2008.

[27] B. Dominguez, R. Felix, and E. Monjaraz, "Upregulation of voltage-gated Na+ channels by long-term activation of the ghrelin-growth hormone secretagogue receptor in clonal GC somatotropes," *American Journal of Physiology-Endocrinology and Metabolism*, vol. 296, no. 5, pp. E1148–E1156, 2009.

[28] X. Han, Y. Zhu, Y. Zhao, and C. Chen, "Ghrelin reduces voltage-gated calcium currents in GH_3 cells via cyclic GMP pathways," *Endocrine*, vol. 40, no. 2, pp. 228–236, 2011.

[29] F. Van Goor, D. Zivadinovic, and S. S. Stojilkovic, "Differential expression of ionic channels in rat anterior pituitary cells," *Molecular Endocrinology*, vol. 15, no. 7, pp. 1222–1236, 2001.

[30] J. Herrington and C. J. Lingle, "Multiple components of voltage-dependent potassium current in normal rat anterior pituitary cells," *Journal of Neurophysiology*, vol. 72, no. 2, pp. 719–729, 1994.

[31] S. S. Stojilkovic, J. Tabak, and R. Bertram, "Ion channels and signaling in the pituitary gland," *Endocrine Reviews*, vol. 31, no. 6, pp. 845–915, 2010.

[32] C. Chen, P. Heyward, J. Zhang, D. Wu, and I. J. Clarke, "Voltage-dependent potassium currents in ovine somatotrophs and their function in growth hormone secretion," *Neuroendocrinology*, vol. 59, no. 1, pp. 1–9, 1994.

[33] S. Sankaranarayanan and S. M. Simasko, "Potassium channel blockers have minimal effect on repolarization of spontaneous action potentials in rat pituitary lactotropes," *Neuroendocrinology*, vol. 68, no. 5, pp. 297–311, 1998.

[34] A. C. Charles, E. T. Piros, C. J. Evans, and T. G. Hales, "L-type Ca^{2+} channels and K^+ channels specifically modulate the frequency and amplitude of spontaneous Ca^{2+} oscillations and have distinct roles in prolactin release in GH_3 cells," *Journal of Biological Chemistry*, vol. 274, no. 11, pp. 7508–7515, 1999.

[35] C. Chen, J. Zhang, J. D. Vincent, and J. M. Israel, "Somatostatin increases voltage-dependent potassium currents in rat somatotropes," *American Journal of Physiology*, vol. 259, no. 6, part 1, pp. C854–C861, 1990.

[36] K. Takano, J. Yasufuku-Takano, A. Teramoto, and T. Fujita, "G(i3) mediates somatostatin-induced activation of an inwardly rectifying K^+ current in human growth hormone-secreting adenoma cells," *Endocrinology*, vol. 138, no. 6, pp. 2405–2409, 1997.

[37] R. Xu, S. G. Roh, K. Loneragan, M. Pullar, and C. Chen, "Human GHRH reduces voltage-gated K^+ currents via a non-cAMP-dependent but PKC-mediated pathway in human GH adenoma cells," *Journal of Physiology*, vol. 520, no. 3, pp. 697–707, 1999.

[38] R. Xu, I. J. Clarke, S. Chen, and C. Chen, "Growth hormone-releasing hormone decreases voltage-gated potassium currents in GH4C1 cells," *Journal of Neuroendocrinology*, vol. 12, no. 2, pp. 147–157, 2000.

[39] J. F. McGurk, S. S. Pong, L. Y. Chaung, M. Gall, B. Butler, and J. P. Arena, "Growth hormone secretagogues modulate potassium currents in rat somatotropes," in *Proceedings of the 23rd Annual Meeting of Society for Neurosciences*, 1993, Abstract 642. 1.

[40] C. Chen, "The effect of two-day treatment of primary cultured ovine somatotropes with GHRP-2 on membrane voltage-gated K^+ currents," *Endocrinology*, vol. 143, no. 7, pp. 2659–2663, 2002.

[41] Z. Peng, Z. Xiaolei, H. Al-Sanaban, X. Chengrui, and Y. Shengyi, "Ghrelin inhibits insulin release by regulating the expression of inwardly rectifying potassium channel 6. 2 in islets," *American Journal Medical Science*, vol. 343, no. 3, pp. 215–219, 2012.

[42] D. Bajic, Q. V. Hoang, S. Nakajima, and Y. Nakajima, "Dissociated histaminergic neuron cultures from the tuberomammillary nucleus of rats: culture methods and ghrelin effects," *Journal of Neuroscience Methods*, vol. 132, no. 2, pp. 177–184, 2004.

[43] M. Korbonits, S. A. Bustin, M. Kojima et al., "The expression of the growth hormone secretagogue receptor ligand ghrelin in normal and abnormal human pituitary and other neuroendocrine tumors," *Journal of Clinical Endocrinology and Metabolism*, vol. 86, no. 2, pp. 881–887, 2001.

[44] C. K. Bauer, I. Davison, I. Kubasov, J. R. Schwarz, and W. T. Mason, "Different G proteins are involved in the biphasic response of clonal rat pituitary cells to thyrotropin-releasing hormone," *Pflugers Archiv European Journal of Physiology*, vol. 428, no. 1, pp. 17–25, 1994.

pH-Dependent Interaction between C-Peptide and Phospholipid Bicelles

Sofia Unnerståle and Lena Mäler

Department of Biochemistry and Biophysics, Center for Biomembrane Research, The Arrhenius Laboratories for Natural Sciences, Stockholm University, 106 91 Stockholm, Sweden

Correspondence should be addressed to Lena Mäler, lena.maler@dbb.su.se

Academic Editor: Andreas Herrmann

C-peptide is the connecting peptide between the A and B chains of insulin in proinsulin. In this paper, we investigate the interaction between C-peptide and phospholipid bicelles, by circular dichroism and nuclear magnetic resonance spectroscopy, and in particular the pH dependence of this interaction. The results demonstrate that C-peptide is largely unstructured independent of pH, but that a weak structural induction towards a short stretch of β-sheet is induced at low pH, corresponding to the isoelectric point of the peptide. Furthermore, it is demonstrated that C-peptide associates with neutral phospholipid bicelles as well as acidic phospholipid bicelles at this low pH. C-peptide does not undergo a large structural rearrangement as a consequence of lipid interaction, which indicates that the folding and binding are uncoupled. *In vivo*, local variations in environment, including pH, may cause C-peptide to associate with lipids, which may affect the aggregation state of the peptide.

1. Introduction

50 years ago, it was discovered that insulin is synthesized as proinsulin, which contains not only the two chains of insulin, A and B, but also a linker peptide, called C-peptide [1, 2]. C-peptide connects the two chains of insulin, which facilitates the disulfide bond formation between them and aids the folding process of insulin [3, 4]. Since the discovery, several biological effects of C-peptide have been demonstrated [5–7].

The primary structure of C-peptide varies significantly between different species, although certain common structural features can be observed. For example, the highly acidic and somewhat conserved N-terminus has properties that appear to be important for C-peptides chaperon-like effects on insulin disaggregation [8]. Further, the C-terminus is somewhat conserved and is likely to be involved in receptor interactions [9–11]. Human C-peptide, which is studied in this paper, consists of 31 aminoacid residues, EAEDLQVGQVELGGGPGAGSLQPLALEGSLQ. It contains many negatively charged amino acid residues and no basic residues resulting in a very low pI (3.5).

Human C-peptide has a random coil structure in buffer, while the N-terminal, third of the C-peptide (residues 1–11), has been demonstrated to be helical in 95% TFE [12]. In H_2O/TFE 1:1, on the other hand, it has been shown that residues A2 through L5 adopt a type I β-turn, while residues E27 through Q31, the so-called pentapeptide, is the most ordered part of C-peptide adopting a type III' β-turn [13]. Further, residues Q9-L12, residues G15-A18 and residues Q22-A25 were all shown to have structural preferences in the NMR-derived ensemble average [13]. It has recently been demonstrated that C-peptide also has the ability, under certain conditions, such as low pH, to form β-sheet structure, resembling amyloid structures [14, 15]. The peptide forms predominantly low-order oligomers [14], but very low concentrations of amyloid-like structures may also form [15]. The formation of amyloid structure can be enhanced in the presence of subcritical micelle concentration (CMC) amounts of SDS (at low pH), while SDS in amounts above the CMC, on the other hand, promote a more α-helical structure [15].

Even though C-peptide appears to only be marginally structured in aqueous solution and in solvents such as TFE, its ability to transiently adopt a variety of structures appears to be of importance for the peptides aggregation propensities. The ability of peptides and proteins to self-associate has been recognized in several diseases, including Alzheimer's disease, amyotrophic lateral sclerosis, and type II diabetes [16, 17]. In many cases, it has been demonstrated that the membrane may serve as a means for peptides to undergo structural rearrangements, which may be important for misfolding events. For the APP $A\beta$ peptide, the composition of the membrane has been shown to be crucial for formation of amyloid structure [18–20]. Due to this feature of the peptide, and its previously demonstrated interaction with SDS, we have in this study examined the interaction between C-peptide and membrane mimetic media composed of isotropic phospholipid bicelles [21–26]. To compare, we have also examined the structure of C-peptide in different large unilamellar vesicle solvents. Further, since it has been demonstrated that pH is an important factor that governs the aggregation state of C-peptide [15], much like for the APP $A\beta$ peptide [27] we have investigated the effect of pH on its structure and lipid interaction properties. In this way, the basic biophysical properties of C-peptide have been deduced, and this study shows that pH affects the ensemble average of the structure of C-peptide, which in turn affects the interaction with membrane-mimicking systems.

2. Materials and Methods

2.1. Materials and Sample Preparation. C-peptide was purchased from PolyPeptide Laboratories, France and used without further purification. 1-palmitoyl-2-oleoyl-*sn*-glycero-3-phosphocholine (POPC) and 1-palmitoyl-2-oleoyl-*sn*-glycero-3-phospho-(1′-rac-glycerol)(POPG) were used to produce large unilamellar vesicles (LUVs). Deuterated lipids, 1,2-dihexanoyl-d$_{22}$-sn-glycero-3-phospho-cho-line (d$_{22}$-DHPC), 1,2-dimyristoyl-d$_{54}$-sn-glycero-3-phosphocholine (d$_{54}$-DMPC) and 1,2-dimyristoyl-d$_{54}$-sn-glycero-3-phospho-(1′-rac-glycerol) (d$_{54}$-DMPG) were used to produce bicelles. All phospholipids were obtained from Avanti Polar Lipids (Alabaster, AL, USA). Different buffers were used to study how the pH affects the C-peptide, three sodium phosphate buffers of pH 5.8, 6.9, and 7.2 and one citrate buffer of pH 3.2.

Large unilamellar vesicles (LUVs) were produced for studying the interaction between C-peptide and bilayers by circular dichroism (CD). First, 20 mM stock solutions of neutral and 50% negatively charged vesicles were prepared by dissolving POPC and POPC/POPG 1 : 1, respectively, in chloroform. The samples were then dried under a flow of N$_2$ gas to create lipid films. To ensure that no chloroform remained, the samples were stored under vacuum overnight. The dried lipid films were then soaked in buffer and vortexed for 10 minutes to obtain a more defined size distribution. The solutions were subsequently subjected to five freeze-thaw cycles to decrease lamellarity. Finally, to obtain uniform samples of LUVs, the samples were extruded around 20 times through a polycarbonate microfilter with 100 nm pore size. The CD samples were then prepared from these stock

solutions and from stock solutions of 100 μM C-peptide to a final concentration in the CD samples of 50 μM C-peptide and 1 mM POPC or POPC/POPG 1 : 1 in 50 mM buffer (citrate buffer at pH 3.2 and sodium phosphate buffer at pH 5.8 or pH 6.9).

Small isotropic bicelles were used to further investigate membrane interaction of C-peptide by diffusion NMR and 2D total correlation spectroscopy (TOCSY). For samples used in pulse field gradient (PFG) diffusion measurements, the peptide was initially dissolved in buffer solution (D$_2$O), and lipids were added to each solution. Buffer solutions of pH 3.2, 5.8, and 7.2, prepared from 50 mM citrate buffer (pH 3.2) or sodium phosphate buffer (pH 5.8 or pH 7.2) were dried and subsequently dissolved in D$_2$O. 250 μM C-peptide was dissolved in each buffer by sonication in a water bath for 1 min. PFG NMR experiments were acquired to measure the self-diffusion of C-peptide, D_{free}. Subsequently d$_{54}$-DMPC lipids and d$_{22}$-DHPC dissolved in D$_2$O were added in a ratio of 1 : 2 to each of the three different samples. This resulted in a final total lipid concentration of 150 mM and a q-ratio (long-chain phospholipids/short-chain phospholipids) of 0.5. After spectra were recorded for these samples, DMPG was added to the three samples, corresponding to 10% of the total long-chain lipids. In this way, bicelles with 10% negative charge were obtained. For 2D TOCSY NMR measurements, 200 μM peptide was dissolved in DMPC/DHPC or in (DMPC/DMPG 9 : 1)/DHPC, respectively. Here, the total lipid concentration was 300 mM, and the q-ratio was 0.25. All bicelle mixtures were vortexed until a clear low-viscous solution was formed.

2.2. Circular Dichroism. The measurements were acquired on a Chirascan CD spectrometer with a 1 mm quartz cell for samples with 50 μM peptide content and 1 mM POPC or POPC/POPG 1 : 1. The temperature was adjusted to 298 K with a TC 125 temperature control. Wavelengths ranging from 190 to 250 nm were measured with a 0.5 nm step resolution. Spectra were collected and averaged over ten measurements. Background spectra of buffers, POPC and POPC/POPG 1 : 1 without any peptide, were also recorded and were subtracted from the peptide spectra.

2.3. NMR Spectroscopy. Translational diffusion experiments were performed on a Bruker Avance spectrometer, equipped with a triple resonance probe-head and operating at a ^{1}H frequency of 600 MHz. A standard sample of 0.01% H$_2$O in D$_2$O, with 1 mg/mL GdCl$_3$ to avoid radiation damping, was used for calibration of the gradient strength. The temperature was adjusted to 298 K using d$_4$-methanol. Diffusion constants were measured using a modified Stejskal-Tanner spin-echo experiment [28–30] using a fixed diffusion time (300 ms) to minimize the influence of relaxation contributions, and a fixed gradient length (2.4 ms in buffer, 3 ms in bicelle solution, and 3.4 ms in acidic bicelle solution) and with a gradient strength varying linearly over 32 steps. The linearity of the gradient was calibrated as described previously [31]. The diffusion coefficient for HDO was measured and compared to the standard diffusion of HDO in D_2O $(1.9 \ 10^{-9})$ [32]. This ratio was then multiplied to all measured diffusion constants to correct for viscosity differences induced by the sample.

2D TOCSY experiments [33] were recorded on Bruker Avance spectrometers operating at ^1H frequencies of 500 MHz or 600 MHz at 298 K. TOCSY spectra were recorded with mixing times of 30 ms. Typically, 24–48 transients were recorded, and the number of increments in the indirect dimension was 256–512. The assignment of the spectra was achieved by the aid of the assignment by Munte et al. with the BMRB accession code 6623 [13], which was made at pH 7.0 and at 283 K. No chemical shift differences larger than ±0.03 ppm were seen between this assignment and our spectra obtained at pH 5.8 and at 298 K.

3. Results

3.1. C-Peptide Shows Overall Random Coil Features. Other studies have shown that C-peptide is predominantly unstructured in aqueous solution and in the presence of lipid vesicles at pH 5 and 7 [12]. This is also seen in our CD spectra (Figures 1(b) and 1(c)), which show random coil features in buffered solutions, POPC and POPC/POPG 1 : 1, respectively. It has previously been shown that SDS induces oligomerization of C-peptide at a pH close to the pI of C-peptide (around 3.5) [15]. Thus, we wanted to study the structure of C-peptide at lower pH. Also at pH 3.2, C-peptide is predominantly in a random coil conformation and our results indicate that no major structural rearrangements are induced as a consequence of adding POPC or POPC/POPG 1 : 1, respectively (Figure 1(a)). The lack of structural induction indicates that there are no tight C-peptide-vesicle complexes formed, or that the peptide does not undergo large structural rearrangements in the presence of bilayers. However, local structural rearrangements or weak ensemble average preferences for the overall structure too small to be detected by far-UV CD may be significant for lipid interactions [34]. Such rearrangements may be detected in the H^N-H^α region in TOCSY spectra, since the H^α chemical shifts are especially sensitive to local environment [35].

3.2. Chemical Shift Changes Reveal That Structural Rearrangements Are Induced by Lowering the pH. Even though the C-peptide is predominantly unstructured at all pH studied in this paper as seen in CD spectra (Figures 1(a)–1(c)), lowering the pH was observed to induce a slightly different population average of structures as evidenced by small shift changes in 2D TOCSY spectra (Figure 2(a)). The peaks were generally observed to become less dispersed at pH 3.2 as compared to at pH 5.8. By comparing the chemical shifts under these two conditions, more detailed information can be gained (Figure 2(b)). The amino acid residues that show the greatest chemical shift differences between pH 3.2 and pH 5.8 are the terminal glutamine Q31, probably due to its greater solvent accessibility, and the negatively charged residues E3, D4, E11, and E27 (the N-terminal E1 is not visible in the spectra). These negatively charged amino acid residues are less likely to be charged at pH 3.2 (close to the pI of the C-peptide) due

to protonation, which most likely is the main reason for the differences in chemical shifts. Other amino acid residues that are significantly affected by the change in pH are A2 and Q6. The H^α chemical shifts for both these residues are likely to be influenced by the change in environment of E1 and E3; nevertheless, significant chemical shift changes are observed for a large part of the N-terminus of C-peptide.

To investigate if these chemical shift differences are of importance for local structure induction, secondary chemical shifts were calculated for the H^α protons (Figure 3) [35]. When comparing the secondary chemical shifts at pH 3.2, 5.8, and 7, two stretches of amino acid residues are seen to move towards higher secondary chemical shift values with decreasing pH, residues A2 through Q6, and residues Q9 through L12, although the secondary chemical shift values are not consistently positive or negative in the first stretch of residues. These regions correspond well with two out of the five local structured regions previously found by Munte et al. in the C-peptide solution structure in H_2O/TFE 1 : 1 [13]. Three amino acid residues not located in this region also have the same pattern, G17, Q22 and E27. Since most of these amino acid residues are located close to acidic residues, these shift changes can be induced by protonating the charged residues and are not necessarily due to a pH-induced structural rearrangement. In Figure 3, three residues in a row with a secondary chemical shift above 0.1 indicates β-sheet structure. Such a stretch is found at the lower pH (3.2), that is, V10-L12, indicating a tendency for β-sheet structure for this small part of the sequence. This corresponds well with previous studies, which identified residues Q9–L12 as being capable of forming β-turns in H_2O/TFE 1 : 1 [13] and showed that lower pH is needed to induce β-structure, but then in the presence of SDS [15]. Here, we show that this small structural arrangement can be induced without any strong structural inducers like TFE or SDS, just by changing the pH. When correcting for nearest-neighbor effects according to Wishart et al. [36] (data not shown) G15, Q22, and L24 were the residues that were most affected, all moving to negative secondary chemical shifts with values larger than −0.1.

In summary, lowering the pH induces small but significant chemical shift changes due to changed preferences in the population average. Some of these changes are most likely due to changes in the protonation state of the acidic residues, while a short N-terminal sequence as well as residues V10–L12 is seen to undergo small conformational changes towards more β-like structures.

3.3. Bicelles Induce Chemical Shift Changes at Low pH. To investigate if the shift changes seen between pH 5.8 and pH 3.2 affect the membrane interaction, we examined changes in the TOCSY spectra when adding DMPC/DHPC bicelles [37, 38] with a q-ratio of 0.25 to 200 μM peptide (giving a total lipid/peptide ratio of 1500 : 1) at pH 5.8 (Figure 4) and at pH 3.2 (Figure 5). No shift changes were seen upon adding bicelles at pH 5.8 (Figure 4(a)), and furthermore, no changes were seen when adding $q = 0.25$ (DMPC/DMPG 9 : 1)/DHPC bicelles either (Figure 4(b)). At this pH, C-peptide has a negative net charge; hence, the association with

(a)

(b)

(c)

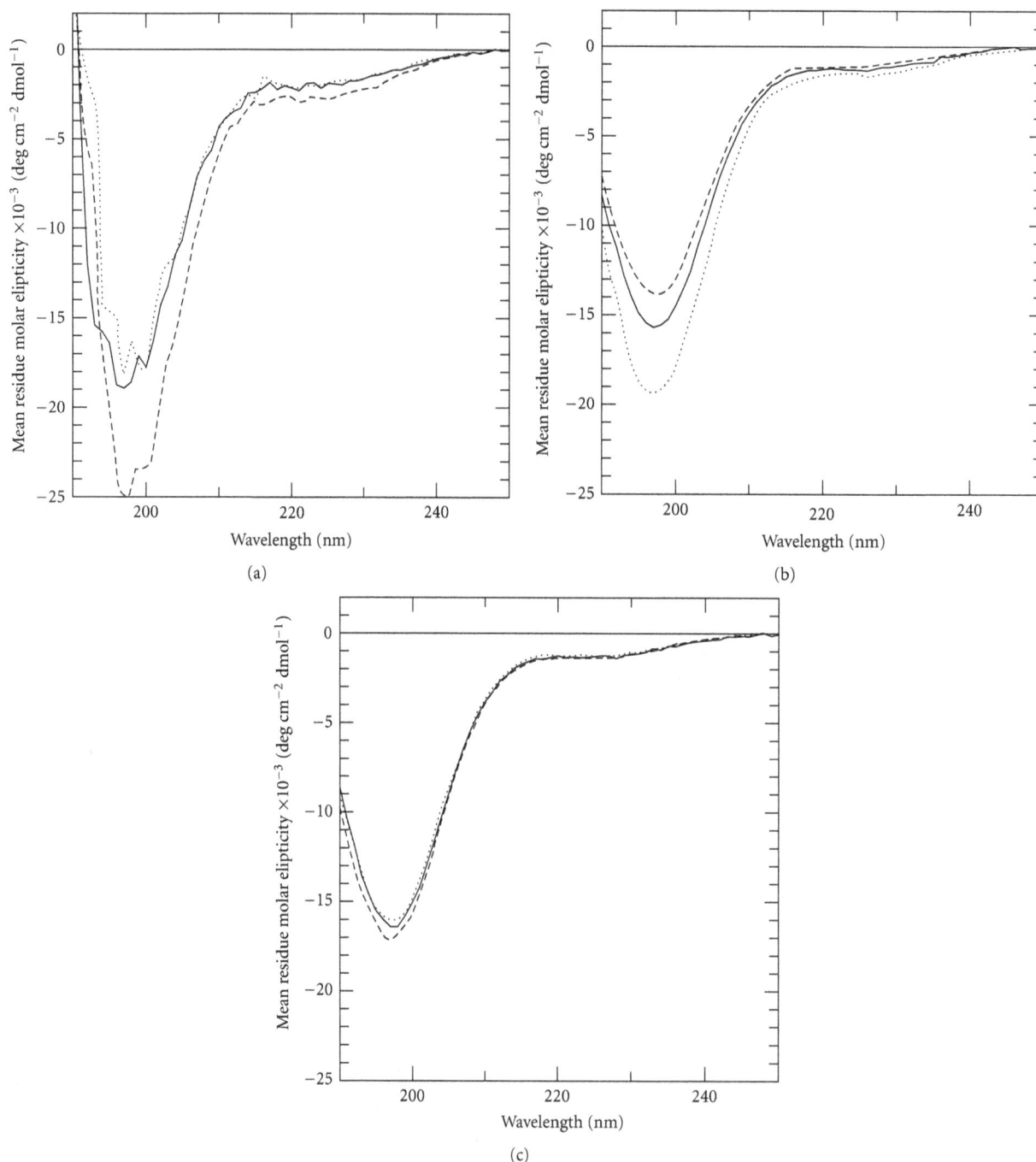

FIGURE 1: Secondary structure of C-peptide in LUVs at different pH as determined by CD. The figures depict CD spectra of 50 μM C-peptide in buffer (dotted line), in 1 mM POPC (continuous line), and in 1 mM POPC/POPG 1 : 1 (dashed line). (a) Shows spectra for C-peptide in 50 mM citrate buffer, pH 3.2, (b) for C-peptide in 50 mM sodium phosphate buffer, pH 5.8, and (c) for C-peptide in 50 mM sodium phosphate buffer, pH 6.9.

bicelles is not expected to increase by adding negative charges to the membrane. At pH 3.2, on the other hand, chemical shift changes are observed (Figure 5(a)). These chemical shift changes are not for the same residues that were identified as pH-sensitive. Rather, the shifts of A2, D4, L5, V7, Q9, E11, G17, G19, and L26 were affected by adding the bicelles (Figure 5(b)). Out of these residues A2, D4, V7, and E11

move to more β-sheet like shifts, while L5, Q9, G17, and L26 move to more α-helical shifts. The small stretch which was observed to adopt β-sheet-like shifts in buffer at pH 3.2 (V10–L12) became even more pronounced when adding bicelles. Hence, the interaction with bicelles stabilizes or favors this structure. Further, this shows that the previously reported structural preferences in C-peptide in TFE are

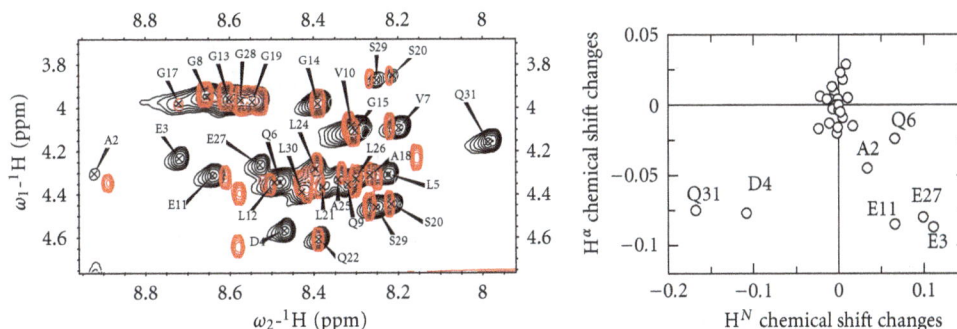

FIGURE 2: (a) 2D TOCSY spectra recorded at 600 MHz for 200 μM C-peptide in 50 mM citrate buffer pH 3.2 (red) and in 50 mM sodium phosphate buffer pH 5.8 (black). (b) Differences in amide proton chemical shifts between pH 3.2 and pH 5.8 (i.e., amide proton shift pH 5.8 and amide proton shift pH 3.2) are plotted against differences in α proton chemical shifts (i.e., α proton shift pH 5.8 and α proton shift pH 3.2).

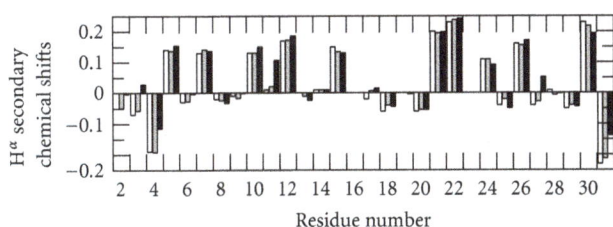

FIGURE 3: (a) Hα secondary chemical shifts for each amino acid residue in C-peptide in pH 7.0 (white), 5.8 (grey), and 3.2 (black).

seen also when using membrane mimetics, suggesting that this stretch may be of relevance *in vivo*, and may at least transiently interact with lipids.

In summary, we observed small but significant bicelle-induced chemical shift changes at pH 3.2 but not at pH 5.8, that stabilize the β-sheet structure of V10–L12.

3.4. PFG Diffusion Data Show That the Bicelle Association Is Greater at Lower pH. To investigate the extent of the C-peptide-lipid interactions, diffusion coefficients were measured in buffer and in the presence of different lipid bicelles. First, the self-diffusion coefficients were measured for C-peptide (250 μM) in 50 mM phosphate buffer of pH 5.8 and 7.2 and in 50 mM citrate buffer of pH 3.2. These values are shown in Table 1. By using the relationship between the hydrodynamic radius (r_H) and the molecular weight (M_r) for nonstructured peptides ($r_H = 0.27M_r^{0.50}$) [39] and by using the theoretical molecular weight of the monomer (3020.3 Da), it became evident that the major population of C-peptide under all these conditions is that of a monomeric peptide ($D = 1.66 \cdot 10^{-10}$ m/s^2). We, therefore, conclude that only changing the pH does not induce a detectable amount of oligomerization of C-peptide monomers. Second, the diffusion coefficients for C-peptide in phospholipid bicelles (150 mM total lipid, $q = 0.5$) with the bilayer part made of DMPC or DMPC/DMPG 9 : 1 were acquired. Measurements were carried out at similar pH values as used in the CD measurements. The diffusion coefficients for C-peptide decreased significantly

when adding the phospholipid bicelles at all pH as seen in Table 1. The populations of free and bicelle-bound molecules can be estimated using a two-state model:

$$xD_{DMPC} + (1 - x)D_{free} = D_{mixture}, \qquad (1)$$

where D_{free} represents the diffusion of free peptide in solution (obtained from measurements in buffer), and D_{DMPC} is the diffusion coefficient for DMPC, that is, the diffusion rate of the bicelles. It has previously been demonstrated that all DMPC molecules are bicelle-bound, and hence it can be safely assumed that the diffusion coefficient for DMPC represents the diffusion of the bicelle [23–26]. Finally, $D_{mixture}$ is the measured diffusion coefficient for C-peptide in the bicelle solution. This calculation suggests that around 65% of the peptide is bound to both DMPC and DMPC/DMPG bicelles at pH 3.2, when the C-peptide is neutral, while at pH 5.8 and 7.2 only 8–34% is bound, depending on the type of bicelle and pH. Hence, pH is important for the degree of association with the bicelles, and negatively charged C-peptide does not associate with the lipids to the same degree as C-peptide at low pH. As seen previously in the chemical shift analyses, charged lipids (DMPG in this case) do not appear to change the interaction.

4. Discussion

In this study, we wished to elucidate the effect of pH on C-peptide structure and lipid interactions. Previous studies have examined the effects of lipids on C-peptide structure at pH 5 or higher, and neither CD spectroscopy or size exclusion chromatography has revealed any lipid interaction [12]. Hence, it was then concluded that stable conformation-dependent interactions of C-peptide with lipid membranes are unlikely to occur. Biological effects of C-peptide, protecting against diabetic complications, are mediated by interaction with insulin or interaction with membrane via specific and/or nonspecific membrane interaction. Most studies support specific interactions with a, yet to be found, GPCR [40, 41]. However, the D-enantiomer of C-peptide has the same biological activity as the L-enantiomer [42], which suggests that other receptor-independent interactions

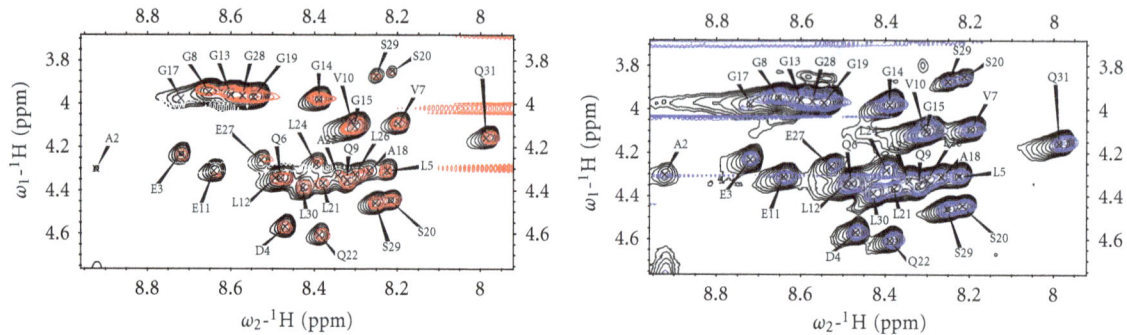

FIGURE 4: (a) 2D TOCSY spectra recorded at 600 MHz for 200 μM C-peptide in bicelles (300 mM total lipid, $q = 0.25$) at pH 5.8. The spectra acquired in buffer are shown in black, while the spectra in bicelle solution are shown in (a) red for DMPC/DHPC bicelles and (b) blue for (DMPC/DMPG 9 : 1)/DHPC bicelles.

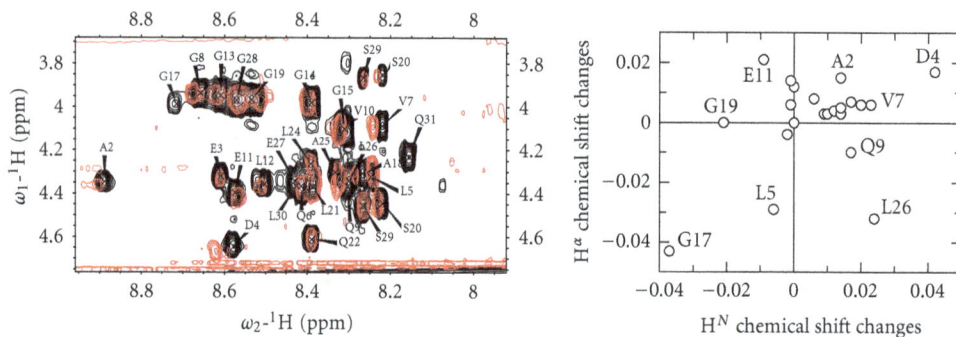

FIGURE 5: (a) 2D TOCSY spectra of 200 μM C-peptide in buffer (black) and in DMPC/DHPC bicelles (300 mM total lipid, $q = 0.25$) (red) at pH 3.2. (b) Differences in amide proton chemical shifts between C-peptide in citrate buffer and C-peptide in DMPC/DHPC bicelles at pH 3.2 are plotted against differences in α proton chemical shifts (i.e., chemical shift in bicelles and chemical shift in buffer).

are important for function. Formation of cation-selective channels in lipid bilayers [43] also suggests a more non-specific interaction. Hence, we find it valuable to investigate nonspecific interactions between C-peptide and the membrane as a part of C-peptides protective function. We have previously demonstrated that, at low pH, C-peptide has the ability to form β-sheet-like aggregates at low detergent concentrations and α-helical structure in SDS micelles [15], indicating that pH is important for structural induction. Thus, the structure and lipid interaction of C-peptide was in the present study also examined at a lower pH close to the pI of the C-peptide. From the results, we see that C-peptide favors a lipid interaction at low pH, when the peptide is neutral, (around 65% of the peptide is associated with bicelles at pH 3.2), suggesting that the relationship between electrostatic and hydrophobic interactions is important for this process (Figure 6). By decreasing the pH, small structural rearrangements in predominately the N-terminal and in the amino acid stretch between V10 and L12 are induced, that facilitate lipid interaction. Upon addition of bicelles, these structural preferences are stabilized. The structural rearrangements of C-peptide, as judged from both CD and NMR spectroscopy are not large, and thus, C-peptide represents a group of membrane interacting peptides that do not appear to undergo large structural changes upon membrane binding. This behavior has previously been observed for, for example, the interaction between the opioid receptor

peptide ligands (dynorphins) and bicelles, which did not cause any structural induction in the peptide ligands [34]. It appears that lack of structure induction is not a conclusive way of demonstrating lack of peptide-membrane interaction. Sometimes local and transient structural preferences in an ensemble of peptides dictate function [44]. This is similar to the recent findings that even protein-protein interactions may not always lead to well-formed secondary or tertiary structures, but indicates a novel mode of action of these intrinsically disordered proteins [45, 46].

Lipid-peptide interaction can further promote aggregation. It is well known that insulin forms oligomeric states and amyloid fibrils as a function of pH and ionic strength [19, 34–37]. However, C-peptide has also been demonstrated to form oligomers under conditions similar to *in vivo* situations, including sub-μM concentrations [12, 13]. Although the interaction between C-peptide and lipids appears to be weak in the present study, the membrane can affect local concentrations of C-peptide and change the local pH, which can shift the equilibrium between C-peptide monomer and membrane bound species. For instance, the interfacial pH of anionic membranes can be much lower than the bulk pH. This may in turn have an effect on the aggregation of C-peptide. Local pH effect has previously been demonstrated to be of importance for, for example, membrane insertion of the pH (low) insertion peptide (pHLIP), where low pH promotes interaction with the membrane[47, 48].

TABLE 1: Diffusion coefficients for the C-peptide.

pH	Buffer	DMPC/DHPC				(DMPC/DMPG 4 : 1)/DHPC		
	D_{free}	$D_{mixture}$	D_{DMPC}	x (%)[b]		$D_{mixture}$	D_{DMPC}	x (%)[b]
3.2	1.5 ± 0.1	0.8 ± 0.1	0.48 ± 0.05	69		0.77 ± 0.1	0.36 ± 0.05	64
5.8	1.6 ± 0.1	1.4 ± 0.1	0.43 ± 0.05	17		1.1 ± 0.1	0.42 ± 0.05	34
7.2	1.5 ± 0.1	1.1 ± 0.1	0.34 ± 0.05	34		1.4 ± 0.1	0.25 ± 0.05	8

Diffusion coefficient (10^{-10} m^2/s)[a]

[a]The diffusion coefficients are normalized according to the diffusion of HDO to account for viscosity differences. [b]Estimation of the percentage of peptide bound to the phospholipid bicelle as calculated by (1).

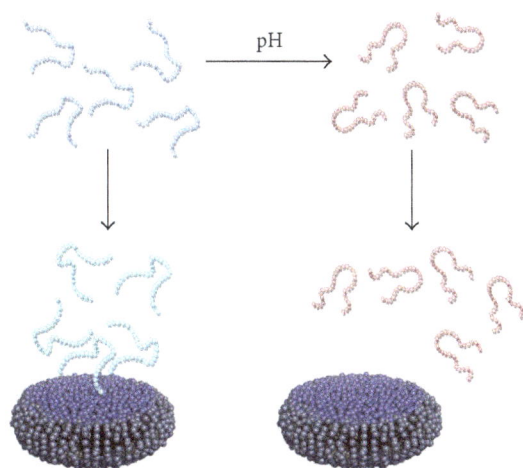

FIGURE 6: When lowering the pH to 3.2 (left panel), small structural rearrangements are induced that facilitate interaction with phospholipid bicelles. The interaction with bicelles further influences the structure. At pH 5.8, on the other hand, (right panel), no structural arrangements are seen upon addition of bicelles and the interaction is much weaker.

Although earlier studies have demonstrated that interactions with lipid vesicle bilayers do not result in any membrane-induced structure conversion in C-peptide [12], local variations in environment, including pH, may cause C-peptide to associate with lipids, as demonstrated here, which may affect the aggregation state of the peptide, and the equilibrium between a dominating population of monomeric peptides and a very small population of oligomers may be shifted [14]. Hence, transient membrane interactions may be of importance *in vivo* for inducing aggregation without any apparent structure intermediate.

Abbreviations

TFE: 2,2,2-Trifluoroethanol
SDS: Sodium dodecyl sulfate
CMC: Critical micelle concentration
APP: Amyloid precursor protein
POPC: 1-palmitoyl-2-oleoyl-sn-glycero-3-phospho-choline
POPG: 1-palmitoyl-2-oleoyl-sn-glycero-3-phospho-(1-rac-glycerol)
LUVs: Large unilamellar vesicles
DHPC: 1,2-dihexanoyl-d22-sn-glycero-3-phosphocholine
DMPC: 1,2-dimyristoyl-d54-sn-glycero-3-phosphocholine
DMPG: 1,2-dimyristoyl-d54-sn-glycero-3-phospho-(1-rac-glycerol)
CD: Circular dichroism
NMR: Nuclear magnetic resonance
TOCSY: Total correlation spectroscopy
PFG: Pulse field gradient
D: Diffusion coefficient
GdCl: Gadolinium(III) chloride
BMRB: Biological Magnetic Resonance Bank
M_r: Molecular weight
r_H: Hydrodynamic radii
GPCR: G protein-coupled receptor.

Funding

This work was supported by the Swedish Research Council.

References

[1] D. F. Steiner, D. Cunningham, L. Spigelman, and B. Aten, "Insulin biosynthesis: evidence for a precursor," *Science*, vol. 157, no. 3789, pp. 697–700, 1967.

[2] D. F. Steiner and P. E. Oyer, "The biosynthesis of insulin and a probable precursor of insulin by a human islet cell adenoma," *Proceedings of the National Academy of Sciences of the United States of America*, vol. 57, no. 2, pp. 473–480, 1967.

[3] R. E. Chance, R. M. Ellis, and W. W. Bromer, "Porcine proinsulin: characterization and amino acid sequence," *Science*, vol. 161, no. 3837, pp. 165–167, 1968.

[4] D. F. Steiner, "Proinsulin and the biosynthesis of insulin," *The New England Journal of Medicine*, vol. 280, no. 20, pp. 1106–1113, 1969.

[5] J. Wahren, B. L. Johansson, and H. Wallberg-Henriksson, "Does C-peptide have a physiological role?" *Diabetologia*, vol. 37, no. 2, pp. S99–S107, 1994.

[6] J. Wahren, K. Ekberg, J. Johansson et al., "Role of C-peptide in human physiology," *American Journal of Physiology*, vol. 278, no. 5, pp. E759–E768, 2000.

[7] J. Wahren, K. Ekberg, and H. Jörnvall, "C-peptide is a bioactive peptide," *Diabetologia*, vol. 50, no. 3, pp. 503–509, 2007.

[8] L. M. Chen, X. W. Yang, and J. G. Tang, "Acidic residues on the N-terminus of proinsulin C-peptide are important for the folding of insulin precursor," *Journal of Biochemistry*, vol. 131, no. 6, pp. 855–859, 2002.

[9] Y. Ohtomo, T. Bergman, B. L. Johansson, H. Jörnvall, and J. Wahren, "Differential effects of proinsulin C-peptide fragments on Na⁺, K⁺- ATPase activity of renal tubule segments," *Diabetologia*, vol. 41, no. 3, pp. 287–291, 1998.

[10] R. Rigler, A. Pramanik, P. Jonasson et al., "Specific binding of proinsulin C-peptide to human cell membranes," *Proceedings of the National Academy of Sciences of the United States of America*, vol. 96, no. 23, pp. 13318–13323, 1999.

[11] A. Pramanik, K. Ekberg, Z. Zhong et al., "C-peptide binding to human cell membranes: importance of Glu27," *Biochemical and Biophysical Research Communications*, vol. 284, no. 1, pp. 94–98, 2001.

[12] M. Henriksson, J. Shafqat, E. Liepinsh et al., "Unordered structure of proinsulin C-peptide in aqueous solution and in the presence of lipid vesicles," *Cellular and Molecular Life Sciences*, vol. 57, no. 2, pp. 337–342, 2000.

[13] C. E. Munte, L. Vilela, H. R. Kalbitzer, and R. C. Garratt, "Solution structure of human proinsulin C-peptide," *FEBS Journal*, vol. 272, no. 16, pp. 4284–4293, 2005.

[14] H. Jörnvall, E. Lindahl, J. Astorga-Wells et al., "Oligomerization and insulin interactions of proinsulin C-peptide: threefold relationships to properties of insulin," *Biochemical and Biophysical Research Communications*, vol. 391, no. 3, pp. 1561–1566, 2010.

[15] J. Lind, E. Lindahl, A. Perálvarez-Marín, A. Holmlund, H. Jörnvall, and L. Mäler, "Structural features of proinsulin C-peptide oligomeric and amyloid states," *FEBS Journal*, vol. 277, no. 18, pp. 3759–3768, 2010.

[16] C. M. Dobson, "Protein misfolding, evolution and disease," *Trends in Biochemical Sciences*, vol. 24, no. 9, pp. 329–332, 1999.

[17] C. M. Dobson, "Protein folding and misfolding," *Nature*, vol. 426, no. 6968, pp. 884–890, 2003.

[18] S. A. Waschuk, E. A. Elton, A. A. Darabie, P. E. Fraser, and J. McLaurin, "Cellular membrane composition defines Aβ-lipid interactions," *Journal of Biological Chemistry*, vol. 276, no. 36, pp. 33561–33568, 2001.

[19] M. Bokvist, F. Lindström, A. Watts, and G. Gröbner, "Two types of Alzheimer's β-amyloid (1–40) peptide membrane interactions: aggregation preventing transmembrane anchoring versus accelerated surface fibril formation," *Journal of Molecular Biology*, vol. 335, no. 4, pp. 1039–1049, 2004.

[20] A. Wahlström, L. Hugonin, A. Perálvarez-Marín, J. Jarvet, and A. Gräslund, "Secondary structure conversions of Alzheimer's Aβ(1–40) peptide induced by membrane-mimicking detergents," *FEBS Journal*, vol. 275, no. 20, pp. 5117–5128, 2008.

[21] C. R. Sanders and R. S. Prosser, "Bicelles: a model membrane system for all seasons?" *Structure*, vol. 6, no. 10, pp. 1227–1234, 1998.

[22] R. R. Vold, R. S. Prosser, and A. J. Deese, "Isotropic solutions of phospholipid bicelles: a new membrane mimetic for high-resolution NMR studies of polypeptides," *Journal of Biomolecular NMR*, vol. 9, no. 3, pp. 329–335, 1997.

[23] K. J. Glover, J. A. Whiles, G. Wu et al., "Structural evaluation of phospholipid bicelles for solution-state studies of membrane-associated biomolecules," *Biophysical Journal*, vol. 81, no. 4, pp. 2163–2171, 2001.

[24] J. J. Chou, J. L. Baber, and A. Bax, "Characterization of phospholipid mixed micelles by translational diffusion," *Journal of Biomolecular NMR*, vol. 29, no. 3, pp. 299–308, 2004.

[25] L. van Dam, G. Karlsson, and K. Edwards, "Direct observation and characterization of DMPC/DHPC aggregates under conditions relevant for biological solution NMR," *Biochimica et Biophysica Acta*, vol. 1664, no. 2, pp. 241–256, 2004.

[26] A. Andersson and L. Mäler, "Magnetic resonance investigations of lipid motion in isotropic bicelles," *Langmuir*, vol. 21, no. 17, pp. 7702–7709, 2005.

[27] A. Perálvarez-Marín, A. Barth, and A. Gräslund, "Time-resolved infrared spectroscopy of pH-induced aggregation of the Alzheimer Aβ1-28 peptide," *Journal of Molecular Biology*, vol. 379, no. 3, pp. 589–596, 2008.

[28] E. O. Stejskal and J. E. Tanner, "Spin diffusion measurements: spin echoes in the presence of a time-dependent field gradient," *The Journal of Chemical Physics*, vol. 42, no. 1, pp. 288–292, 1965.

[29] E. Von Meerwall and M. Kamat, "Effect of residual field gradients on pulsed-gradient NMR diffusion measurements," *Journal of Magnetic Resonance*, vol. 83, no. 2, pp. 309–323, 1989.

[30] P. T. Callaghan, M. E. Komlosh, and M. Nyden, "High magnetic field gradient PGSE NMR in the presence of a large polarizing field," *Journal of Magnetic Resonance*, vol. 133, no. 1, pp. 177–182, 1998.

[31] P. Damberg, J. Jarvet, and A. Gräslund, "Accurate measurement of translational diffusion coefficients: a practical method to account for nonlinear gradients," *Journal of Magnetic Resonance*, vol. 148, no. 2, pp. 343–348, 2001.

[32] L. G. Longsworth, "The mutual diffusion of light and heavy water," *Journal of Physical Chemistry*, vol. 64, no. 12, pp. 1914–1917, 1961.

[33] L. Braunschweiler and R. R. Ernst, "Coherence transfer by isotropic mixing: application to proton correlation spectroscopy," *Journal of Magnetic Resonance*, vol. 53, no. 3, pp. 521–528, 1983.

[34] J. Lind, A. Gräslund, and L. Mäler, "Membrane interactions of dynorphins," *Biochemistry*, vol. 45, no. 51, pp. 15931–15940, 2006.

[35] D. S. Wishart, B. D. Sykes, and F. M. Richards, "The chemical shift index: a fast and simple method for the assignment of protein secondary structure through NMR spectroscopy," *Biochemistry*, vol. 31, no. 6, pp. 1647–1651, 1992.

[36] D. S. Wishart, C. G. Bigam, A. Holm, R. S. Hodges, and B. D. Sykes, "¹H, ¹³C and ¹⁵N random coil NMR chemical shifts of the common amino acids. I. Investigations of nearest-neighbor effects," *Journal of Biomolecular NMR*, vol. 5, no. 1, pp. 67–81, 1995.

[37] I. Marcotte and M. Auger, "Bicelles as model membranes for solid- and solution-state NMR studies of membrane peptides and proteins," *Concepts in Magnetic Resonance A*, vol. 24, no. 1, pp. 17–35, 2005.

[38] R. S. Prosser, F. Evanics, J. L. Kitevski, and M. S. Al-Abdul-Wahid, "Current applications of bicelles in NMR studies of membrane-associated amphiphiles and proteins," *Biochemistry*, vol. 45, no. 28, pp. 8453–8465, 2006.

[39] J. Danielsson, J. Jarvet, P. Damberg, and A. Gräslund, "Translational diffusion measured by PFG-NMR on full length and fragments of the Alzheimer Aβ(1–40) peptide. Determination of hydrodynamic radii of random coil peptides of varying length," *Magnetic Resonance in Chemistry*, vol. 40, pp. S89–S97, 2002.

[40] J. Shafqat, E. Melles, K. Sigmundsson et al., "Proinsulin C-peptide elicits disaggregation of insulin resulting in enhanced physiological insulin effects," *Cellular and Molecular Life Sciences*, vol. 63, no. 15, pp. 1805–1811, 2006.

[41] Y. Ohtomo, A. Aperia, B. Sahlgren, B. L. Johansson, and J. Wahren, "C-peptide stimulates rat renal tubular Na⁺, K⁺-ATPase activity in synergism with neuropeptide Y," *Diabetologia*, vol. 39, no. 2, pp. 199–205, 1996.

[42] Y. Ido, A. Vindigni, K. Chang et al., "Prevention of vascular and neural dysfunction in diabetic rats by C- peptide," *Science*, vol. 277, no. 5325, pp. 563–566, 1997.

[43] P. Schlesinger, Y. Ido, and J. Williamsson, "Conductive channel properties of human C-peptide in-corporated into planar lipid bilayers," *Diabetes*, vol. 47, supplement 1, p. A29, 1998.

[44] P. Tompa, "Unstructural biology coming of age," *Current Opinion in Structural Biology*, vol. 21, no. 3, pp. 419–425, 2011.

[45] T. Mittag, J. Marsh, A. Grishaev et al., "Structure/function implications in a dynamic complex of the intrinsically disordered Sic1 with the Cdc4 subunit of an SCF ubiquitin ligase," *Structure*, vol. 18, no. 4, pp. 494–506, 2010.

[46] V. N. Uversky, "Intrinsically disordered proteins from A to Z," *International Journal of Biochemistry and Cell Biology*, vol. 43, no. 8, pp. 1090–1103, 2011.

[47] O. A. Andreev, A. D. Dupuy, M. Segala et al., "Mechanism and uses of a membrane peptide that targets tumors and other acidic tissues in vivo," *Proceedings of the National Academy of Sciences of the United States of America*, vol. 104, no. 19, pp. 7893–7898, 2007.

[48] D. Wijesinghe, D. M. Engelman, O. A. Andreev, and Y. K. Reshetnyak, "Tuning a polar molecule for selective cytoplasmic delivery by a pH (Low) Insertion Peptide," *Biochemistry*, vol. 50, no. 47, pp. 10215–10222, 2011.

14

Inhibitory Effects of Arginine on the Aggregation of Bovine Insulin

Michael M. Varughese and Jay Newman

The Department of Physics and Astronomy, Union College, Schenectady, NY 12308, USA

Correspondence should be addressed to Jay Newman, newmanj@union.edu

Academic Editor: Peter Schuck

Static and dynamic light scattering were used to investigate the effects of L-arginine, commonly used to inhibit protein aggregation, on the initial aggregation kinetics of solutions of bovine insulin in 20% acetic acid and 0.1 M NaCl as a model system for amyloidosis. Measurements were made as a function of insulin concentration (0.5–2.0 mM), quench temperature (60–85°C), and arginine concentration (10–500 mM). Aggregation kinetics under all conditions had a lag phase, whose duration decreased with increasing temperature and with increasing insulin concentration but which increased by up to a factor of 8 with increasing added arginine. Further, the initial growth rate after the lag phase also slowed by up to a factor of about 20 in the presence of increasing concentrations of arginine. From the temperature dependence of the lag phase duration, we find that the nucleation activation energy doubles from 17 ± 5 to 36 ± 3 kcal/mol in the presence of 500 mM arginine.

1. Introduction

Conformational misfolding under destabilizing conditions and subsequent beta-sheet amyloid fibril formation has been shown to be a very common, perhaps generic, property of proteins [1, 2]. In a number of proteins, such aggregation is pathological leading to over 40 different neurodegenerative or systemic diseases [3, 4] such as Alzheimer's, Parkinson's, Huntington's, and type II diabetes. It is becoming more apparent that the toxic species of many proteins is formed in the early stages of soluble aggregation. Such proteins include $A\beta$, for which an annular protofibril form has recently been isolated [5], and insulin [6, 7]. Much recent effort has been devoted to understanding the details of aggregate formation and also to find ways to control or inhibit this process. In the case of insulin, the motivation for such studies includes improving its pharmacological use in treating diabetes, as well as using it as a model system for studying amyloid aggregation.

Insulin, a key component in glucose metabolism, is a 51-residue protein consisting of two chains linked by disulfide bonds. Since the pioneering studies of David Waugh in the 1940's and 50's [8, 9], it has been known that, at elevated temperatures at low pH, insulin aggregates to form fibers that precipitate and/or form gels following a nucleation and elongation reaction. Despite this early finding and many subsequent studies [10–23], the molecular mechanism of insulin aggregation is not fully understood.

Aggregation of insulin has been studied under a variety of solvent conditions and with added cosolutes. Added ethanol [24], metalloporphyrins [25, 26], targeted peptides [27], and proteins such as α-crystallin [28], β-casein [29], or heparin [30] have all been shown to have inhibitory effects on insulin aggregation. The addition of amino acids to insulin solutions also has been shown to inhibit aggregation and, of 15 amino acids studied, arginine was shown to be most effective [31]. Arginine has been studied as a suppressor of protein aggregation in a variety of protein systems [32–34]. The role of the guanidinium group of arginine binding to aromatic groups of insulin when these are exposed in misfolded monomers leading to inhibition of aggregation has been suggested by Lyutova et al. [35]. A recent paper by Shah et al. [36] reports that arginine can also act to promote aggregation of proteins and that the key as to whether it inhibits or promotes aggregation appears to be whether the guanidinium group binds to aromatic or acidic residues, respectively.

In this work, we use dynamic and static light scattering to examine the effects of arginine on the early stages of thermally-induced insulin aggregation at low pH. We study the nucleation and early growth stages of insulin solutions in 20% acetic acid plus 0.1 M NaCl at varying concentrations of insulin and arginine and at various quench temperatures. From the temperature dependence of the nucleation lag time we are able to find the activation energy for nucleation of insulin in the absence and abundant presence (500 : 1 molar ratio) of arginine. We note explicitly that we have used light scattering rather than the more commonly used fluorescence of Thioflavin T to monitor aggregation since light scattering is more directly related to the weight average mass of the aggregates.

2. Experimental Procedures

2.1. Materials. Solutions of bovine insulin (Sigma-Aldrich #I5500, stored at $-20°C$) were prepared in 20% (v/v) acetic acid and 0.1 M NaCl using SuperQ water. Protein powder was dissolved at the appropriate concentration and spun at 5000 rpm in a JA-25.50 rotor of an Avanti Model J-25 centrifuge at 2°C for 5 minutes and then filtered, using a glass syringe and 13 mm diameter 0.22 micron pore-sized filters (Millex-GV) directly into precleaned 1 cm path length square optical glass cuvettes. These were stored at 2°C until used (within a few days).

Filtered stock solutions of L-arginine (Sigma-Aldrich #A5006) were made in the same solvent as the insulin and stored at 2°C. Some insulin samples had appropriate volumetric additions of arginine to result in arginine concentrations ranging from 10 to 500 mM just prior to quenching at elevated temperatures in the light scattering thermal bath.

2.2. Static and Dynamic Light Scattering. Static and dynamic light scattering measurements were done using a home-made optical arrangement described previously [37], a Lexel argon ion laser operating in TEM_{00} mode at 514.5 nm and a Brookhaven BI-9000 correlator. A series of sequential two-minute experiments at a 90° scattering angle of both the average scattered intensity and the intensity autocorrelation function were made after insulin samples were placed in the thermal optical bath and quenched to the measurement temperature. Temperature equilibration occurred within two minutes at all quench temperatures. The intensity autocorrelation functions were analyzed by the method of cumulants to give an average hydrodynamic diameter of the scatterers in each experiment based on the Stokes-Einstein relation. Sequential experiments allowed us to follow the time dependence of the scattered intensity and the hydrodynamic diameter as the insulin aggregated. Control experiments with monodisperse polystyrene latex spheres assured that intensity autocorrelation function measurements correctly determined scatterer diameters.

3. Results

A series of light scattering measurements at a fixed insulin concentration (1 mM) were done at varying temperatures in the range of 60–85°C to study the thermal-induced aggregation of insulin. Note that the solvent was 20% acetic acid plus 0.1 M NaCl throughout this study and that it differed from the solvent in a recent previous study co-authored by one of us [12] by the addition of the salt. The presence of salt speeded up the kinetics dramatically and allowed us to study the inhibition effects of arginine on more reasonable time scales. At each constant temperature, static and dynamic light scattering measurements were made as a function of time (see Figure 1). The results from these measurements are characterized by a temperature-dependent lag time, followed by an exponential increase in scattered intensity and in hydrodynamic diameter (data not shown), obtained from simultaneous static and dynamic measurements, respectively. To obtain parameters describing these two phases of the aggregation, we fit data at each temperature to an exponential growth function of the form:

$$I = \exp[a(t - t_o)], \tag{1}$$

where I is the adjusted count rate and t is the time from quench, and the data were fit to two parameters: lag time, t_o, and growth rate, a. The count rates were adjusted by subtracting the mean value of the initial count rate during the lag phase. Figure 2 shows the dependence of these two parameters on quench temperature. At higher temperatures, the lag time becomes significantly shorter (by more than a factor of 8) with a linear dependence on quench temperature and the exponential growth rate becomes much faster (by close to a factor of 20).

We point out that the data presented here are for the early stages in the aggregation process. Samples monitored for longer periods of time became inhomogeneous with large random fluctuations in intensity and hydrodynamic diameter, and upon removal of the cuvette from the thermal bath, it was apparent that a gel had formed. For the purposes of this study we only included the initial growth phase data and samples remained fairly homogeneous and in solution during all reported results. We performed all measurements two to four times, with resulting lag times and growth rates reproducing within about 10%.

To study the inhibitory effects of L-arginine on thermally-induced insulin aggregation, arginine was added to 1 mM insulin solutions in varying molar ratios from 10 : 1 to 500 : 1 and the time dependence of both the scattered intensity and hydrodynamic diameter were obtained at a fixed quench temperature. Figure 3 shows the results of these measurements for the scattered intensity when samples were quenched to 70°C. Note the significantly lengthened time axis in this figure as compared to that in Figure 1, indicating that, in the presence of increasing amounts of arginine, the lag time increases. While less apparent from this figure, it can also be seen that the growth rate decreases with increasing arginine. This is made clearer in Figure 4 in which the lag time and the growth rate are shown, each normalized to the values in the absence of arginine, as functions of the arginine

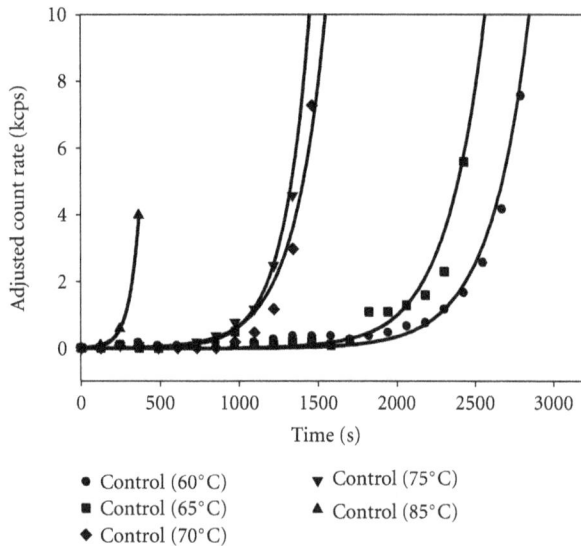

FIGURE 1: Average scattered intensity measurements versus time after quenching at varying temperatures for 1 mM insulin in 20% acetic acid + 0.1 M NaCl solutions. The count rates were measured in sequential two-minute dynamic light scattering experiments and, for purposes of comparison, the count rates were adjusted by subtracting the average initial count rates (typically several kcps). The smooth curves are fits to the data using (1) as described in the text. Similar data were obtained from plots of hydrodynamic diameter versus time after quenching to different temperatures, with the lag times typically somewhat shorter (10–30%) from these data.

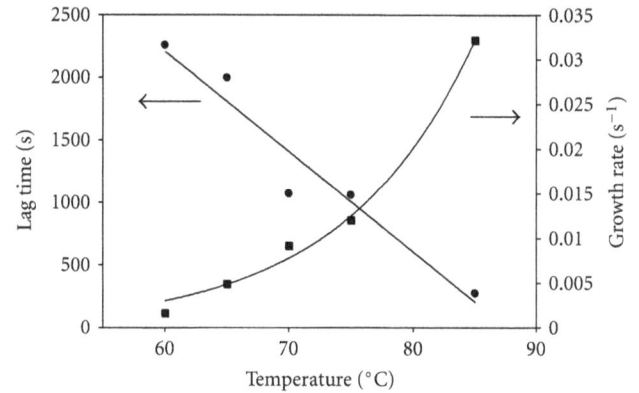

FIGURE 2: Data from Figure 1 were fit to two parameters, lag time and growth rate (using (1) as described in the text), and these are shown here as functions of the quench temperature. The smooth curves are linear and exponential fits to the data.

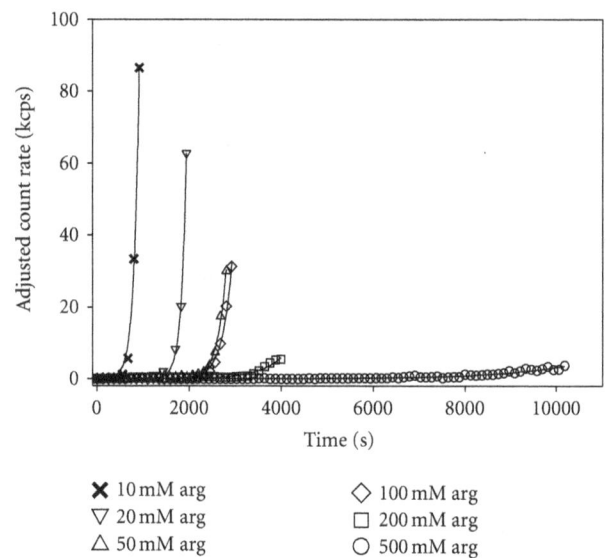

FIGURE 3: Adjusted (as in Figure 1 caption) average scattered intensity versus time after quenching to 70°C for 1 mM insulin with varying amounts of L-arginine added. Note the longer time scale (compared to Figure 1), the longer lag times, and the apparent slower growth rates at increasing arginine concentrations.

concentration. There is a linear growth in lag time and a nonlinear decrease in the growth rate with increasing arginine concentration. At the lowest arginine concentration used the lag time increases by 40% while the growth rate only slightly decreases by a few percent, while at 500 mM arginine the lag time increases by nearly a factor of 8 and the growth rate slows to only about 6.5% of its rate in the absence of arginine.

A more limited set of results at both 0.5 mM and 2 mM insulin concentrations (not shown) show similar behavior with lag times that increase linearly with increasing arginine concentration. As a function of insulin concentration at a fixed arginine concentration, there is a monotonic decrease of approximately a factor of two in lag time with increasing insulin concentration from 0.5 to 2 mM.

In the presence of arginine, we observed two distinct behaviors at later times (times beyond those reported in Figure 3) in the growth phase of the insulin aggregation. At arginine concentrations below 100 mM, we observed, similar to those samples in the absence of arginine, inhomogeneous scattering at these later times and a final gel-like appearance when examined by eye after removal from the temperature bath. For arginine concentrations of 100 mM and above, after the initial growth phase reported in Figure 3, precipitates formed and both sedimented to the bottom of the cuvette and collected at the meniscus due to surface tension. Also these samples had not formed a gel when examined after

removal from the thermal bath. We reiterate that these late-stage behaviors are not presented further in this study.

The temperature dependence of the inhibitory effect of L-arginine on insulin (1 mM) aggregation was studied at the fixed molar ratio of arginine to insulin of 500 : 1. As shown in Figure 5, as monitored by hydrodynamic diameter, at progressively lower quench temperatures the lag time increases and the growth rate decreases (as evidenced by the decreasing rate of slope change after the lag time). Again, we can quantitate this process by fitting the time dependence data to (1), with the intensity replaced by the hydrodynamic diameter, and extracting lag times and growth rates, and, in Figure 6, we show the normalized lag time and growth rate as functions of quench temperature. At this arginine

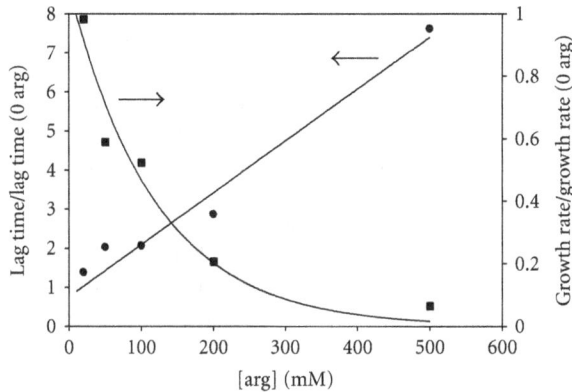

FIGURE 4: The lag time and growth rate, each normalized to the values with no added arginine, as functions of the amount of added arginine. The smooth curves are linear and exponential fits to the data.

concentration (500 mM), we see that the ratio of lag time with arginine to that without arginine decreases linearly with increasing temperature. At 60°C the lag time is roughly 7 times longer than that in the absence of arginine, while at 85°C the lag time is only about twice as long as that in the absence of arginine. Linear extrapolation to 37°C would predict a lag time that is over 12 times longer than in the absence of arginine at that temperature; these measurements were not attempted because of the extremely slow kinetics.

The growth rates in the presence of 500 mM arginine increase monotonically with increasing temperature, as would be expected, by about a factor of 5 over our range of quench temperatures. On the other hand, the growth rate in the absence of arginine (see Figure 2) increases by about a factor of 20 over the same temperature range. Therefore, when plotted as the normalized growth rate (the ratio of growth rate with arginine to that in its absence) as in Figure 6, the normalized growth rate decreases nonlinearly from its value at 60°C and remains fairly flat at higher temperatures. In the presence of 500 mM arginine, the growth rate is roughly ten times slower than without arginine for most temperatures.

From our data for the lag times at various quench temperatures in the absence and presence of arginine, we can make an Arrhenius plot using the inverse lag times as indicative of the nucleation rate:

$$\ln\left(\frac{1}{\tau}\right) = \ln A - \left(\frac{\Delta G}{R}\right)\left(\frac{1}{T}\right), \qquad (2)$$

where τ is the lag time, ΔG is the nucleation free energy, R is the molar gas constant, and T is the absolute temperature, to determine ΔG. In Figure 7 we show this plot both for insulin alone and for insulin with 500 mM arginine. The negative slope of this plot is equal to the nucleation activation energy divided by the molar gas constant. We obtain an activation free energy in the absence of arginine of 17 ± 5 kcal/mol, and an activation free energy in the presence of 500 mM arginine of 36 ± 3 kcal/mol, an increase of more than a factor of two.

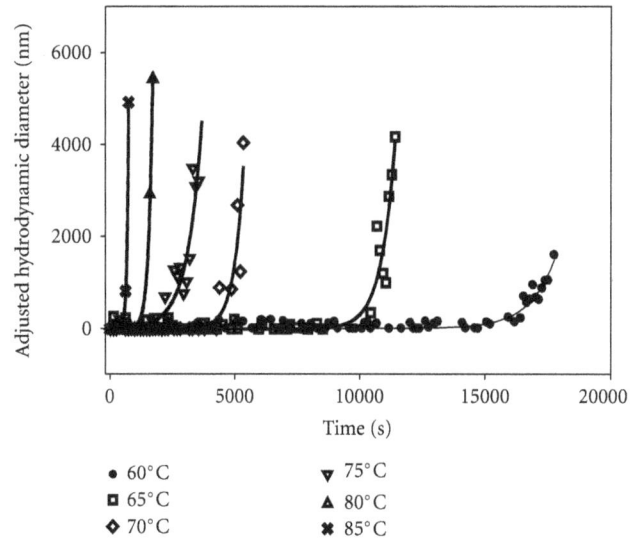

FIGURE 5: Hydrodynamic diameters of scattering species versus time for solutions of 1 mM insulin with 500 mM arginine at varying quench temperatures. The data are adjusted by subtracting the mean diameter during the lag time period in order to fit to the same two-parameter expression in (1). The smooth curves are fits to the same functional dependence as in (1) replacing intensity with hydrodynamic diameter.

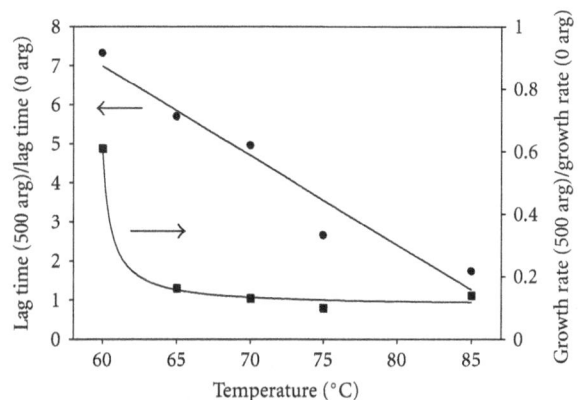

FIGURE 6: Lag times and growth rates of 1 mM insulin in the presence of 500 mM arginine, each normalized by the corresponding value in the absence of arginine, as function of the quench temperature. The smooth curves are linear and exponential fits to the data.

4. Discussion

Under the proper destabilizing conditions, essentially all proteins are now believed capable of forming amyloid fibrils through a similar mechanism of aggregation leading to cross-β structure formation. Through this process, proteins may not only lose their biological functioning but may also become toxic leading to a variety of diseases. Insulin has long been known to form amyloid fibrils and has been extensively studied because of problems in its production, delivery, and storage for use in treating type II diabetes and also as a model system.

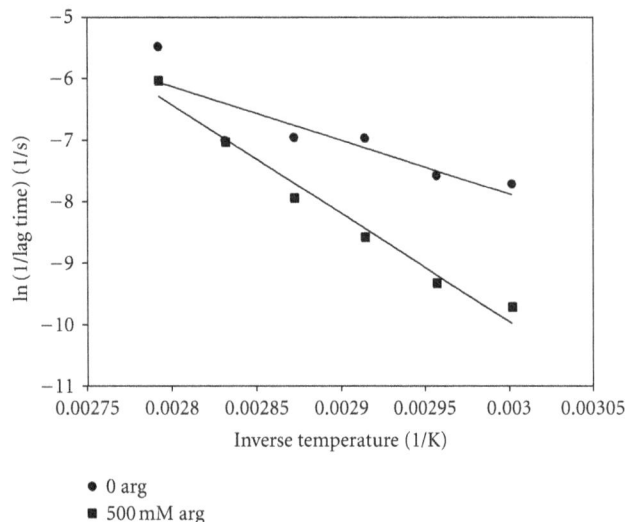

FIGURE 7: Arrhenius plots of the natural logarithm of the initial phase rate (taken as the inverse of the corresponding lag time) versus inverse absolute temperature for 1 mM insulin in the presence of and absence of 500 mM arginine.

The aggregation of insulin follows a nucleation-elongation process in which there is a lag phase during which monomers undergo a conformational change exposing hydrophobic residues and forming nuclei. It is known that in 20% acetic acid at ambient temperatures insulin is preferentially in the monomeric form, most susceptible to subsequent amyloidosis upon quenching to elevated temperature [16]. Recently the kinetics of amyloid growth has been modeled in terms of secondary, as well as primary, nucleation in which fragmentation and other secondary processes contribute to nuclei formation and to the kinetics [38]. This study found a strong correlation between the inverse lag time and growth rate based on their model. As we saw in Figure 2, for our insulin data we found this strong correlation: at increasing quench temperatures, lag time decreases and growth rate increases. After the lag phase, fibril growth has been shown to be exponential followed by gel-like formation and/or floccules formation, depending upon the conditions [11, 13].

In this study we have primarily studied the inhibitory effects of arginine on insulin aggregation. We chose, in these experiments, to add 0.1 M NaCl to the 20% acetic acid solvent in order to accelerate the aggregation kinetics, but not to agitate the samples (also known to accelerate the aggregation kinetics). Static and dynamic light scattering experiments were performed, in which the insulin concentration was varied between 0.5 and 2 mM, the arginine concentration varied from 10–500 mM, and the quench temperature ranged from 60 to 85°C.

Control experiments in the absence of arginine (see Figures 1 and 2 for 1 mM insulin) show that the lag time decreases with increasing quench temperature and that the

elongation phase has exponential growth with time, in qualitative agreement with previous studies which used a different solvent [12]. We also found that the elongation growth rate itself increases exponentially with quench temperature as shown in Figure 2. This same behavior was observed in a more limited data set at both 0.5 mM and 2 mM insulin with the lag time decreasing with increasing insulin concentration, in agreement with previous work [13]. Again we point out that the data reported here are only for the early elongation stage. Samples left longer at the quench temperatures showed large fluctuations in count rate and apparent hydrodynamic diameter and, when examined after removal from the thermal bath, had formed clear gels that did not flow at the higher insulin concentrations.

Our data (see Figures 3 and 4) show that arginine acts as an effective inhibitor of insulin aggregation with lag times increasing linearly with increasing arginine concentration. The lag time increases by nearly a factor of 8 and the growth rate slows by about a factor of 15 at the highest arginine concentration compared to kinetics in the absence of arginine. We note that the average scattered intensity versus time and the apparent hydrodynamic diameter versus time (not shown) results are quite similar and give independent confirmation of the role of arginine in inhibiting insulin aggregation.

The time course kinetics of both the average scattered intensity (not shown) and effective hydrodynamic diameter of insulin samples in the presence of 500 mM arginine (see Figures 5 and 6) show a characteristic temperature dependence. The normalized lag times (always longer than in the absence of arginine) decrease linearly with increasing temperature, with the ratio of lag time with arginine to that without arginine decreasing from a factor of nearly 8 at 60°C to only 1.7 at 85°C. These various experiments at different quench temperatures were further analyzed using an Arrhenius plot to obtain values for the nucleation free energy in the absence and presence of 500 mM arginine. We found about a two-fold increase in the activation energy for nucleation in the presence versus absence of arginine, with values of 36 and 17 kcal/mol, respectively. In the similar amyloid aggregating system of $A\beta$ there are two studies of the elongation kinetics that find values for the free energy associated with the fibril elongation process of 7 kcal/mol [39] and of 11 kcal/mol [40]. Assuming similar thermodynamics for the cross-β formation during aggregation, our values are somewhat larger than those, as might be expected, for the rate-limiting step of nucleation vs that of elongation. The significant increase in nucleation free energy in the presence of arginine emphasizes the inhibitory effect of arginine.

Our results show the inhibitory effects of arginine on the initial aggregation kinetics of bovine insulin over a range of insulin concentrations and quench temperatures. While the mechanism of arginine's interaction with insulin and with other proteins is not yet well understood [32–36, 41, 42] it appears to be related to its binding to aromatic groups in partially unfolded states of proteins and may therefore be an effective inhibitory agent for a wide variety of amyloid forming proteins.

Acknowledgments

The authors thank M. Manno for critical reading of the manuscript and helpful comments, and C. Wong and N. Poulan for collaborations.

References

[1] C. M. Dobson, "Protein folding and misfolding," *Nature*, vol. 426, no. 6968, pp. 884–890, 2003.

[2] R. M. Murphy and B. S. Kendrick, "Protein misfolding and aggregation," *Biotechnology Progress*, vol. 23, no. 3, pp. 548–552, 2007.

[3] F. Chiti and C. M. Dobson, "Protein misfolding, functional amyloid, and human disease," *Annual Review of Biochemistry*, vol. 75, pp. 333–366, 2006.

[4] C. G. Glabe, "Structural classification of toxic amyloid oligomers," *The Journal of Biological Chemistry*, vol. 283, no. 44, pp. 29639–29643, 2008.

[5] C. A. Lasagna-Reeves, C. G. Glabe, and R. Kayed, "Amyloid-β annular protofibrils evade fibrillar fate in Alzheimer disease brain," *The Journal of Biological Chemistry*, vol. 286, no. 25, pp. 22122–22130, 2011.

[6] T. Zako, M. Sakono, N. Hashimoto, M. Ihara, and M. Maeda, "Bovine insulin filaments induced by reducing disulfide bonds show a different morphology, secondary structure, and cell toxicity from intact insulin amyloid fibrils," *Biophysical Journal*, vol. 96, no. 8, pp. 3331–3340, 2009.

[7] C. L. Heldt, D. Kurouski, M. Sorci, E. Grafeld, I. K. Lednev, and G. Belfort, "Isolating toxic insulin amyloid reactive species that lack B-sheets and have wide pH stability," *Biophysical Journal*, vol. 100, no. 11, pp. 2792–2800, 2011.

[8] D. F. Waugh, "A fibrous modification of insulin. I. The heat precipitate of insulin," *Journal of the American Chemical Society*, vol. 68, no. 2, pp. 247–250, 1946.

[9] D. F. Waugh, D. F. Wilhelmson, S. L. Commerford, and M. L. Sackler, "Studies of the nucleation and growth reactions of selected types of insulin fibrils," *Journal of the American Chemical Society*, vol. 75, no. 11, pp. 2592–2600, 1953.

[10] M. Manno, E. F. Craparo, V. Martorana, D. Bulone, and P. L. San Biagio, "Kinetics of insulin aggregation: disentanglement of amyloid fibrillation from large-size cluster formation," *Biophysical Journal*, vol. 90, no. 12, pp. 4585–4591, 2006.

[11] M. Mauro, E. F. Craparo, A. Podestà et al., "Kinetics of different processes in human insulin amyloid formation," *Journal of Molecular Biology*, vol. 366, no. 1, pp. 258–274, 2007.

[12] M. Manno, D. Giacomazza, J. Newman, V. Martorana, and P. L. San Biagio, "Amyloid gels: precocious appearance of elastic properties during the formation of an insulin fibrillar network," *Langmuir*, vol. 26, no. 3, pp. 1424–1426, 2010.

[13] L. Nielsen, R. Khurana, A. Coats et al., "Effect of environmental factors on the kinetics of insulin fibril formation: elucidation of the molecular mechanism," *Biochemistry*, vol. 40, no. 20, pp. 6036–6046, 2001.

[14] J. L. Jiménez, E. J. Nettleton, M. Bouchard, C. V. Robinson, C. M. Dobson, and H. R. Saibil, "The protofilament structure of insulin amyloid fibrils," *Proceedings of the National Academy of Sciences of the United States of America*, vol. 99, no. 14, pp. 9196–9201, 2002.

[15] Q. Hua and M. A. Weiss, "Mechanism of insulin fibrillation: the structure of insulin under amyloidogenic conditions resembles a protein-folding intermediate," *The Journal of Biological Chemistry*, vol. 279, no. 20, pp. 21449–21460, 2004.

[16] A. Ahmad, V. N. Uversky, D. Hong, and A. L. Fink, "Early events in the fibrillation of monomeric insulin," *The Journal of Biological Chemistry*, vol. 280, no. 52, pp. 42669–42675, 2005.

[17] M. R. H. Krebs, C. E. MacPhee, A. Miller, I. E. Dunlop, C. M. Dobson, and A. M. Donald, "The formation of spherulites by amyloid fibrils of bovine insulin," *Proceedings of the National Academy of Sciences of the United States of America*, vol. 101, no. 40, pp. 14420–14424, 2004.

[18] M. R. H. Krebs, E. H. C. Bromley, S. S. Rogers, and A. M. Donald, "The mechanism of amyloid spherulite formation by bovine insulin," *Biophysical Journal*, vol. 88, no. 3, pp. 2013–2021, 2005.

[19] F. Librizzi and C. Rischel, "The kinetic behavior of insulin fibrillation is determined by heterogeneous nucleation pathways," *Protein Science*, vol. 14, no. 12, pp. 3129–3134, 2005.

[20] A. Podestà, G. Tiana, P. Milani, and M. Manno, "Early events in insulin fibrillization studied by time-lapse atomic force microscopy," *Biophysical Journal*, vol. 90, no. 2, pp. 589–597, 2006.

[21] S. Grudzielanek, A. Velkova, A. Shukla et al., "Cytotoxicity of insulin within its self-assembly and amyloidogenic pathways," *Journal of Molecular Biology*, vol. 370, no. 2, pp. 372–384, 2007.

[22] M. I. Smith, J. S. Sharp, and C. J. Roberts, "Insulin fibril nucleation: the role of prefibrillar aggregates," *Biophysical Journal*, vol. 95, no. 7, pp. 3400–3406, 2008.

[23] L. F. Pease III, M. Sorci, S. Guha et al., "Probing the nucleus model for oligomer formation during insulin amyloid fibrillogenesis," *Biophysical Journal*, vol. 99, no. 12, pp. 3979–3985, 2010.

[24] W. Dzwolak, S. Grudzielanek, V. Smirnovas et al., "Ethanol-perturbed amyloidogenic self-assembly of insulin: looking for origins of amyloid strains," *Biochemistry*, vol. 44, no. 25, pp. 8948–8958, 2005.

[25] R. F. Pasternack, E. J. Gibbs, S. Sibley et al., "Formation kinetics of insulin-based amyloid gels and the effect of added metalloporphyrins," *Biophysical Journal*, vol. 90, no. 3, pp. 1033–1042, 2006.

[26] S. P. Sibley, K. Sosinsky, L. E. Gulian, E. J. Gibbs, and R. F. Pasternack, "Probing the mechanism of insulin aggregation with added metalloporphyrins," *Biochemistry*, vol. 47, no. 9, pp. 2858–2865, 2008.

[27] T. J. Gibson and R. M. Murphy, "Inhibition of insulin fibrillogenesis with targeted peptides," *Protein Science*, vol. 15, no. 5, pp. 1133–1141, 2006.

[28] T. Rasmussen, M. R. Kasimova, W. Jiskoot, and M. van de Weert, "The chaperone-like protein α-crystallin dissociates insulin dimers and hexamers," *Biochemistry*, vol. 48, no. 39, pp. 9313–9320, 2009.

[29] X. Zhang, X. Fu, H. Zhang, C. Liu, W. Jiao, and Z. Chang, "Chaperone-like activity of β-casein," *International Journal of Biochemistry & Cell Biology*, vol. 37, no. 6, pp. 1232–1240, 2005.

[30] K. Giger, R. P. Vanam, E. Seyrek, and P. L. Dubin, "Suppression of insulin aggregation by heparin," *Biomacromolecules*, vol. 9, no. 9, pp. 2338–2344, 2008.

[31] K. Shiraki, M. Kudou, S. Fujiwara, T. Imanaka, and M. Takagi, "Biophysical effect of amino acids on the prevention of protein aggregation," *Journal of Biochemistry*, vol. 132, no. 4, pp. 591–595, 2002.

[32] R. Ghosh, S. Sharma, and K. Chattopadhyay, "Effect of arginine on protein aggregation studied by fluorescence correlation spectroscopy and other biophysical methods," *Biochemistry*, vol. 48, no. 5, pp. 1135–1143, 2009.

[33] B. M. Baynes, D. I. C. Wang, and B. L. Trout, "Role of arginine in the stabilization of proteins against aggregation," *Biochemistry*, vol. 44, no. 12, pp. 4919–4925, 2005.

[34] J. Chen, Y. Liu, Y. Wang, H. Ding, and Z. Su, "Different effects of L-arginine on protein refolding: suppressing aggregates of hydrophobic interaction, not covalent binding," *Biotechnology Progress*, vol. 24, no. 6, pp. 1365–1372, 2008.

[35] E. M. Lyutova, A. S. Kasakov, and B. Y. Gurvits, "Effects of arginine on kinetics of protein aggregation studied by dynamic laser light scattering and tubidimetry techniques," *Biotechnology Progress*, vol. 23, no. 6, pp. 1411–1416, 2007.

[36] D. Shah, A. R. Shaikh, X. Peng, and R. Rajagopalan, "Effects of arginine on heat-induced aggregation of concentrated protein solutions," *Biotechnology Progress*, vol. 27, no. 2, pp. 513–520, 2011.

[37] J. Newman, L. A. Day, G. W. Dalack, and D. Eden, "Hydrodynamic determination of molecular weight, dimensions, and structural parameters of Pf3 virus," *Biochemistry*, vol. 21, no. 14, pp. 3352–3358, 1982.

[38] T. P. Knowles, C. A. Waudby, G. L. Devlin et al., "An analytical solution to the kinetics of breakable filament assembly," *Science*, vol. 326, no. 5959, pp. 1533–1537, 2009.

[39] Y. Kusumoto, A. Lomakin, D. B. Teplow, and G. B. Benedek, "Temperature dependence of amyloid β-protein fibrillization," *Proceedings of the National Academy of Sciences of the United States of America*, vol. 95, no. 21, pp. 12277–12282, 1998.

[40] R. Carrotta, M. Manno, D. Bulone, V. Martorana, and P. L. San Biagio, "Protofibril formation of amyloid β-protein at low pH via a non-cooperative elongation mechanism," *The Journal of Biological Chemistry*, vol. 280, no. 34, pp. 30001–30008, 2005.

[41] K. Tsumoto, M. Umetsu, I. Kumagai, D. Ejima, J. S. Philo, and T. Arakawa, "Role of arginine in protein refolding, solubilization, and purification," *Biotechnology Progress*, vol. 20, no. 5, pp. 1301–1308, 2004.

[42] U. Das, G. Hariprasad, A. S. Ethayathulla et al., "Inhibition of protein aggregation: supramolecular assemblies of Arginine hold the key," *PLoS ONE*, vol. 2, no. 11, Article ID e1176, 2007.

The Aggregation of Huntingtin and α-Synuclein

María Elena Chánez-Cárdenas¹ and Edgar Vázquez-Contreras²

¹ *Laboratorio de Patología Vascular Cerebral, Instituto Nacional de Neurología y Neurocirugía, insurgentes sur 3877 col. la fama,*
  14269 Mexico, DF, Mexico
² *Departamento de Ciencias Naturales, CNI, Universidad Autónoma Metropolitana Cuajimalpa,*
  Pedro Antonio de los Santos 84 Col. Sn. Miguel Chapultepec Deleg, Miguel Hidalgo, 11851 México, DF, Mexico

Correspondence should be addressed to Edgar Vázquez-Contreras, evazquez@correo.cua.uam.mx

Academic Editor: Valeria Militello

Huntington's and Parkinson's diseases are neurodegenerative disorders associated with unusual protein interactions. Although the origin and evolution of these diseases are completely different, characteristic deposits of protein aggregates (huntingtin and α-synuclein resp.), are a common feature in both diseases. After these observations, many studies are performed with both proteins. Some of them try to understand the nature and driving forces of the aggregation process; others try to find a correlation between the genetic and failure in protein function. Finally with the combination of both approaches, it was proposed that possible strategies deal with pathologic aggregation. Unfortunately, if protein aggregation is a cause or a consequence of the neurodegeneration observed in these pathologies, it is still debatable. This paper describes the process of aggregation of two proteins: huntingtin and α synuclein. The characteristics of the aggregation reaction of these proteins have been followed with novel methods both *in vivo* and *in vitro*; these studies include both the combination with other proteins and the presence of various chemical compounds. The ultimate goal of this study was to summarize recent findings on protein aggregation and its possible role as a therapeutic target in neurodegenerative diseases and their role in biomaterial science.

1. Introduction

At present, it is evident that if proteins adopt their correct three-dimensional structure, that is, acquire their native structure; they will be able to perform multiple biological functions including macromolecular recognition, catalysis, and interaction with biomolecular targets among other possibilities. This conclusion was reached after studying protein denaturation, where the loss of the different levels of protein structure can be followed by many biophysical techniques. In some cases, when recovery of the structure and biological function of proteins were studied *in vitro*, it was observed that the process of nonspecific aggregation appears frequently; this situation is obtained using several types of proteins and perturbing with many denaturing conditions [1]. At the end of the twentieth century protein aggregation was observed *in vivo* in some pathological conditions. It was found that several phenomena may be involved in protein aggregation process. For example, some thermodynamic or kinetic situations could lead to the stabilization and association of partially folded intermediates [2], and it has been observed that some denaturing conditions induce the presence of nonnative conformations [3]. *In vivo*, the accumulation of nonfunctional proteins or their incorrect compartmentalization, is related to a failure to remove misfolded or nonfolded proteins, and with the disruption in the posttranslational control (inhibition of glycosylation or the reduction of disulfide linkages, among others). Also a genetic origin can be related with aggregation; on the one hand, it was found that mutations alter the protein stability, and secondly there are mutations that modify the rate of interconversion between the native state and the aggregate.

The reported pathological conditions may be the result of a loss of the original protein function, or by the gain of some toxic function. For a few human alterations such as cystic fibrosis, sickle cell anemia, and familial pulmonary emphysema, there is a clear relationship between an error in the protein folding and a cellular dysfunction. Moreover, there are cases where the proteins do not work or are not secreted in suitable amounts [4]. There are also a large number of pathologies, under the name of amyloidosis, in which the errors of protein folding lead to the irreversible formation of insoluble fibrillar aggregates called amyloids. This group includes several diseases, for example Alzheimer's disease, Creutzfeldt-Jakob disease, diabetes type II, and bovine spongiform encephalopathy. Currently, it is not known whether the presence of amyloid fibers in these pathologies is the cause or effect of cellular dysfunction. In fact, the causes of the formation of these fibrils are also debatable. Even though Huntington's (HD) and Parkinson's diseases (PD) have not been considered as amyloidoses, they involve the presence of a distinctive protein prone to aggregate. The protein aggregation associated with these disorders will be reviewed in the following sections.

2. Aims and Scope of the Chapter

Huntington and Parkinson diseases (HD and PD resp.) are two major neurodegenerative disorders pathologically characterized by the accumulation of the aggregate-prone proteins: mutant huntingtin (Htt) in HD and α-synuclein (α-syn) in PD. HD is the most common of nine inherited neurological disorders caused by expanded polyglutamine (polyQ) sequences, which confer propensity to self-aggregate and toxicity to their corresponding mutant proteins. On the other hand, α-syn is likely to be a major contributor to PD since overexpression of this protein from genetic triplication is sufficient to cause human forms of PD. Due to the importance of these diseases, researches around the Htt and α-syn aggregation are numerous. On the one hand, there are studies related to thermodynamic factors that govern the processes of stabilization of certain conformers that are involved in oligomerization; on the other hand, there is the study of the rate at which energy barriers are overcome to produce the three-dimensional species involved in the formation of the aggregates. Thermodynamic and kinetic processes are studied primarily in vitro; however, many experiments are carried out in vivo, using all the models and techniques available since cell lines to transgenic mice. Resent research has found that could be a relationship between the aggregation of these two proteins in vivo. To try to understand which are the molecular processes involved in the aggregation of these proteins, molecular biology is a very useful tool for dissecting the involvement of certain genes and the modification of its resulting proteins. Pharmacology is involved in the generation of molecules that could eventually be used as a probe, as well as inhibitors for the formation of the aggregates. Even it should be noted that all these molecular studies are related to research at the metabolic, cytological, and behavioral level because of the significance of the implications that these diseases have on the quality of human life. Also, some research in materials science have focused on the formation of aggregates of α-syn and HTT, as the seeds used to try to produce biomaterials. In this review, is described the proteins Htt and α-syn in HD and PD, respectively, as an approach to understand these pathologies as protein aggregation alterations. In vivo and in vitro aggregation studies of these proteins are reviewed in order to summarize recent findings on protein aggregation and its possible role as a target in neurodegenerative diseases.

3. Huntington's Disease (HD) and Huntingtin Protein

HD is an autosomal dominant neurodegenerative disease characterized by movement disorders and cognitive impairment. Psychiatric manifestations such as depression, irritability, and dementia are also observed. Brains with HD show a massive loss of striatum neurons (especially the medium-sized spiny neurons), although in advanced stages, the hippocampus, hypothalamus, cerebellum, amygdala, and thalamic nuclei are also affected. The symptoms of HD appear in midlife; however, it can range from childhood up to 70 years of age and over. In general patients die 10 to 15 years after the onset of symptoms, and no treatment or cure exists to date.

In HD a mutation occurs in the exon 1 of the huntingtin gene (*HTT*). It results in an expanded allele with a variable length of repetitive trinucleotides "CAG." In wild-type alleles, the length of CAG trinucleotides is between 6 and 37 and produces a 350 kDa protein-denominated huntingtin (Htt); on the other hand, in expanded alleles, the length of CAG trinucleotides ranges from 35 to 121 and produces proteins with a polyglutamine (polyQ) tract, variable in length. This characteristic includes HD in the family of polyQ diseases. Notably, the severity of the disease and the earlier age of onset are correlated with the length of the polyQ tract.

In humans Htt is a ubiquitously expressed protein; however, its precise role in metabolism is unclear, even so the importance of this protein is relevant because Htt knockout mice show cell degeneration and embryonic lethality [5]. It has been proposed that Htt is involved in different processes including neurogenesis, apoptosis, vesicle trafficking, caspases activation, signaling pathways, proteasomal function, and gene expression. Some consequences of Htt mutation have been reported; between them it is the loss of regulation of transcription of several kind of genes, for example, those involved with encoding neurotransmitter receptors, stress response, and enzymes and proteins involved in neuron structure. It was also reported a significant interaction of this protein with basal transcription factors such as TBP (TATA binding transcription factor), the transcription factor II F, and the cyclic AMP-response element binding (CREB) protein (CBP) amongst others (reviewed in [6]). It is known that transcriptional alterations are involved in HD and that the mutation of Htt could be part of this deregulation. Chromatin immunoprecipitation experiments have demonstrated that mutant Htt modifies the normal expression of specific mRNA species and then the gene expression is modulated through abnormal interactions of this protein with genomic DNA. Finally this interaction modifies the DNA conformation, producing alterations in the binding of transcription factors that changes the expression of those proteins [7].

Between the most relevant regulations produced by Htt is that of the brain-derived neurotrophic factor (BDNF). Cortical neurons produce this prosurvival factor that is necessary for the maintenance of striatal neurons in the brain. In the wild type cells Htt upregulates the transcription of BDNF; on the other hand in mutant Htt cells, the production of cortical BDNF decreases, then the neurotrophic support for striatal neurons is insufficient, causing cell death [8]. Mutant Htt affects not only BDNF transcription but it is known that wild-type Htt also promotes BDNF vesicular transport along microtubules, interacting with Htt-associated protein-1 (HAP1) and the p150 (Glued) subunit of dynactin, an essential component in molecular motors. The deficient mutant Htt transport of BDNF results in the loss of neurotrophic support and neuronal toxicity [9]. Finally it was found that the presence of Htt mutant protein diminishes the serum levels of BDNF in HD patients [10]. The participation of Htt in axonal transport, vesicle secretion, and trafficking has been studied using fusion proteins as the markers for fast axonal transport, and it was studied in presymptomatic homozygous mutant mice carrying 150 Q Htt knock-in mutations. It was found that the absence of Htt function also disrupts axonal transport, whereas the overexpression of the wild type (wt) but not of the mutant Htt enhances it [11]. On the other hand, the role of Htt in post-Golgi trafficking of secreted proteins was recently described. It was shown that mutant Htt perturbs post-Golgi trafficking to lysosomal compartments by delocalizing the optineurin/Rab8 complex, which in turn, affects the lysosomal function [12]. It was reported that mutant Htt has also been involved in excitotoxicity; the presence of this specie produces an abnormal increase of receptors for excitatory amino acids, especially those for N-methyl-D-aspartic acid (NMDA). Finally, observations concerning mutant Htt interactions with scaffold protein PSD 95 and transgenic mice overexpressing human exon 1 of Htt, show that the trafficking of NMDA receptor to the cell surface alters the presence of this receptor at the membrane level which in turn causes hypersensitivity [13, 14].

4. Huntingtin Aggregation

In vivo and in cell cultures it was observed that mutant Htt forms intranuclear inclusions bodies and cytoplasmic aggregates in neurons. Those forms are composed by N-terminal Htt fragments, which have been located as both neuronal intranuclear inclusions and dystrophic neurites in the cortex and striatum of HD patients [15]. It was found that the distribution of Htt inclusions is different throughout the different types of neurons; for example, two days after an adenovirus infection, it was observed that cytoplasmic inclusions become dominant in cortical and striatal neurons, whereas the opposite occurs on day 4 of the infection; that is, the ratio of nuclear inclusions overtakes that of cytoplasmic inclusions [16]. PolyQ proteins produce spherical and annular oligomeric structures; it is believed that they are reminiscent forms of those constituted by amyloid β in Alzheimer's disease and α-syn in PD. These

annular structures have been observed in aggregates of polyQ containing proteins and other proteins associated with diseases of the brain. They have also been observed in large systems of polyQ peptide model simulations at a molecular-level [17]. Htt aggregates are colocalized with several proteins including those involved in proteolysis, transcription regulation, vesicle trafficking, and ubiquitin-proteasome system components (proteasome subunits and chaperones). It has also been suggested that these proteins are sequestered by mutant Htt, contributing to the pathogenesis of the disease [18], and that the cleavage of Htt by caspases and calpains plays an important role in the existence of inclusion bodies and aggregates. The cleavage of Htt for caspase 3 produces N-terminal fragments of the polyQ tract; consequently, the increase in these fragments promotes the formation of intranuclear inclusions; it was also observed that the inhibition of caspase activity reduces their formation. In order to understand the progression and pathogenesis of HD, several metabolic factors have been studied; it was found that both the proteolytic processing of the mutant Htt and the abnormal calcium signaling could play a critical role and that the mutant Htt is also a substrate of the calcium-dependent proteases and calpains. Several references indicate that calpains may participate in the increased and/or altered patterns of Htt proteolysis, leading to the selective toxicity observed in the HD striatum. It was reported that the mutation of two calpain cleavage sites in Htt produces a polyQ-expanded Htt, and this protein is less susceptible to proteolysis and aggregation than the wt forms, and produces a decreased toxicity in a cell-culture model [19]. Moreover, it has been suggested that not all Htt proteolytic fragments contribute to toxicity and therefore pathogenic as well as a nonpathogenic role has been suggested for the Htt cleavage by different caspases; for example, using mass spectrometry and site-directed mutagenesis, it has been recently demonstrated that the generation of a cp-2 fragment (specifically cleaved at the level of arginine 167) produces a pathogenic fragment; this study reveals that specific sites of Htt are involved in the toxicity produced by the mutant Htt in HD [20].

5. The Effect of PolyQ Length and Its Concentration in Htt Aggregation

From some observations, it may be argued that Htt aggregation has a pathogenic role in HD. The characterization of aggregates by agarose gel electrophoresis has shown that cytoplasmic (but not nuclear) Htt aggregates become larger as the disease progresses in the brain of transgenic mice; it has demonstrated the formation of these aggregates in primary striatal neurons and in the brains of R6/2 and Htt Q150 mice; it was also observed that the presence of aggregates preceded the initiation of any other functional deficits in the cell [21]. On the other hand, when purified Htt fragments were used to follow the Htt aggregation *in vitro*, it was observed by electron and atomic force microscopy that globular and protofibrilar states appear as intermediates in the mature of fibril aggregation process. This observation supports a strong correlation between structural changes of mutant Htt and the consequential appearance of mature fibrils and neurotoxicity. The observation of similar intermediates in aggregates such as amyloid β and α-syn, suggests that the presence of these conformers plays an important role in the neurotoxicity observed in HD and other neurodegenerative diseases [22]. Moreover, when a set of Htt with various glutamine repetitions (between 23 and 150 units) is expressed in cell culture, the aggregation of the protein was not induced when the repetition was ≤ 49 units; however, the process was promoted by repetitions ≥ 120 units, suggesting that for the reaction of aggregation the length of the polyQ tail is important [23]. This observation correlates with the HD age of onset, the longer the CAG expansion in the *HTT* gene, the earlier the HD symptoms appear. Furthermore, it was reported that Htt, with polyQ tracts in a pathological range (>37 Q), form sodium dodecyl sulfate (SDS) resistant aggregates with fibrilar morphology, while wt Htt does not. In fact, the study of coaggregation of mutant and wt Htt fragments showed that mutant Htt promotes the aggregation of wt aggregates with the characteristic fibrilar morphology; then it was proposed that this observation could be related with the loss of normal functions in neuronal cells of HD patients [24]. The formation of Htt aggregates not only depends on polyQ repetition length but also on protein concentration and time. For example, when the aggregation process is followed in transfected COS cells as a function of time, it was found that the formation of Htt aggregates follows a kinetic mechanism, in which nucleation is the rate-limiting step [25]. It was proposed that sufficiently long polyQ peptides first undergo a unimolecular conformational change, in which they form a nucleus that eventually seeds aggregation. This proposed nucleation mechanism has been used to derive a stochastic mathematical model to describe the probability of aggregate formation in cells, as a function of time and mutant Htt protein concentration [26]. The development of this kind of models has important implications in the design of therapeutic strategies for HD since it has been shown that modest reductions in Htt expression delay the aggregate formation. Theoretical simulations of Htt protein aggregation have confirmed important characteristics involved in the process, for example, the differential hydrophobicity between glutamine versus nonglutamine segments (which was originally proposed as the driving force of the association process), the protein concentration, and the length of the polyQ tract. These studies also demonstrate that the physicochemical behavior of the exon 1 fragment not only depends on the polyQ properties, but also on the rest of the sequence [27].

In Htt the C terminus of the polyQ (aggregation domain) could be flanked by polyproline (polyP) tracts regions; it was proposed that polyP regions could be "protective" against aggregation and cytotoxicity. For example, it was synthesized polyQ peptides with 3–15 glutamine residues and a corresponding set of polyQ peptides flanked on the C terminus by 11 proline residues (poly(Q)-poly(P)), as occurs in the Htt sequence [28]; it was observed that the shorter soluble polyQ peptides (three or six glutamine residues) present both a polyproline type II-like-(PPII) helix

conformation (obtained by circular dichroism spectroscopy) and a monomeric organization (verified by size-exclusion chromatography); on the other hand, the longer poly(Q) peptides (nine or fifteen glutamine residues) showed a beta-sheet conformation and defined oligomers.

6. Proteasomal Dysfunction and Chaperones in Htt Aggregation

The generation of Htt aggregates is strongly determined proteasomal Htt degradation. Several studies show that in the presence of mutant Htt protein the accurate proteasome function is disrupted, increasing the level of Htt fragments and facilitating the formation of inclusions. As mentioned before, in these inclusions of Htt, chaperones and proteasomal subunits are "sequestered." A few components of the ubiquitin system have shown to be potential modulators of Htt aggregation, and it was suggested that failures in ubiquitination could play an important role in the pathology of polyglutamine diseases. On the other hand, as mutant Htt forms insoluble aggregates associated with components of the ubiquitin system, including ubiquitin, ubiquitin-like proteins, and proteins that bind ubiquitin, it has been demonstrated that isolated ubiquitin-interacting motifs might serve as potential inhibitors of polyQ aggregation *in vivo* [29]. Besides, it was reported that the expression of single and isolated ubiquitin-interacting motifs inhibited aggregation of the mutant Htt. There is also information that E2-25K (Hip2), an ubiquitin-conjugating enzyme, which interacts directly with Htt, may mediate the ubiquitination of the neuronal intranuclear inclusions of Htt and that is involved in the aggregate formation, modulating aggregation, and toxicity of the expanded Htt [30]. It was also reported that several situations produces an increase in the aggregation of mutant Htt. For example a decrease in the proteasome activity as result of cellular conditions that promote proteasomal malfunction such as oxidative stress [31] and the presence of specific compounds [32] was observed.

It was proposed that the aminoterminal segment of the protein could be involved in the accumulation and aggregation process; in order to study this possibility, mutant Htt fragments (smaller than the first 508 amino acids) were generated in *HTT*-transfected cells and *HTT* knock in mouse brains. It was found that these fragments constituted neuronal nuclear inclusions, which appeared before the neurological symptoms. It was also observed that their accumulation and aggregation were associated with an age-dependent decrease in proteasome activity and that these process were promoted by the inhibition of proteasome activity. After these facts it was suggested that the decrease in proteasome activity contributes to a late onset of Htt toxicity and ultimately restore the ability to remove NH_2-terminal fragments. These observations suggest a route to find a more effective therapy for HD rather than only the inhibition of their production [33].

On the other hand, a relevant role for chaperones in HD aggregation has been proposed. It has been shown that the aggregation of proteins with polyQ is not a consequence of proteolytic failure of the 20S proteasome. Rather and independent of proteolysis, aggregation is elicited by chaperone subunits of the 19S particle. Apparently, polyQ aggregation requires the transition between a benign conformation and an aggregation-prone form; it was proposed that proteasomal chaperones could facilitate the presence of this conformation [34]. Moreover, it has been reported that two molecular chaperones (HSP70 and HSP40) modulate the polyglutamine aggregation reactions by partitioning monomeric conformations and disfavoring the accretion of spherical and annular oligomers [35]. For example, *in vitro* and in an ATP-dependent process, the presence of these chaperones produces the suppression of the assembly of Htt into detergent-insoluble amyloid-like fibrils and, however, caused the formation of amorphous, detergent-soluble aggregates. Besides, it was observed that these chaperones were most active in preventing fibrillization when added during the lag phase of the polymerization reaction. Finally, when the system was repeated in yeast cells, similar inhibition detergent-insoluble polyQ aggregates were observed [36].

7. Chaperon-Like Proteins That Decrease Htt Aggregation

Recently, the use of novel chaperones, nonmammal chaperones, and proteins with chaperon-like activity has shown that it can be useful in the decrease of Htt aggregation. For example, the chaperonin TRiC, TCP-1 Ring Complex, also called CCT for chaperonin containing TCP-1, is a cytosolic protein that can alter the course of aggregation and cytotoxicity of Htt in mammalian cells. The aggregation states of the Htt in the presence and absence of TRiC were analyzed by fluorescence correlation spectroscopy. These studies revealed that the depletion of TRiC results in the appearance of soluble Htt aggregates. It was found that the eukaryotic chaperonin is composed of eight different subunits, each thought to be represented once per eight-membered ring. It was found that the overexpression of all eight subunits of CCT suppressed Htt aggregation and neuronal cell death [37]. The chaperone p97 plays a protective role in neurodegenerative disorders suppressing the protein aggregation. It was found that, in *Caenorhabditis elegans*, two homologues of chaperone p97, CDC-49.1 and CDC-48.2, partially suppress aggregation of the *HTT* exon1 fragment. It has been proved that CDC-48.1 and CDC-48.2 bind to the *HTT* exon1 fragment directly, suppressing the formation of SDS-insoluble aggregates of Htt fragments and also modulating the oligomeric states during the aggregate formation [38]. Recently, it has been reported that Htt interacts with the cue domain of gp78 and inhibits gp78 binding to ubiquitin and p97 [39]. Htt-interacting protein (HYPK) alters the number and distribution of aggregates formed by N-terminal Htt with 40 Q in Neuro2A cells. Even though HYPK does not possess any sequence similarity with the known chaperones it was found that it also modify the kinetics of mutated N-terminal Htt-mediated aggregate formation. Also it was reported that both *in vitro* and *in*

vivo HYPK possesses a novel chaperone-like activity; then it was suggested that this protein play an effective role in protecting neuronal cells against apoptosis induced by mutated N-terminal of Htt; it was proposed that this process occurs by modulating the formation of aggregates [40]. Moreover, HYPK acts together with NatA protein (N(alpha)-terminal-acetyltransferase complex) in cotranslational N-terminal acetylation and prevention of Htt aggregation [41]. Finally it was found that the yeast chaperone HSP104 reduces the aggregation of polyglutamine protein. This observation is supported by both the generation of a new transgenic mouse, which overexpresses the yeast chaperone HSP104, as well as the result of crossing of *HSP104* transgenic mice with mutant mice that express only the first 171 residues of mutant Htt. The overexpression of HSP104 reduces the aggregation process and prolongs the survival of a transgenic mouse model of HD. Then it was proposed that the reduction and protection processes are apparently due to a change in the conformation of a putative toxic monomer of the protein, reducing oligomerization or aggregation or the combination of both possibilities [42].

8. Could Aggregation in HD Be a Therapeutic Target?

Preventing the formation of insoluble Htt aggregates in neurons may represent an attractive therapeutic strategy to ameliorate HD symptoms. Different strategies to decrease Htt aggregation have been developed, such as the use of antibodies, oligonucleotides, and different compounds that modify Htt fibrillogenesis. The advance of novel methods for testing new drugs that are able to reduce the aggregation and toxicity of Htt has become a relevant issue; for example, the fluorescence resonance energy transfer-(FRET) based screening assay is a recent improvement in the screening of libraries of small active biological molecules, even in mammalian cells and *Drosophila* model. Recently, the use of these models has been found to be very effective to test drugs against Htt aggregation, providing new therapeutic tools for HD [43]. Other methods, such as the automated filter retardation assay, have been used for the identification of chemical compounds, which prevent the aggregation of *HTT* exon 1 product *in vitro*. This method was used in the screening of molecules that suppress the self-assembly of Htt, in order to find compounds with clinical and research applications. For example, twenty-five benzothiazol derivatives, which inhibit Htt fibrillogenesis in a dose-dependent manner, were discovered in a library of approximately 184,000 small molecules. Furthermore, cell culture studies revealed that 2-amino-4,7-dimethyl-benzothiazol-6-ol, a chemical compound similar to riluzol, significantly inhibits the aggregation *in vivo* of the *HTT* exon 1 product [44]. Other compound, the rapamycin, an immunosuppressant drug used to prevent rejection in organ transplantation, also reduces the amount of soluble polyQ protein via the inhibition of protein synthesis, which in turn significantly reduces the formation of insoluble polyQ protein and the formation of inclusion bodies. Therefore, a modest reduction in Htt synthesis by rapamycin may lead to a substantial decrease in probability of reaching the critical concentration required for a nucleation event and the subsequent toxic polyQ aggregation [45]. The use of compounds such as rapamycin has been proposed as promising therapeutic tools. For example, *in vitro* Htt studies have shown that reaching a critical concentration of these molecules is crucial for a nucleation event; only after this concentration is reached, the aggregation of toxic polyQ proceeds. In mammalian cells other compounds, such as pharmacologically improved derivatives of geldanamycin (17-DMAG and 17-AAG), have been proved to inhibit protein aggregation apparently through the induction of heat-shock response. It was observed that 17-DMAG induces the expression of HSP40, HSP70, and HSP105 chaperones. It was also found that this molecule inhibits the formation of mutant Htt aggregates with a greater efficiency than 17-AAG or even geldanamycin [46]. Efforts to prevent the misfolding and aggregation of Htt also involve the use of intracellular antibodies against Htt. It was observed that high levels of intrabody expression have been required to obtain even a limited number of reductions in aggregation. On the other hand, it was found that engineered single-domain intracellular antibody against Htt inhibits aggregation at low expression levels; it is believed that this process occurs by increasing its affinity in the absence of a disulfide bond [47]. Since the rate limiting of the nucleation process of polyQ aggregation involves the folding of mutated Htt monomers, a monoclonal antibody which selectively recognizes expanded pathogenic and aggregate-prone polyQ repetitions has been used. A direct correlation between polyQ tract aggregation, lag times and the intensity of the anti-1C2 signal in soluble monomers of Htt was found [48]. Finally, following the same logic, the use of short synthetic oligonucleotides has also been proposed as a therapeutic approach in blocking Htt nucleation and elongation phases of aggregation. It was found that introducing modified, single-stranded oligonucleotides into the neuronal cell line, the inclusion formation, is retarded [49]. In addition, it was shown that short guanosine single-stranded DNA oligonucleotides are capable to adopt a G-quartet structure and then be effective inhibitors of aggregation; this inhibition was tested in a cell-based assay and revealed that this oligomer is an effective inhibitor of the aggregation of both, the mutant Htt fragments and a green fluorescent protein fusion [49].

9. α-Synuclein in Parkinson's Disease (PD)

PD is the second most frequent progressive neurodegenerative disorder. Clinical manifestations of PD include motor disorders such as a resting tremor, rigidity, and bradykinesia (abnormally slow movements). Pathologically, PD is characterized by the progressive and selective degeneration of the dopaminergic neurons in the *substantia nigra (SN)*. A variety of cellular and biochemical alterations are observed in PD: (1) Mitochondrial dysfunction caused by complex I inhibition; (2) excitotoxicity; (3) oxidative stress and subsequently oxidative damage to protein, lipid, and nucleic acid molecules; (4) inflammatory response by microglia activation and a raise in cytokine levels; (5) proteaseome dysfunction and (6) proapoptotic and autophagic pathways. In addition, it has been demonstrated that the neurodegeneration of the *SN pars compacta* is accompanied by protein aggregation in the Lewy bodies; these proteinaceous inclusions are localized in the cytoplasm and axon of the dopaminergic neurons of the *SN* although other areas are also affected. Sporadic or idiopathic PD accounts for a large number of cases. However, <5–10% of them are considered to be familiar PD, with a genetic component. The genes associated to familial forms of PD mediate autosomal dominant or recessive forms and include α-syn (SNCA, also known as PARK1), Ubiquitin C-terminal hydrolase L1 (UCH-L1, also known as PARK5), Leucine-rich repetition kinase 2 (LRRK2, also known as PARK8), Parkin (PARKIN, also known as PARK2), DJ-1 (also known as PARK7), and PTEN-induced kinase 1 (PINK1, also known as PARK6). The study of familiar PD genes and their mutation is useful to understand the underlying mechanisms of the disease. Protein aggregation is a key factor in the pathogenesis of both the familial and sporadic disease. The Lewy bodies are the pathological hallmarks of PD and they appear in both PD forms. Sporadic PD shows evidence of extensive protein damage. However, it has been proposed that the impaired degradation of these misfolded proteins is a consequence of a pathological process such as oxidative stress, mitochondrial as well as proteasomal dysfunction, inflammation, and excitotoxicity. In familial PD on the other hand, related mutations of α-synuclein (α-syn), as well as the duplication or triplication of the gene correlate with the possibility of the misfolding and aggregation of this protein. These observations suggest that the overexpression of α-syn is responsible of aggregation. This section of the text is focused on the protein product of the *PARK1* gene, the α-syn protein, due to its unfolded nature and the increasing interest in the folding studies and its proposed role as a therapeutic target for PD. However, it is interesting to note that the other familial PD genes (*PARKIN, UCHL-1, DJ-1, PINK1,* and *DARDARIN/LRRK2*) are also related with the cellular unfolding protein response. PARKIN is an E-3 ubiquitin ligase, which labels proteins with ubiquitin, tagging them for their degradation in the proteasome. Mutations in the Parkin gene lead to a deficiency in ubiquitination. UCH-L1 is a deubiquitination enzyme, and its function is to cleave ubiquitin. Defects in this protein could impair proteasomal clearance of misfolded proteins and reduce the availability of ubiquitin monomers. DARDARIN/LRRK2 mutations suggest that it promotes abnormal phosphorylation and aggregation of target proteins [50]. The involvement of DJ-1 and PINK-1 in protein aggregation is not clear; however, it has been observed that mutations in DJ-1 could promote damage to the free-radical-mediated protein and PINK-1 mutations could reduce the ability to protect cells from proteasome inhibition. Recently, it has been reported that PINK-1 exists as a dimer in mitochondrial protein complexes that comigrate with respiratory chain complexes in sucrose gradients. In cell culture systems, mutant PINK-1 or PINK-1 knockdown caused deficits in mitochondrial respiration and ATP synthesis. Furthermore, proteasome function is impaired with the loss of PINK-1, and deficits are accompanied by an increase in α-syn aggregation [51]. On the other hand, DJ-1 shows antioxidant and chaperone-like activities. With the development of a cellular model of oxidative stress to address the subject of whether α-syn and DJ-1 interact functionally, it has been demonstrated that the inactivation of DJ-1 may promote the aggregation of α-syn, also proving that HSP70 is involved in the antioxidant response and in the regulation of α-syn fibril formation [52]. Mutations in several of these familial PD genes have also been sporadically identified in PD, suggesting that their study might provide insight into both familial and sporadic forms.

10. α-Synuclein Aggregation

synuclein family is formed by the α-, β-, and γ-isoforms all of them are proteins between 15 and 20 kDa. α- and β-synuclein colocalize in presynaptic nerve terminals, close to synaptic vesicles, while γ-synuclein appears to be axonal and cytosolic. In addition to synucleins, a wide range of proteins, neurofilaments, and ubiquitin-proteasome system proteins are present in Lewy bodies. α-syn is the main component of these structures and could be present in its full-length, oligomeric form and in aggregates. It was also observed that these forms present several posttranslational modifications including phosphorylation, gycosylation, nitration, or ubiquitination [53]. As α-syn aggregates are present in PD, it was proposed that the study of the process could be an approach for therapeutics. For example, a new and useful tool for study aggregation is bimolecular fluorescence complementation, which allows for the direct visualization of α-syn oligomeric species in living cells, facilitating the study of initial events and leading to determine how manipulations affect the process [54].

α-syn is a monomeric protein considered to be "unstructured," because of its lack of rigid structure. It is described as a protein with a remarkable conformational plasticity since it can be found in several forms including unfolded protein and amyloidogenic partially folded conformations [55]; it has also been reported that it can fold into α-helical or β-structure. In addition, α-syn can form different kind of aggregates, including oligomers, amorphous aggregates, or amyloid-like fibrils [56]. Experimental evidence has shown that the causes of α-syn assembly into fibers could involve several situations including an increase in concentration, inappropriate interaction with other proteins, posttranslational modifications (such as phosphorylation, partial or inappropriate cleavage of the flexible and highly negatively charged C terminus), and oxidative posttranslational modifications; eventually all of them result in the formation of oxidized and nitrated α-syn form [56]. Numerous efforts have been made to understand the early events of α-syn aggregation such as oligomerization and nucleation. For example, it has been reported that the photoinduced cross-linking of unmodified α-syn produces species with different oligomeric nature in solution, including monomers, dimers, and trimers [57]. It was proposed that it is probable that

the seeds for α-syn aggregation are the dimeric and trimeric species, and that the N-terminal amphipathic region of the protein is necessary for the oligomerization process. Using fluorescence lifetime imaging, two forms of α-syn homomeric interactions have been detected: (1) an antiparallel interaction between the amino and carboxy terminus of α-syn molecules and (2) a close amino terminus-carboxy terminus interaction within single α-syn molecules [58]. The central hydrophobic region of α-syn is critical for the β-sheet interaction [59]. Furthermore, conformational heterogeneity of the monomeric α-syn has been characterized at a single-molecule level. It was found that, in this condition, there is equilibrium between a random coil, a mechanically weak fold, and β-like structures. It was also proposed that this conformational equilibrium controls the population of the specific class of monomeric α-syn conformers. Besides, these species are positively correlated with the conditions previously used to promote the formation of aggregates, such as the presence of Cu^{2+}, the pathogenic A30P mutation, and high ionic strength [60]. Finally, PD-associated mutations in α-syn have been used to determine the presence of intermediates in the fibrillization process of the wt protein. For example results from fluorescence and CD experiments combined with singular value decomposition method demonstrate that wt α-syn presents a multistep process in the formation of fibrils. It was found that there are at least five intermediates present in the early stages of fibrillation [61]. Even though unstructured form of α-syn in solution is capable to form β-sheet-rich fibrils that accumulate as intracytoplasmic inclusions, a distinct protein conformation of the protein is observed when bound to phospholipid membranes. It was observed that in the presence of acidic phospholipid vesicles, the N terminus of α-syn folds into an amphipathic helical conformation that mediates membrane binding [62]. This fact suggests that lipid-dependent protein interactions might stabilize the folded structure of α-syn, and that this stabilization could decrease the oligomerization process. *In vivo*, α-syn is in dynamic equilibrium between membrane-bound form and soluble cytoplasmic specie. Therefore, understanding the role of lipid binding in modulating both the α-syn conformations and its aggregation is an important issue to propose PD therapeutics. In order to contribute to solve this questions, the thermodynamic characterization of the monomeric α-syn folding was performed. The experiments were carried out in the presence of SDS and followed by far-UV CD spectroscopy in order to detect the conformational transitions induced by the detergent, temperature, and pH. It was observed that the α-syn folding is a multistate process and that the structure of two of the intermediates stabilized is predominantly α-helical, however, they are only partially folded [63]. Moreover, to understand how the α-syn aggregation state affects its lipid binding activity, surface plasmon resonance, fluorescence anisotropy, and CD spectroscopy experiments were performed [64]. It was identified that during the early aggregation phase in the formation of fibrils, an intermediate with high affinity lipid binding appear [63]. On the other hand, to study the binding of α-syn to proteins in its lipid-associated helical conformation, a new *in vitro* approach was developed [65]. This procedure uses bacteriophage display to screen proteins that selectively bind to α-syn in this induced conformation. Interestingly, it was identified that 20 different human brain proteins specifically interact with phospholipid-bound α-syn; this observation confirms that the helical N terminus of α-syn can mediate specific interactions with other proteins and suggests that membrane binding may regulate their physiological activity *in vivo*.

It was argued that oxidative stress has been implicated in a number of neurodegenerative disorders including PD. An increasing amount of evidence shows that the α-syn aggregation process can be altered by oxidative and nitrative damage. It is widely known that reactive oxygen species induce lipoperoxidation, and that one of the most important products of this reaction, 4-hydroxy-2-nonenal (HNE), is implicated in the pathogenesis of PD. In order to know if there is an interaction between HNE and α-syn, the protein was incubated in the presence of the compound [66]; it was found that the process produces covalent modifications of the protein, with up to six HNE molecules being attached. The characterization of the modified protein by Fourier transform, infrared and CD spectra, indicated that HNE modification of α-syn resulted in a major conformational change involving an increase in the β-sheet content. It was also observed that the HNE-modified α-syn does not undergo fibrillation and is argued that this is due to the formation of stable soluble oligomers in a HNE concentration-dependent manner. In addition, HNE-induced oligomers showed to be toxic for dopaminergic neurons in cell cultures [66].

11. α-Synuclein Aggregation and Degradative Dopamine Metabolism

It has been proposed in the literature that the dopamine (DA) metabolism can be related with both α-syn aggregation and DA neuron loss in PD. Experiments *in vitro* demonstrated that the DA interaction with α-syn inhibits the fibril formation by stabilizing oligomeric kinetic intermediates [67]. Therefore, extensive research has been done in order to understand the effect of DA concentration, oxidation and metabolism. First, it was reported that the presence of DA causes the aggregation of the wt α-syn. On the other hand, it was observed that the A53T mutation in the α-syn gene might predispose the mutant protein to aggregation. Then it was found that the presence of DA causes the aggregation of both wt and mutated proteins in a temperature-dependent manner; however, the pathogenic A53T mutant shows a greater propensity to aggregate. It was suggested that this point mutation results in a structural change of the protein that drastically increases its propensity to aggregate in the presence of DA, contributing to PD pathogenesis [68]. Finally an *in vivo* cellular model, which regulates the intracellular steady-state levels of DA by expressing different forms of tyrosine hydroxylase, was developed in order to study aggregation of α-syn [69]. In these conditions, the formation of α-syn aggregates was inhibited and

that the formation of innocuous oligomeric intermediates was induced. These results support the observations that intraneuronal DA levels can be a major modulator of the α-syn aggregation and the inclusion formation; both processes have important implications on the selective degeneration of these neurons in PD [70]. Moreover, the formation of α-syn aggregates is inhibited by oxidized catechols, stabilizing their soluble oligomeric intermediates. For example, it was reported that some of the oxidative modifications to α-syn are produced by the reactive orthoquinone and aminochrome species generated by catechols oxidation. To characterize the mechanism of DA-induced alterations in α-syn aggregation, the same cellular model that regulates DA concentration mentioned above was used [69]. In this condition it was found that the wt protein was modified in five residues; it was proposed that they compose a catechol interaction domain, responsible for inhibiting α-syn aggregation. Furthermore, it was demonstrated that DA and other catechols modulate the progression of PD through interactions with the C-terminal of α-syn [71]. It was also shown that aberrant C-terminal processing of α-syn might lead to selective vulnerability of dopaminergic neurons by accelerating the formation of insoluble inclusions. To understand the link between the α-syn aggregation and DA metabolism, the monoamine oxidase (MAO) metabolite of DA, 3,4-dihydroxyphenylacetaldehyde (DOPAL), was tested using several techniques including Western blot, fluorescent confocal microscopy, and immunohistochemistry in a cell-free system; the effect of DOPAL was tested *in vitro* (in DA neuron cultures) and *in vivo* (with stereotactic injections into the *SN* of Sprague-Dawley rats). It was found that in the cell-free system *in vitro* and in the cell cultures, the use of DOPAL in physiologically relevant concentrations triggers the formation of toxic α-syn oligomers that finally aggregate. On the other hand, the results in the *in vivo* system showed that the use of DOPAL resulted in DA neuron loss and the accumulation of high molecular weight oligomers of α-syn [72].

12. Therapeutic Research in α-Synuclein Aggregation

As mentioned before, α-syn aggregation has become an important target in PD-therapeutic studies. Several strategies are developed in order to propose a treatment to prevent the aggregation of α-syn, including those that act directly with the protein and those that modify certain metabolic parameters that ultimately cause inhibition of the oligomerization. For example, it was proposed that some antibodies could inhibit the formation of the particularly toxic small oligomeric aggregates of α-syn; the dopamine (DA), nicotine, curcumin, and heat shock proteins as well as several NSAIDs regulate the α-syn fibril formation. On the other hand, it was found that oxidative stress and the presence of cholesterol results in mechanisms that can influence the formation of stable α-syn fibrils; the use of agonists and inhibitors for enzymes related with oxidative stress has been proposed as an antioxidant therapeutic strategy that could be involved in the inhibition of oligomeric forms of α-syn. Finally the regulation of cholesterol concentrations, with specific inhibitors as the statins, might be a novel approach for the treatment of PD. In order to find a therapeutic result against α-syn aggregation, several cell lines, mutant variants of the protein, and experimental approaches have been used.

Since the DA-induced oxidative stress has been related with PD pathology, the use of DA agonists, as well as monoamine oxidase β inhibitors has been proposed as an antioxidant therapeutic strategy. The effect of anti-Parkinsonian agents was studied such as selegiline, DA, pergolide, bromocriptine, and trihexyphenidyl in the formation of α-syn fibrils and in preformed fibrils by several methodologies including the fluorescence of thioflavin S, electron microscopy, and atomic force microscopy experiments. The results demonstrate that in a dose-dependent response and except for trihexyphenidyl, all the molecules tested inhibited the formation of the fibrils and destabilized the preformed α-syn fibrils [73]. It was proposed that the use of antibodies to neutralize the particularly toxic small oligomeric aggregates of α-syn could be used in therapeutics against PD [74]. For example, a single-chain variable fragment (scFv) antibody against the nonamyloid component of α-syn was produced; the presence of this antibody reduces both the intracellular aggregation and the toxicity produced by the protein [75]. Other antibodies that recognize the different morphologies of α-syn have been obtained. For example using a phage-displayed antibody library, an antibody was found that binds to the oligomeric form of α-syn; it was shown that this *in vitro* antibody decreases both aggregation and toxicity [76]. Moreover, a second antibody, which binds to a larger but latter stage oligomeric form of α-syn was also reported; this antibody inhibits aggregation *in vitro*, blocking extracellular α-syn and inducing toxicity in both an undifferentiated and differentiated human neuroblastoma cell lines. Interestingly, it was found that this antibody specifically recognizes naturally occurring aggregates in PD, but not in healthy human brain tissue [77]. On the other hand it has been proposed that interactions among closely related homologous proteins, included in the same family, might regulate the state of aggregation in PD. For example, it is known that α-syn adopts β-sheet-rich conformations that promote the oligomerization and neurotoxicity; on the other hand, it has been observed that β synuclein lacks the central hydrophobic cluster observed in α-syn. This observation suggests that β synuclein could be a negative regulator for the amyloid formation of α-syn. In order to understand the molecular mechanism through which β-synuclein exerts its anti-amyloidogenic effect, the study of the structural and dynamic features of the protein has been performed characterizing the ensemble of β-synuclein conformations in a transgenic mice line [78]. It was observed that this model presents both a marked reduction in α-syn expression in the cortex and an overexpressing β-synuclein. This overexpression produces a reduction of α-syn protein expression; however it was not accompanied by a decrease in α-syn mRNA. In addition to this observation, retardation in the progression of impaired motor performance, reduction in aggregation of both α-syn (α-syn A53T mouse model) and β-synuclein was obtained,

suggesting relevant implications of β-synuclein in the development of therapies for PD [79]. Moreover, the use of the cholesterol synthesis inhibitors (statins) lovastatin [80], simvastatin, and pravastatin was tested in α-syn aggregation in a transfected neuronal cell line as well as in primary human neurons. It was observed that the use of statins reduced the levels of α-syn accumulation in the detergent insoluble fraction of the transfected cells. This observation was accompanied by several effects including redistribution of α-syn in caveolar fractions, reduction in oxidized α-syn, and enhanced neurite outgrowth. In these experiments the addition of cholesterol to the media increased α-syn aggregation in detergent insoluble fractions of transfected cells. The cholesterol also reduces neurite outgrowth, suggesting that regulation of cholesterol levels, with specific inhibitors, might be a novel approach for the treatment of PD [81]. Based on epidemiological studies, it was proposed that the therapeutic use of nonsteroidal anti-inflammatory drugs (NSAIDs) could reduce the risk to develop PD. The effects of NSAIDs on the formation and destabilization of α-syn fibrils were studied using fluorescence spectroscopy in presence of thioflavin S and electron microscopy. The effect of several NSAIDs molecules including ibuprofen, aspirin, acetaminophen, meclofenamic acid sodium salt, sulindac sulfide, ketoprofen, flurbiprofen, diclofenac sodium salt, naproxen, and indomethacin, was tested. It was found that except for naproxen and indomethacin, the rest of the molecules inhibited the formation of α-syn fibrils in a dose-dependent response. It was proposed that NSAIDS also could destabilize preformed fibers [82]. In addition, it was reported that nicotine in a dose-dependent response inhibits the α-syn fibril formation in both wt and A53T forms. In addition, it was found that nicotine could destabilize preformed fibrillar α-syn in a dose-dependent manner, and that both effects could be attributed to the N-methylpyrrolidine moieties of the nicotine. These observations are a relevant issue in PD therapeutics since they correlate in retrospective and prospective epidemiological studies with an association between cigarette smoking and PD [83, 84]. Curcumin is a constituent of the Indian curry spice Turmeric, a herb of the ginger family, which is structurally similar to Congo red. It is known that curcumin is involved in the prevention of the oligomerization of Aβ-amyloid protein; considering this observation, the effect of this molecule in α-syn aggregation was tested. The addition of curcumin to an *in vitro* model of α-syn decreases the aggregation in a dose-dependent manner and increases its solubility. The aggregation-inhibiting effect of curcumin was investigated in the catecholaminergic SH-SY5Y cell line. Images captured using a high-throughput cell-based screening microscope showed that curcumin inhibits α-syn oligomerization into higher molecular weight aggregates. These observations suggest that curcumin could be a potential therapeutic compound for PD [85]. Finally, it was reported that some chaperones, specifically HSP70, are involved in regulate the α-syn fibril formation [52]. The assembly of α-syn in the presence of HSP70 was studied, and it was found that *in vitro* the assembly of α-syn was efficiently inhibited by substoichiometric concentrations of purified HSP70 in the absence of cofactors [86]. In fact, the use of α-syn deletion mutants indicated that interactions between the HSP70 substrate binding domain and the α-syn core hydrophobic region underlie assembly inhibition. Finally, it was found that the mammalian protein quality control system increases by the introduction of a nonmetazoan HSP104 in a rat lentiviral model of PD. This observation showed that both the dopaminergic neurodegeneration and the phosphorylated α-syn inclusion formation are reduced. HSP104 prevented fibrillization of α-syn and PD-linked variants (A30P, A53T, and E46K) and might have therapeutic potential in PD [87].

13. Cross Talk between α-Synuclein and Huntingtin

After observing that HD and PD have characteristics in common, some research groups have studied the possible relationships between Htt and α-syn *in vivo*. For example, it was found that polyQ inclusions in HD and other polyQ disorders are immunopositive for α-syn; after this observation it was speculated that α-syn might be recruited as an additional mediator of polyQ toxicity; the hypothesis was confirmed because the accumulation of α-syn in polyQ inclusions was found in HD postmortem brains and in the R6/1 mouse model of HD. It was also found that N-terminal mutant huntingtin (N-mutHtt) and α-syn form independent filamentous microaggregates in R6/1 mouse brain as well as in the inducible HD94 mouse model and that N-mutHtt expression increases the load of α-syn filaments [88]. In the same direction, however, using a bimolecular fluorescence complementation assay, it was found that the initial steps of the coaggregation of Htt and α-syn it was found that HTT (exon 1) oligomerized with α-syn and sequestered it in the cytosol. In turn, α-syn increased the number of cells displaying aggregates, decreased the number of aggregates per cell, and increased the average size of the aggregates [88]. Finally, it was demonstrated that wild-type α-syn overexpression impairs macroautophagy in mammalian cells and in transgenic mice [89]. Then, and considering that the overexpression of human wild-type α-syn in cells and Drosophila models of HD worsens the disease phenotype it was examined whether α-syn overexpression also worsens the HD phenotype in a mammalian system using two widely used N-terminal HD mouse models (R6/1 and N171-82Q). The effects of α-syn deletion in the same N-terminal HD mouse models, as well as assessed the effects of α-syn deletion on macroautophagy in mouse brains were also studied and it was found that overexpression of wild-type α-syn in both mouse models of HD enhances the onset of tremors and has some influence on the rate of weight loss. On the other hand, α-syn deletion in both HD models increases autophagosome numbers, and this is associated with a delayed onset of tremors and weight loss, two of the most prominent endophenotypes of the HD-like disease in mice [90].

These observations therefore suggest for one hand that α-syn could be a modifier of polyQ toxicity *in vivo*, or in other words maybe in mammals there is a functional link between these two aggregate-prone proteins. On the other hand, these observations raise the possibility that potential PD-related therapies aimed to counteract α-syn toxicity might help to slow HD. Finally, these results support the idea that coaggregation of these aggregation-prone proteins can contribute to the histopathology of neurodegenerative disorders, helping with diagnosis.

14. Protein Aggregation and Biomaterials

Proteins are large, complex molecules, which play many critical roles in the body. They do work in cells and are essential for the structure, function, and regulation of the body's tissues and organs. In addition to their natural functions, proteins can also act as biocatalysts and biomaterials. A biomaterial is a synthetic or natural material used to replace part of a living system or to function in intimate contact with living tissue. The study of the biophysical properties of specific biomaterials constitutes a first approach and a critical role for applications in tissue engineering, reconstructive surgery and regenerative medicine. The design of biomaterials that elicit specific cellular behavior constitutes a major challenge for the fields of tissue engineering and materials science [91]. Therefore, the rational design technique is often used in order to produce biomaterials from proteins; this method functions through protein engineering and requires an understanding of the molecular interactions that stabilize proteins in specific folded configurations [92]. However, naturally programmed self-assembly proteins offer significant promise for the generation of new types of well-defined multifunctional materials to provide structural and functional supports. Anyway, the information on the biophysical properties of proteins, during and after their aggregation, can be used to manipulate the aggregates in different forms (gel, film, or solid matrix), depending on the final use and application of the new biomaterial. Collagen, fibrin, and albumin are popular proteins for making biological scaffolds for tissue engineering because of their biocompatibility, biodegradability, and availability. Considering that a major drawback of biological protein-based biomaterials is the limited control over their biodegradation, the properties of the naturally occurring aggregates have been studied. For example, the stability and robustness of amyloid-like fibrils, their nanoscale dimensions, their suitability for chemical modification, and their inherent biological compatibility, are generating increasing interest in the use of these fibrils as new biomaterials. In this sense, as α-syn turned into two morphologically distinctive amyloid fibrils—"curly" (CAF) versus "straight" (SAF)—depending on its fibrillation processes. The differences in secondary structures of CAF and SAF have been studied and suggested to be responsible for their morphological uniqueness with structural flexibility and mechanical strength. Accumulation of CAF produced amyloid hydrogel composed of fine nanoscaled three-dimensional protein fibrillar network resulting in a nanomatrix for enzyme entrapment; then nano-scaled fibrillar network of CAF is a potential application in the promising areas of nanobiotechnology including tissue engineering, drug delivery, nanofiltration and biosensor development [93]. Also a nanomaterial fabrication was performed utilizing two peptide fragments of key region for α-synuclein fibrillation. These peptides modify the fibril nanostructure of full-length α-synuclein, and the effects depend on the peptide sequences. It was proposed that the combination of amyloid-forming protein, and its partial peptide fragments with some mutations have a potential for novel nanomaterial fabrication [94]. Then, the cases of Htt and α-syn are studied in several ways that will open the access to the future understanding of these and other proteins that could be related with the generation of innovative materials with various purposes.

Acknowledgments

E. Vázquez-Contreras Grants nos. 47310106 and 168177 and M. E. Chánez-Cárdenas received Grant no. 90720 from CONACyT and UAM (agreements 11 and 13/07 of the General Rector).

References

[1] E. Vázquez-Contreras, P. Ibarra Rodríguez, V. Castillo-Sánchez, and M. E. Chánez-Cárdenas, "The unfolding of proteins induced by different denaturants," in *Advances in Protein Physical Chemistry*, E. García-Hernández and D. A. Fernández-Velasco, Eds., pp. 169–192, Transworld Research Network, Kerala, India, 2008.

[2] T. R. Jahn and S. E. Radford, "The Yin and Yang of protein folding," *FEBS Journal*, vol. 272, no. 23, pp. 5962–5970, 2005.

[3] F. Chiti, P. Webster, N. Taddei et al., "Designing conditions for *in vitro* formation of amyloid protofilaments and fibrils," *Proceedings of the National Academy of Sciences of the United States of America*, vol. 96, no. 7, pp. 3590–3594, 1999.

[4] C. M. Dobson, "Protein misfolding, evolution and disease," *Trends in Biochemical Sciences*, vol. 24, no. 9, pp. 329–332, 1999.

[5] I. Dragatsis, M. S. Levine, and S. Zeitlin, "Inactivation of Hdh in the brain and testis results in progressive neurodegeneration and sterility in mice," *Nature Genetics*, vol. 26, no. 3, pp. 300–306, 2000.

[6] E. Roze, F. Saudou, and J. Caboche, "Pathophysiology of Huntington's disease: from huntingtin functions to potential treatments," *Current Opinion in Neurology*, vol. 21, no. 4, pp. 497–503, 2008.

[7] C. L. Benn, T. Sun, G. Sadri-Vakili et al., "Huntingtin modulates transcription, occupies gene promoters *in vivo*, and binds directly to DNA in a polyglutamine-dependent manner," *Journal of Neuroscience*, vol. 28, no. 42, pp. 10720–10733, 2008.

[8] C. Zuccato, A. Ciammola, D. Rigamonti et al., "Loss of huntingtin-mediated BDNF gene transcription in Huntington's disease," *Science*, vol. 293, no. 5529, pp. 493–498, 2001.

[9] L. R. Gauthier, B. C. Charrin, M. Borrell-Pagès et al., "Huntingtin controls neurotrophic support and survival of neurons by enhancing BDNF vesicular transport along microtubules," *Cell*, vol. 118, no. 1, pp. 127–138, 2004.

[10] A. Ciammola, J. Sassone, M. Cannella et al., "Low brain-derived neurotrophic factor (BDNF) levels in serum of Huntington's disease patients," *American Journal of Medical Genetics, Part B*, vol. 144, no. 4, pp. 574–577, 2007.

[11] L. S. Her and L. S. B. Goldstein, "Enhanced sensitivity of striatal neurons to axonal transport defects induced by mutant huntingtin," *Journal of Neuroscience*, vol. 28, no. 50, pp. 13662–13672, 2008.

[12] D. D. Toro, J. Alberch, F. Lazaro-Dieguez et al., "Mutant huntingtin impairs post-golgi trafficking to lysosomes by delocalizing optineurin/rab8 complex from the golgi apparatus," *Molecular Biology of the Cell*, vol. 20, no. 5, pp. 1478–1492, 2009.

[13] Y. Sun, A. Savanenin, P. H. Reddy, and Y. F. Liu, "Polyglutamine-expanded huntingtin promotes sensitization of N-methyl-D-aspartate receptors via post-synaptic density 95," *The Journal of Biological Chemistry*, vol. 276, no. 27, pp. 24713–24718, 2001.

[14] M. M. Y. Fan, H. B. Fernandes, L. Y. J. Zhang, M. R. Hayden, and L. A. Raymond, "Altered NMDA receptor trafficking in a yeast artificial chromosome transgenic mouse model of Huntington's disease," *Journal of Neuroscience*, vol. 27, no. 14, pp. 3768–3779, 2007.

[15] M. DiFiglia, E. Sapp, K. O. Chase et al., "Aggregation of huntingtin in neuronal intranuclear inclusions and dystrophic neurites in brain," *Science*, vol. 277, no. 5334, pp. 1990–1993, 1997.

[16] K. Tagawa, M. Hoshino, T. Okuda et al., "Distinct aggregation and cell death patterns among different types of primary neurons induced by mutant huntingtin protein," *Journal of Neurochemistry*, vol. 89, no. 4, pp. 974–987, 2004.

[17] A. J. Marchut and C. K. Hall, "Spontaneous formation of annular structures observed in molecular dynamics simulations of polyglutamine peptides," *Computational Biology and Chemistry*, vol. 30, no. 3, pp. 215–218, 2006.

[18] Z. H. Qin and Z. L. Gu, "Huntingtin processing in pathogenesis of Huntington disease," *Acta Pharmacologica Sinica*, vol. 25, no. 10, pp. 1243–1249, 2004.

[19] J. Gafni, E. Hermel, J. E. Young, C. L. Wellington, M. R. Hayden, and L. M. Ellerby, "Inhibition of calpain cleavage of Huntingtin reduces toxicity: accumulation of calpain/caspase fragments in the nucleus," *The Journal of Biological Chemistry*, vol. 279, no. 19, pp. 20211–20220, 2004.

[20] T. Ratovitski, M. Gucek, H. Jiang et al., "Mutant huntingtin N-terminal fragments of specific size mediate aggregation and toxicity in neuronal cells," *The Journal of Biological Chemistry*, vol. 284, no. 16, pp. 10855–10867, 2009.

[21] A. Weiss, C. Klein, B. Woodman et al., "Sensitive biochemical aggregate detection reveals aggregation onset before symptom development in cellular and murine models of Huntington's disease," *Journal of Neurochemistry*, vol. 104, no. 3, pp. 846–858, 2008.

[22] M. A. Poirier, H. Li, J. Macosko, S. Cai, M. Amzel, and C. A. Ross, "Huntingtin spheroids and protofibrils as precursors in polyglutamine fibrilization," *The Journal of Biological Chemistry*, vol. 277, no. 43, pp. 41032–41037, 2002.

[23] S. H. Li and X. J. Li, "Aggregation of N-terminal huntingtin is dependent on the length of its glutamine repeats," *Human Molecular Genetics*, vol. 7, no. 5, pp. 777–782, 1998.

[24] A. Busch, S. Engemann, R. Lurz, H. Okazawa, H. Lehrach, and E. E. Wanker, "Mutant huntingtin promotes the fibrillogenesis of wild-type huntingtin: a potential mechanism for loss of huntingtin function in Huntington's disease," *The Journal of Biological Chemistry*, vol. 278, no. 42, pp. 41452–41461, 2003.

[25] E. Scherzinger, A. Sittler, K. Schweiger et al., "Self-assembly of polyglutamine-containing huntingtin fragments into amyloid-like fibrils: implications for Huntington's disease pathology," *Proceedings of the National Academy of Sciences of the United States of America*, vol. 96, no. 8, pp. 4604–4609, 1999.

[26] D. W. Colby, J. P. Cassady, G. C. Lin, V. M. Ingram, and K. D. Wittrup, "Stochastic kinetics of intracellular huntingtin aggregate formation," *Nature Chemical Biology*, vol. 2, no. 6, pp. 319–323, 2006.

[27] M. G. Burke, R. Woscholski, and S. N. Yaliraki, "Differential hydrophobicity drives self-assembly in Huntington's disease," *Proceedings of the National Academy of Sciences of the United States of America*, vol. 100, no. 2, pp. 13928–13933, 2003.

[28] G. Darnell, J. P. R. O. Orgel, R. Pahl, and S. C. Meredith, "Flanking polyproline sequences inhibit β-sheet Structure in polyglutamine segments by inducing PPII-like helix structure," *Journal of Molecular Biology*, vol. 374, no. 3, pp. 688–704, 2007.

[29] S. L. H. Miller, E. L. Scappini, and J. O'Bryan, "Ubiquitin-interacting motifs inhibit aggregation of polyQ-expanded Huntingtin," *The Journal of Biological Chemistry*, vol. 282, no. 13, pp. 10096–10103, 2007.

[30] R. de Pril, D. F. Fischer, R. A. C. Roos, and F. W. van Leeuwen, "Ubiquitin-conjugating enzyme E2-25K increases aggregate formation and cell death in polyglutamine diseases," *Molecular and Cellular Neuroscience*, vol. 34, no. 1, pp. 10–19, 2007.

[31] A. Goswami, P. Dikshit, A. Mishra, S. Mulherkar, N. Nukina, and N. R. Jana, "Oxidative stress promotes mutant huntingtin aggregation and mutant huntingtin-dependent cell death by mimicking proteasomal malfunction," *Biochemical and Biophysical Research Communications*, vol. 342, no. 1, pp. 184–190, 2006.

[32] P. Dikshit, A. Goswami, A. Mishra, N. Nukina, and N. R. Jana, "Curcumin enhances the polyglutamine-expanded truncated N-terminal huntingtin-induced cell death by promoting proteasomal malfunction," *Biochemical and Biophysical Research Communications*, vol. 342, no. 4, pp. 1323–1328, 2006.

[33] H. Zhou, F. Cao, Z. Wang et al., "Huntingtin forms toxic NH_2-terminal fragment complexes that are promoted by the age-dependent decrease in proteasome activity," *Journal of Cell Biology*, vol. 163, no. 1, pp. 109–118, 2003.

[34] E. Rousseau, R. Kojima, G. Hoffner, P. Djian, and A. Bertolotti, "Misfolding of proteins with a polyglutamine expansion is facilitated by proteasomal chaperones," *The Journal of Biological Chemistry*, vol. 284, no. 3, pp. 1917–1929, 2009.

[35] J. L. Wacker, M. H. Zareie, H. Fong, M. Sarikaya, and P. J. Muchowski, "Hsp70 and Hsp40 attenuate formation of spherical and annular polyglutamine oligomers by partitioning monomer," *Nature structural & molecular biology*, vol. 11, no. 12, pp. 1215–1222, 2004.

[36] P. J. Muchowski, G. Schaffar, A. Sittler, E. E. Wanker, M. K. Hayer-Hartl, and F. U. Hartl, "Hsp70 and Hsp40 chaperones can inhibit self-assembly of polyglutamine proteins into amyloid-like fibrils," *Proceedings of the National Academy of Sciences of the United States of America*, vol. 97, no. 14, pp. 7841–7846, 2000.

[37] A. Kitamura, H. Kubota, C. G. Pack et al., "Cytosolic chaperonin prevents polyglutamine toxicity with altering the aggregation state," *Nature Cell Biology*, vol. 8, no. 10, pp. 1163–1170, 2006.

[38] S. Nishikori, K. Yamanaka, T. Sakurai, M. Esaki, and T. Ogura, "p97 homologs from *Caenorhabditis elegans*, CDC-48.1 and CDC-48.2, suppress the aggregate formation of huntingtin exon1 containing expanded polyQ repeat," *Genes to Cells*, vol. 13, no. 8, pp. 827–838, 2008.

[39] H. Yang, C. Liu, Y. Zhong, S. Luo, M. J. Monteiro, and S. Fang, "Huntingtin interacts with the cue domain of gp78 and inhibits gp78 binding to ubiquitin and p97/VCP," *PLoS ONE*, vol. 5, no. 1, article e8905, 2010.

[40] S. Raychaudhuri, M. Sinha, D. Mukhopadhyay, and N. P. Bhattacharyya, "HYPK, a Huntingtin interacting protein, reduces aggregates and apoptosis induced by N-terminal Huntingtin with 40 glutamines in Neuro2a cells and exhibits chaperone-like activity," *Human Molecular Genetics*, vol. 17, no. 2, pp. 240–255, 2008.

[41] T. Arnesen, K. K. Starheim, P. Van Damme et al., "The chaperone-like protein HYPK acts together with natA in cotranslational N-terminal acetylation and prevention of Huntingtin aggregation," *Molecular and Cellular Biology*, vol. 30, no. 8, pp. 1898–1909, 2010.

[42] C. Vacher, L. Garcia-Oroz, and D. C. Rubinsztein, "Overexpression of yeast hsp104 reduces polyglutamine aggregation and prolongs survival of a transgenic mouse model of Huntington's disease," *Human Molecular Genetics*, vol. 14, no. 22, pp. 3425–3433, 2005.

[43] U. A. Desai, J. Pallos, A. A. K. Ma et al., "Biologically active molecules that reduce polyglutamine aggregation and toxicity," *Human Molecular Genetics*, vol. 15, no. 13, pp. 2114–2124, 2006.

[44] V. Heiser, S. Engemann, W. Bröcker et al., "Identification of benzothiazoles as potential polyglutamine aggregation inhibitors of Huntington's disease by using an automated filter retardation assay," *Proceedings of the National Academy of Sciences of the United States of America*, vol. 99, no. 4, pp. 16400–16406, 2002.

[45] M. A. King, S. Hands, F. Hafiz, N. Mizushima, A. M. Tolkovsky, and A. Wyttenbach, "Rapamycin inhibits polyglutamine aggregation independently of autophagy by reducing protein synthesis," *Molecular Pharmacology*, vol. 73, no. 4, pp. 1052–1063, 2008.

[46] M. Herbst and E. E. Wanker, "Small molecule inducers of heat-shock response reduce polyQ-mediated huntingtin aggregation: a possible therapeutic strategy," *Neurodegenerative Diseases*, vol. 4, no. 2-3, pp. 254–260, 2007.

[47] D. W. Colby, Y. Chu, J. P. Cassady et al., "Potent inhibition of huntingtin aggregation and cytotoxicity by a disulfide bond-free single-domain intracellular antibody," *Proceedings of the National Academy of Sciences of the United States of America*, vol. 101, no. 51, pp. 17616–17621, 2004.

[48] K. Sugaya, S. Matsubara, Y. Kagamihara, A. Kawata, and H. Hayashi, "Polyglutamine expansion mutation yields a pathological epitope linked to nucleation of protein aggregate: determinant of Huntington's disease onset," *PLoS ONE*, vol. 2, no. 7, article e635, 2007.

[49] H. Parekh-Olmedo, J. Wang, J. F. Gusella, and E. B. Kmiec, "Modified single-stranded oligonucleotides inhibit aggregate formation and toxicity induced by expanded polyglutamine," *Journal of Molecular Neuroscience*, vol. 24, no. 2, pp. 257–267, 2004.

[50] K. S. P. McNaught and C. W. Olanow, "Protein aggregation in the pathogenesis of familial and sporadic Parkinson's disease," *Neurobiology of Aging*, vol. 27, no. 4, pp. 530–545, 2006.

[51] W. Liu, C. Vives-Bauza, R. Acín-Peréz- et al., "PINK1 defect causes mitochondrial dysfunction, proteasomal deficit and

alpha-synuclein aggregation in cell culture models of Parkinson's disease," *PLoS ONE*, vol. 4, no. 2, article e4597, 2009.

[52] S. Batelli, D. Albani, R. Rametta et al., "DJ-1 modulates α-synuclein aggregation state in a cellular model of oxidative stress: relevance for Parkinson's Disease and involvement of HSP70," *PLoS ONE*, vol. 3, no. 4, article e1884, 2008.

[53] M. Skogen, J. Roth, S. Yerkes, H. Parekh-Olmedo, and E. Kmiec, "Short G-rich oligonucleotides as a potential therapeutic for Huntington's Disease," *BMC Neuroscience*, vol. 7, article 65, 2006.

[54] T. F. Outeiro, P. Putcha, J. E. Tetzlaff et al., "Formation of toxic oligomeric α-synuclein species in living cells," *PLoS ONE*, vol. 3, no. 4, article e1867, 2008.

[55] V. N. Uversky, H. J. Lee, J. Li, A. L. Fink, and S. J. Lee, "Stabilization of partially folded conformation during α-synuclein oligomerization in both purified and cytosolic preparations," *The Journal of Biological Chemistry*, vol. 276, no. 47, pp. 43495–43498, 2001.

[56] V. N. Uversky, "A protein-chameleon: conformational plasticity of α-synuclein, a disordered protein involved in neurodegenerative disorders," *Journal of Biomolecular Structure and Dynamics*, vol. 21, no. 2, pp. 211–234, 2003.

[57] H. T. Li, X. J. Lin, Y. Y. Xie, and H. Y. Hu, "The early events of α-synuclein oligomerization revealed by photo-induced cross-linking," *Protein and Peptide Letters*, vol. 13, no. 4, pp. 385–390, 2006.

[58] J. Klucken, T. F. Outeiro, P. Nguyen, P. J. McLean, and B. T. Hyman, "Detection of novel intracellular α-synuclein oligomeric species by fluorescence lifetime imaging," *FASEB Journal*, vol. 20, no. 12, pp. 2050–2057, 2006.

[59] X. J. Lin, F. Zhang, Y. Y. Xie, W. J. Bao, J. H. He, and H. Y. Hu, "Secondary structural formation of α-synuclein amyloids as revealed by g-factor of solid-state circular dichroism," *Biopolymers*, vol. 83, no. 3, pp. 226–232, 2006.

[60] M. Sandal, F. Valle, I. Tessari et al., "Conformational equilibria in monomeric alpha-synuclein at the single-molecule level," *PLoS Biology*, vol. 6, no. 1, article e6, 2008.

[61] T. Kamiyoshihara, M. Kojima, K. Uéda, M. Tashiro, and S. Shimotakahara, "Observation of multiple intermediates in α-synuclein fibril formation by singular value decomposition analysis," *Biochemical and Biophysical Research Communications*, vol. 355, no. 2, pp. 398–403, 2007.

[62] D. Eliezer, E. Kutluay, R. Bussell, and G. Browne, "Conformational properties of α-synuclein in its free and lipid-associated states," *Journal of Molecular Biology*, vol. 307, no. 4, pp. 1061–1073, 2001.

[63] A. C. M. Ferreon and A. A. Deniz, "α-Synuclein multistate folding thermodynamics: implications for protein misfolding and aggregation," *Biochemistry*, vol. 46, no. 15, pp. 4499–4509, 2007.

[64] D. P. Smith, D. J. Tew, A. F. Hill et al., "Formation of a high affinity lipid-binding intermediate during the early aggregation phase of α-synuclein," *Biochemistry*, vol. 47, no. 5, pp. 1425–1434, 2008.

[65] W. S. Woods, J. M. Boettcher, D. H. Zhou et al., "Conformation-specific binding of α-synuclein to novel protein partners detected by phage display and NMR spectroscopy," *The Journal of Biological Chemistry*, vol. 282, no. 47, pp. 34555–34567, 2007.

[66] Z. Qin, D. Hu, S. Han, S. H. Reaney, D. A. Di Monte, and A. L. Fink, "Effect of 4-hydroxy-2-nonenal modification on α-synuclein aggregation," *The Journal of Biological Chemistry*, vol. 282, no. 8, pp. 5862–5870, 2007.

[67] E. H. Norris, B. I. Giasson, H. Ischiropoulos, and V. M. Y. Lee, "Effects of oxidative and nitrative challenges on α-synuclein fibrillogenesis involve distinct mechanisms of protein modifications," *The Journal of Biological Chemistry*, vol. 278, no. 29, pp. 27230–27240, 2003.

[68] C. E. H. Moussa, F. Mahmoodian, Y. Tomita, and A. Sidhu, "Dopamine differentially induces aggregation of A53T mutant and wild type α-synuclein: insights into the protein chemistry of Parkinson's disease," *Biochemical and Biophysical Research Communications*, vol. 365, no. 4, pp. 833–839, 2008.

[69] J. R. Mazzulli, A. J. Mishizen, B. I. Giasson et al., "Cytosolic catechols inhibit α-synuclein aggregation and facilitate the formation of intracellular soluble oligomeric intermediates," *Journal of Neuroscience*, vol. 26, no. 39, pp. 10068–10078, 2006.

[70] S. L. Leong, R. Cappai, K. J. Barnham, and C. L. L. Pham, "Modulation of α-synuclein aggregation by dopamine: a review," *Neurochemical Research*, vol. 34, no. 10, pp. 1838–1846, 2009.

[71] J. R. Mazzulli, M. Armakola, M. Dumoulin, I. Parastatidis, and H. Ischiropoulos, "Cellular oligomerization of α-synuclein is determined by the interaction of oxidized catechols with a C-terminal sequence," *The Journal of Biological Chemistry*, vol. 282, no. 43, pp. 31621–31630, 2007.

[72] W. J. Burke, V. B. Kumar, N. Pandey et al., "Aggregation of α-synuclein by DOPAL, the monoamine oxidase metabolite of dopamine," *Acta Neuropathologica*, vol. 115, no. 2, pp. 193–203, 2008.

[73] K. Ono, M. Hirohata, and M. Yamada, "Anti-fibrillogenic and fibril-destabilizing activities of anti-Parkinsonian agents for α-synuclein fibrils *in vitro*," *Journal of Neuroscience Research*, vol. 85, no. 7, pp. 1547–1557, 2007.

[74] T. Näsström, S. Gonçalves, C. Sahlin et al., "Antibodies against alpha-synuclein reduce oligomerization in living cells," *PLoS One*, vol. 6, no. 10, article e27230, 2011.

[75] S. M. Lynch, C. Zhou, and A. Messer, "An scFv intrabody against the nonamyloid component of α-synuclein reduces intracellular aggregation and toxicity," *Journal of Molecular Biology*, vol. 377, no. 1, pp. 136–147, 2008.

[76] S. Emadi, H. Barkhordarian, M. S. Wang, P. Schulz, and M. R. Sierks, "Isolation of a human single chain antibody fragment against oligomeric α-synuclein that inhibits aggregation and prevents α-synuclein-induced toxicity," *Journal of Molecular Biology*, vol. 368, no. 4, pp. 1132–1144, 2007.

[77] S. Emadi, S. Kasturirangan, M. S. Wang, P. Schulz, and M. R. Sierks, "Detecting morphologically distinct oligomeric forms of α-synuclein," *The Journal of Biological Chemistry*, vol. 284, no. 17, pp. 11048–11058, 2009.

[78] C. W. Bertoncini, R. M. Rasia, G. R. Lamberto et al., "Structural characterization of the intrinsically unfolded protein β-synuclein, a natural negative regulator of α-synuclein aggregation," *Journal of Molecular Biology*, vol. 372, no. 3, pp. 708–722, 2007.

[79] Y. Fan, P. Limprasert, I. V. J. Murray et al., "β-synuclein modulates α-synuclein neurotoxicity by reducing α-synuclein protein expression," *Human Molecular Genetics*, vol. 15, no. 20, pp. 3002–3011, 2006.

[80] A. O. Koob, K. Ubhi, J. F. Paulsson et al., "Lovastatin ameliorates α-synuclein accumulation and oxidation in transgenic mouse models of α-synucleinopathies," *Experimental Neurology*, vol. 221, no. 2, pp. 267–274, 2010.

[81] P. Bar-On, L. Crews, A. O. Koob et al., "Statins reduce neuronal α-synuclein aggregation in *in vitro* models of Parkinson's disease," *Journal of Neurochemistry*, vol. 105, no. 5, pp. 1656–1667, 2008.

[82] M. Hirohata, K. Ono, A. Morinaga, and M. Yamada, "Non-steroidal anti-inflammatory drugs have potent anti-fibrillogenic and fibril-destabilizing effects for α-synuclein fibrils *in vitro*," *Neuropharmacology*, vol. 54, no. 3, pp. 620–627, 2008.

[83] K. Ono, M. Hirohata, and M. Yamada, "Anti-fibrillogenic and fibril-destabilizing activity of nicotine *in vitro*: implications for the prevention and therapeutics of Lewy body diseases," *Experimental Neurology*, vol. 205, no. 2, pp. 414–424, 2007.

[84] D. P. Hong, A. L. Fink, and V. N. Uversky, "Smoking and Parkinson's disease: does nicotine affect α-synuclein fibrillation?" *Biochimica et Biophysica Acta*, vol. 1794, no. 2, pp. 282–290, 2009.

[85] N. Pandey, J. Strider, W. C. Nolan, S. X. Yan, and J. E. Galvin, "Curcumin inhibits aggregation of α-synuclein," *Acta Neuropathologica*, vol. 115, no. 4, pp. 479–489, 2008.

[86] K. C. Luk, I. P. Mills, J. Q. Trojanowski, and V. M. Y. Lee, "Interactions between Hsp70 and the hydrophobic core of α-synuclein inhibit fibril assembly," *Biochemistry*, vol. 47, no. 47, pp. 12614–12625, 2008.

[87] C. Lo Bianco, J. Shorter, E. Régulier et al., "Hsp104 antagonizes α-synuclein aggregation and reduces dopaminergic degeneration in a rat model of Parkinson disease," *Journal of Clinical Investigation*, vol. 118, no. 9, pp. 3087–3097, 2008.

[88] C. Tomás-Zapico, M. Díez-Zaera, I. Ferrer et al., "α-Synuclein accumulates in huntingtin inclusions but forms independent filaments and its deficiency attenuates early phenotype in a mouse model of Huntington's disease," *Human Molecular Genetics*, vol. 21, no. 3, pp. 495–510, 2012.

[89] S. Corrochano, M. Renna, C. Tomas-Zapico et al., "α-Synuclein levels affect autophagosome numbers *in vivo* and modulate Huntington disease pathology," *Autophagy*, vol. 8, no. 3, pp. 431–432, 2012.

[90] S. Corrochano, M. Renna, S. Carter et al., "α-Synuclein levels modulate Huntington's disease in mice," *Human Molecular Genetics*, vol. 21, pp. 485–494, 2012.

[91] S. A. Maskarinec and D. A. Tirrell, "Protein engineering approaches to biomaterials design," *Current Opinion in Biotechnology*, vol. 16, no. 4, pp. 422–426, 2005.

[92] A. Saghatellan, Y. Yokobayashi, K. Soltani, and M. R. Ghadiri, "A chiroselective peptide replicator," *Nature*, vol. 409, no. 6822, pp. 797–801, 2001.

[93] G. Bhak, S. Lee, J. W. Park, S. Cho, and S. R. Paik s, "Amyloid hydrogel derived from curly protein fibrils of α-synuclein," *Biomaterials*, vol. 31, no. 23, pp. 5986–5995, 2010.

[94] N. Kobayashi, S. Han, C. Nakamura, and K. Sode, "Nanostructure fabrication based on engineered α-synuclein," *Nanobiotechnology*, vol. 4, no. 1–4, pp. 50–55, 2008.

16

A Molecular Dynamics Approach to Ligand-Receptor Interaction in the Aspirin-Human Serum Albumin Complex

H. Ariel Alvarez, Andrés N. McCarthy, and J. Raúl Grigera

Instituto de Física de Líquidos y Sistemas Biológicos (IFLYSIB), CONICET y Departamento de Ciencias Biológicas, Facultad de Ciencias Exactas, Universidad Nacional de La Plata, 49-789, cc 565, B1900BTE La Plata, Argentina

Correspondence should be addressed to J. Raúl Grigera, grigera@iflysib.unlp.edu.ar

Academic Editor: Claudio M. Soares

In this work, we present a study of the interaction between human serum albumin (HSA) and acetylsalicylic acid (ASA, $C_9H_8O_4$) by molecular dynamics simulations (MD). Starting from an experimentally resolved structure of the complex, we performed the extraction of the ligand by means of the application of an external force. After stabilization of the system, we quantified the force used to remove the ASA from its specific site of binding to HSA and calculated the mechanical nonequilibrium external work done during this process. We obtain a reasonable value for the upper boundary of the Gibbs free energy difference (an equilibrium thermodynamic potential) between the complexed and noncomplexed states. To achieve this goal, we used the finite sampling estimator of the average work, calculated from the Jarzynski Equality. To evaluate the effect of the solvent, we calculated the so-called "viscous work," that is, the work done to move the aspirin in the same trajectory through the solvent in absence of the protein, so as to assess the relevance of its contribution to the total work. The results are in good agreement with the available experimental data for the albumin affinity constant for aspirin, obtained through quenching fluorescence methods.

1. Introduction

Human serum albumin (HSA) is the most abundant plasma protein in the human body and plays an important role in drug transport and metabolism. Generally regarded as a nonspecific transport protein, HSA has been assigned a number of enzymatic properties [1–3]. Additionally, the enzymatic activity of HSA on different substrates and drugs has also been studied and documented. Nevertheless, the structural mechanism of this activity is yet unknown.

To assess the structural basis of binding mechanisms, we evaluated the interaction between HSA and acetylsalicylic acid (ASA, $C_9H_8O_4$) by means of molecular dynamics simulations (MD).

Starting from an experimentally resolved structure of the complex, we extracted the ligand by means of the application of an external force, under near quasistatic conditions, thus evaluating the work involved in breaking the interactions present in the protein-ligand complex, and hence obtaining

an upper boundary for the free energy of binding of the complex.

We quantified the force used to remove the ASA from its specific site of binding to HSA and calculated the mechanical nonequilibrium external work done during this process.

The aim of the present study is to calculate an upper boundary for the Gibbs free energy difference associated to that process, through the average work obtained using the finite sampling estimator from the Jarzynski equality.

To evaluate the effect of the solvent (treated explicitly), we calculated the "viscous work." This magnitude represents the work done to move the aspirin in the same trajectory through the solvent in absence of the protein. Its principal sense follows from the fact that experimental techniques to obtain the protein-ligand affinity (e.g., quenching fluorescence, microcalorimetry) do not take into account the contribution of the solvent to this out of equilibrium process (since they analyze equilibrium states).

Finally, we compared our computational results with the experimental data available for the affinity constant of the HSA-ASA complex, obtained by quenching fluorescence methods [4], showing reasonable agreement.

2. General Procedure

We carried out all molecular dynamics (MD) simulations using the GROMACS 4.0.1 package [5] in which the equations of motion are solved using a leap-frog integration step. We used GROMOS96 (43a1) force field [6] for the minimization process, as well as for all the MD simulation steps, and kept all protein bond lengths constrained using the LINCS algorithm [7]. Water molecules were constrained using the SETTLE algorithm [8]. For the calculation of long-range coulombic interactions, we applied the reaction field method, with a 1.4 nm cut-off radius. Likewise, we calculated Lennard-Jones interactions within a cut-off radius of 1.4 nm.

For all the simulation runs, we have used a Xeon-based, dual-processor cluster, running under GNU/Linux, and for all plots and graphics MS Windows or GNU/Linux, using the reference Visual Molecular Dynamics package, Swiss PDB Viewer, or XGrace software [9–11].

As starting configuration we used the human serum albumin-acetylsalicylic acid complex (PDB ID: 2I30-Resolution 2.9Å) that contains five myristic acid molecules (MYR, $C_{14}H_{28}O_2$) forming part of the albumin structure [1]. Since the original structure was incomplete, we reconstructed both the N-terminus and the C-terminus. Thus, ASP1, ALA2, and LEU585 had to be added *ad hoc* to the experimental structure using Swiss PDB Viewer [12], using standard $\phi - \Psi$ angles for the reconstructed peptidic planes. Moreover, the experimental structure corresponds to the complex in the post reaction state; that is, the protein was acetylated in residue LYS199 and the aspirin hydrolyzed. Thus, we removed the acetyl group from the protein and reconstructed the modified aspirin.

Having focused on the interaction (and dissociation) between HSA and its most internal (and more tightly attached) aspirin ligand, such process will be expected to occur in absence of the less tightly bonded (and external) aspirin ligand. Consequently, the external aspirin was removed from the initial configuration.

The starting system consisted of a truncated dodecahedral simulation box of dimension parameter $d = 12.9702$ nm and a total volume of 1540.65 nm³, containing one HSA molecule, one ASA molecule, five myristic acid molecules, and 47.803 water molecules. We used the SPC/E [13] water model. We aligned the principal axes of the protein and the box.

We constructed the topologies for aspirin and myristic acid and added them to GROMOS96 43a1 force field residue data base, using all the corresponding atom types, charges, bonds, angles, proper, and improper dihedral angles from that database. The values used for all atom charges in both aspirin and myristic acid topologies are summarized in Tables I(a) and I(b). A detailed schematic representation for both molecules is additionally provided as supplementary material (see Supplementary Material available online at

TABLE 1: Charge distribution used for every atom in the construction of the Aspirin and Myristic Acid topologies, respectively.

(a)

Atom number	Atom Type	Name	Charge group	Partial Charge	Mass
1	OM	OL	1	−0.635	15.9994
2	C	CK	1	0.27	12.011
3	OM	OM	1	−0.635	15.9994
4	C	CJ	2	0	12.011
5	CR1	CI	2	0	13.019
6	CR1	CH	2	0	13.019
7	CR1	CG	3	0	13.019
8	CR1	CF	3	0	13.019
9	C	CE	3	0	12.011
10	OA	OD	4	0	15.9994
11	C	CB	4	0.38	12.011
12	O	OC	4	−0.38	15.9994
13	CH3	CA	5	0	15.035

(b)

Atom number	Atom Type	Name	Charge group	Partial Charge	Mass
1	OM	OB	1	−0.635	15.9994
2	C	CA	1	0.27	12.011
3	OM	OC	1	−0.635	15.9994
4	CH2	CD	2	0	14.027
5	CH2	CE	2	0	14.027
6	CH2	CF	2	0	14.027
7	CH2	CG	2	0	14.027
8	CH2	CH	2	0	14.027
9	CH2	CI	2	0	14.027
10	CH2	CJ	2	0	14.027
11	CH2	CK	2	0	14.027
12	CH2	CL	2	0	14.027
13	CH2	CM	2	0	14.027
14	CH2	CN	2	0	14.027
15	CH2	CO	2	0	14.027
16	CH3	CP	2	0	15.035

doi:10.1155/2012/642745). As previously validated in recent literature [14], myristic acid was modeled in the unprotonated state and described using parameters derived from the lipid force field [15]. The carboxylic acid group was based on the parameters of glutamic acid, which were available from the corresponding GROMOS96 43a1 force field.

We generated the topology of the total system, with standard protonation states for all amino acids (pH 7).

In every step, we removed the motion of the center of mass of the system.

The production system was weakly coupled to a thermal and hydrostatic bath, in order to work in the isothermal-isobaric ensemble [16] at $T = 300$ K and $P = 1$ bar.

3. Equilibration

We minimized the energy of the system using firstly the steepest descent method, converging to machine precision. Secondly, we applied the conjugated gradient method, converging in less than 20 cycles.

The equilibration of the complete system proved to be nontrivial, because we had to complete the original structure and chemically modify it, and we had to perform a nonstandard series of computational steps to release the exogenous tensions introduced due to the structural additions and modifications mentioned previously. Namely, we performed an MD run *in vacuo* (in absence of solvent), using a 0.0001 fs time step and for a total time of 2 ps lightly coupled to a 100 K temperature bath [16], after energy minimization in order to let the incorporated residues explore more stable configurations. During this run, we applied position restrains on all of the backbone atoms (except for those incorporated *ad hoc*). In the same conditions, we subsequently performed a 10 ps run, using a 0.5 fs time step. Next and for another 10 ps (using the same time step), we incorporated the LINCS algorithm to add restrictions in bond lengths and angles for all molecules belonging to the complex.

Once the previous procedure was completed, we gradually increased the time step up to 2 fs, likewise increasing temperature up to the final value of 300 K.

At this point, we solvated the system adding 47.803 SPC/E explicit water molecules. In these conditions, we allowed the solvated system to relax (keeping all solute heavy atoms restrained to their corresponding crystallographic positions) during 200 ps, at 300 K and 1bar. At the same temperature and pressure, we subsequently decreased the position restraints force constants during a 200 ps MD run. Finally, we released all position restraints and allowed the system to relax freely for a total simulation time of 20 ns, at 300 K and 1bar, after which we considered the system to be in conformational equilibrium.

The stability of MYR molecules in their binding sites throughout the total 20 ns of the equilibration run indicates that the conformational equilibrium was correctly achieved. Together with this, we observed a relative stability of the HSA backbone (evaluated by the convergence of HSA alpha-carbon root mean square displacement (RMSD), (see Supplementary Material available online at doi:10.1155/2012/642745)). These facts are in notable agreement with the widely known role of fatty acids in the stabilization of a HSA structure [17].

4. AFM Pulling Simulation

Once the system was effectively equilibrated, we performed 30 MD quasistatic runs, applying an external force on the center of mass of the aspirin, in order to extract this molecule from its binding site. We coupled all 30 production simulation runs to a weak temperature and pressure bath of 300 K and 1 bar, in order to work in the isothermal-isobaric ensemble. We randomly generated the initial velocities, using a unique seed for every production run. The total simulation time of every production run was 500 ps, with a time step of 2 fs. The constant of the harmonical potential applied at the center of mass of the aspirin was of $k = 1,000 \, \text{kJ/(mol·nm}^2)$, and the pulling rate (i.e., the velocity used to move the free end of the "*spring*" in the pulling direction) was $v = 0.01 \, \text{nm/ps}$.

The pulling direction was established as the line connecting the center of mass of the ligand and the center of the triangle determined by the centers of mass of the R groups of residues GLU 292, HIS 440, and CYS 448 (i.e., the binding pocket threshold).

To calculate the work done during this process, we plotted each component of the force (F_x, F_y, F_z) as a function of the respective position component of the center of mass of the acetylsalicylic acid molecule ($x_{ASA}, y_{ASA}, z_{ASA}$). This is shown in Figure 1. These graphics show resemblance to the experimental data obtained from Atomic Force Microscopy [18].

5. The Jarzynski Relationship

The process of removing a ligand from the site of binding to its receptor can be done by applying an external force to the ligand, hence performing a corresponding external mechanical work. If this work is done quasistatically, the value obtained for the applied external work is equal to the free energy variation associated to that process.

When the work done is not quasistatic, the Jarzynski equality [19, 20] gives a relationship between the variation in free energy values that describes the state changes in the system under study, and the correctly averaged ensemble of the measurements of the external work done to change that state in a finite time (out of equilibrium). In the case of a system in contact to an external thermal bath (in the Canonical ensemble), the correct description of the state changes of the system is through Helmholtz free energy, whereas for a system in the isothermal-isobaric ensemble like the one we are studying, the suitable descriptor of the equilibrium states is Gibbs free energy.

If the actual distribution is approximated by performing a finite sample of work performed through an applied external force, a superior value to the free energy variation can be obtained from the average work W^a of the finite sampling, taken as usual:

$$W^a \equiv \frac{1}{N_s} \sum_{i=1}^{N_s} W_i, \qquad (1)$$

where N_s is the number of measurements and W_i the external work done in the *i*th process. We can think of W_i as a random sampling from the distribution $\rho(W)$ that satisfies the so-called Jarzynski equality at any time t_s:

$$\left\langle e^{-W/k_B T} \right\rangle \equiv \int W \rho(W, t_s) e^{-W/k_B T} dW = e^{-\Delta G/k_B T}, \qquad (2)$$

where T is the temperature, k_B the Boltzmann constant, t_s is the time spent to do the described process, and ΔG is the Gibbs free energy variation.

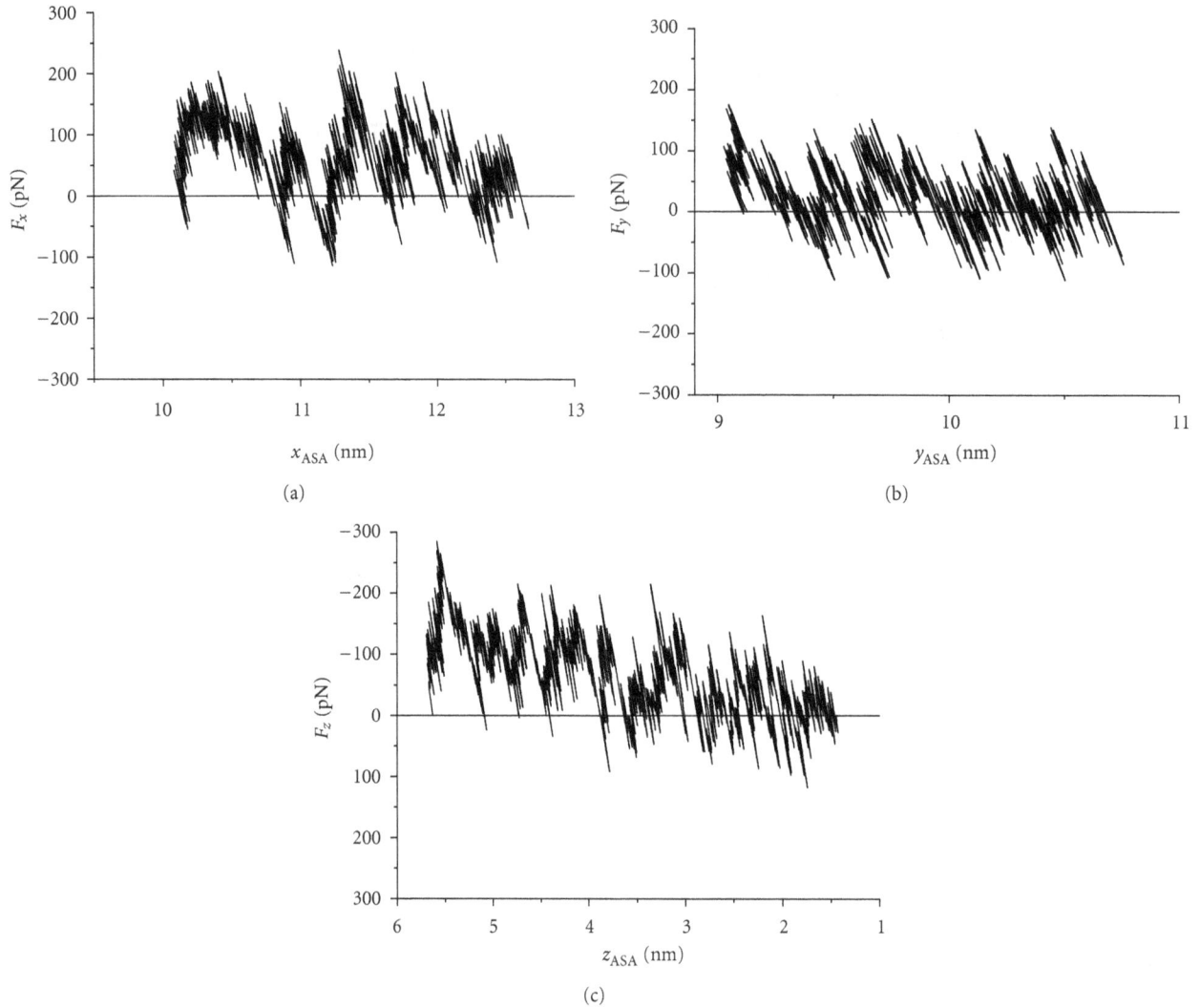

FIGURE 1: Example of the components x, y, and z of the applied force, as a function of the respective position component of the acetylsalicylic acid molecule center of mass (Figures (a), (b), and (c), resp.). The displacement in the z axis was in the negative direction, while for x and y axes the displacement was in the positive direction.

In this context, the expectation value of W^a, $\langle\langle W^a\rangle\rangle$, will give an upper boundary to the Gibbs free energy variation:

$$\langle\langle W^a\rangle\rangle = \langle W\rangle \equiv \int W\rho(W,t_s)dW \geq \Delta G. \tag{3}$$

The Jarzynski equality suggests that W^a is not the correct quantity to estimate ΔG. A better upper boundary to this magnitude is given by the so-called finite sampling estimator of the average from the Jarzynski equality [21]:

$$W^x \equiv -k_B T \cdot \ln\left[\frac{1}{N_s}\sum_{i=1}^{N_s}e^{-W_i/k_B T}\right]. \tag{4}$$

For $N_s = 1$, W^x and W^a will be equal and the expectation value for both will be $\langle W\rangle$, and for $N_s \rightarrow \infty$, W^x will tend to ΔG, and W^a will tend to $\langle W\rangle$.

For intermediate values of N_s,

$$\Delta G \leq \langle\langle W^x\rangle\rangle \leq \langle\langle W^a\rangle\rangle, \tag{5}$$

is satisfied.

6. The Viscous Work

The data obtained through nonequilibrium molecular dynamics pulling of a ligand may reasonably contain a nondesired contribution due to the work that is necessary to move a molecule through the solvent under such conditions, when compared to experimental data obtained in conditions of equilibrium. Hence, we have undertaken the evaluation of the relevance of the so-called "viscous work," in comparison with the total work.

With this purpose, we prepared a system of the same shape and dimensions, consisting only of one acetylsalicylic

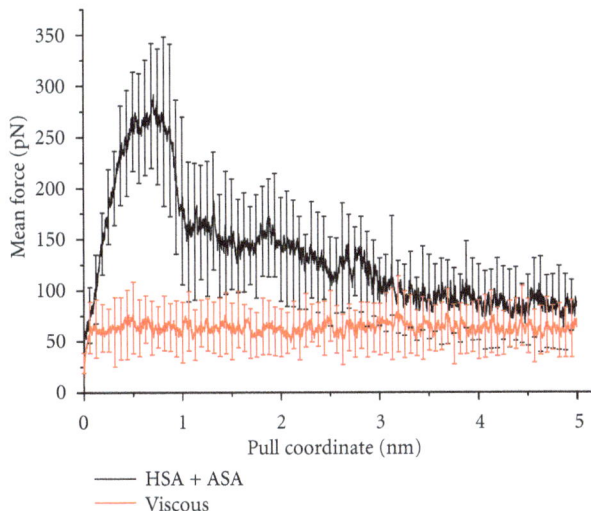

FIGURE 2: Dependence of the value of the magnitude of the mean force applied on the center of mass of ASA as a function of the magnitude of the displacement of the spring (pull coordinate), both for the system in presence (black lines) and absence (red lines) of the HSA complex. These values represent the mean force at each step, averaged over the thirty MD runs for both systems. The error bars show the standard deviation of these mean values (only 80 values are here presented in order to enhance graphical clearness).

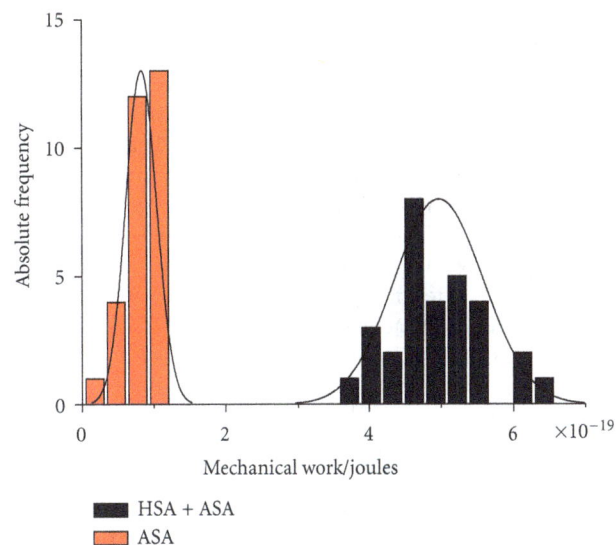

FIGURE 3: Histogram representation of the mechanical work, both for the system in presence (black bars) and absence (red bars) of the HSA complex. The continuous lines are the Gaussian distributions for the corresponding values of standard deviation and mean value for each histogram population.

acid molecule in the original position, solely surrounded by water molecules.

Firstly, we minimized the energy of this system as described in the previous section and, subsequently, performed short 200 ps stabilization. Once temperature and pressure reached equilibrium values of 300 K and 1 bar, we coupled the system to a weak thermostatic bath and fixed the box dimensions. After stabilization, we proceeded as with the system in presence of HSA, performing 30 MD runs with the same general setup.

7. Results and Discussion

The data presented in Figure 2 corresponds to the mean value of the magnitude of the force applied on the center of mass of the acetylsalicylic acid molecule, as a function of the magnitude of the displacement of the free end of the ideal spring, both for the system in presence (black lines) and absence (red lines) of the HSA complex. Error bars show the standard deviation of those values. In order to obtain greater graphical clearness, only 80 of these bars are shown. In this Figure, a marked difference between curves may be observed up to 2 nm. After this point the error bars cross each other, finally reaching statistical convergence. The duplication of this distance (the point of error bar crossing, that is, 4 nm) was considered as a reasonable criterion for convergence and was thus chosen as the limit value for production data. After this point, the magnitude of the applied force becomes comparable for both systems, which allows us to consider the acetylsalicylic acid molecule to be out of the binding site.

In Figure 3, we present a histogram representation of the distribution of the performed mechanical work, both for the system in presence (red bars) and absence (black bars) of the HSA complex. From this study, we may interestingly observe that the studied samples of both systems are large enough to clearly distinguish between populations.

The obtained values for W^a and W^x, using (1) and (4), respectively, are

$$W^a = (5.0 \pm 0.6) \times 10^{-19} \text{ J} \quad \text{(usual average)},$$
$$W^x = (4.0 \pm 0.6) \times 10^{-19} \text{ J} \quad \text{(Jarzynski's average)}. \tag{6}$$

From these results we can observe that W^x establishes a better highest value for ΔG than W^a, in agreement with (5).

Likewise, the average value of the viscous work obtained using the usual average is

$$W^a{}_{\text{viscous}} = 8 \pm 2 \times 10^{-20} \text{ J}. \tag{7}$$

And for the finite sampling estimator of the average value from Jarzynski equality,

$$W^x{}_{\text{viscous}} = 4 \pm 2 \times 10^{-20} \text{ J}. \tag{8}$$

For the present case, the obtained value of the viscous work falls within the order of magnitude of the standard deviation of the calculated binding work. However, this may not be the general situation. Additionally, as such computational methods are refined and sampling capacities increased, this "viscous work" may reasonably become a more important source of discrepancy with experimental observation, therefore rendering its evaluation mandatory.

Taking these into account, the obtained value for W^x would be

$$W^x{}_{\text{corrected}} = (3.6 \pm 0.6) \times 10^{-19} \text{ J}. \tag{9}$$

Regarding the ligand trajectory in this process, we can distinguish two regions: an interior region that corresponds to the movement within the binding pocket, where the pulling direction is univocally determined by the shortest path between the original position of the ligand and the binding pocket threshold and a second exterior one.

In the first region, where the unbinding path is conformationally determined, we can assume that the contribution to the total average work is close to the optimal one, whereas in the exterior region it is expected that better path optimization may be achieved.

The separation between both regions was established as the point in which the center of mass of the ligand crosses the plane formed by residues GLU 292, HIS 440, and CYS 448 (i.e., the binding pocket threshold).

The first region represents 29.4% of the total trajectory, with a corresponding pulling coordinate of 1.47 nm, and its contribution to the corrected work is of $(1.9 \pm 0.6) \times 10^{-19}$ J.

The experimental affinity constant of the complex HSA-aspirin, obtained by Bojko and coworkers from quenching fluorescence measurements [4], is $K_a = 18.79 \times 10^3 \, \mathrm{M}^{-1}$. From this value, we obtain the experimental ΔG as

$$\Delta G = k_B T \ln K_a = 4.07 \times 10^{-20} \, \mathrm{J}. \tag{10}$$

The upper boundary obtained using the finite sampling estimator of the average work gives a better value than that obtained from the usual average work. Albeit more recent comments on Jarzynski's work [22], his response is clear and categorical [23], and this result comes in line with what he classically proposes.

8. Conclusion

In the present study, we have succeeded in reconstructing, stabilizing, and equilibrating a protein ligand complex which proceeded from incomplete data provided from experiment. This achieved equilibrium state enabled us to study the HSA-ASA-MYR complex through nonequilibrium molecular dynamics, in near quasistatic conditions. The fact that complex biological structures, that are not totally resolved through experiment, may be eligible targets for molecular dynamic studies reinforces the importance and potentiality of this technique.

We have likewise obtained an average upper boundary for the Gibbs free energy of binding of the HSA-ASA complex that is less than one order of magnitude above the experimental value.

As regards the assessment of the relevance of the viscous work, the value obtained is low compared to the upper boundary for the Gibbs free energy difference of the HSA-ASA complex. Nevertheless, its difference with the experimental free energy of binding is not negligible and should therefore be generally taken into account in such studies. Moreover, the relevance of this effect, and the need for its correct evaluation may reasonably increase as computational methods are refined and sampling capacities increased.

Finally, the promising level of agreement of the data presented in this study with that available experimentally

calls for further development of such studies as well as encourages the analysis of other similar systems.

Acknowledgments

The authors thank Dr. O. Chara for useful comments and remarks. This work was partially supported by the National Council of Scientific and Technical Research (CONICET), the National University of La Plata (UNLP), and the National Agency for Scientific and Technological Promotion (ANPCyT). H. A. Alvarez is a Doctoral fellow of CONICET. A. N. McCarthy. is member of the Researcher Career of the Scientific Research Committee of the Buenos Aires Province (CICPBA). J. R. Grigera is member of the Researcher Career of CONICET.

References

[1] F. Yang, C. Bian, L. Zhu, G. Zhao, Z. Huang, and M. Huang, "Effect of human serum albumin on drug metabolism: structural evidence of esterase activity of human serum albumin," *Journal of Structural Biology*, vol. 157, no. 2, pp. 348–355, 2007.

[2] N. Ahmed, D. Dobler, M. Dean, and P. J. Thornalley, "Peptide mapping identifies hotspot site of modification in human serum albumin by methylglyoxal involved in ligand binding and esterase activity," *Journal of Biological Chemistry*, vol. 280, no. 7, pp. 5724–5732, 2005.

[3] N. Dubois-Presle, F. Lapicque, M. H. Maurice et al., "Stereoselective esterase activity of human serum albumin toward ketoprofen glucuronide," *Molecular Pharmacology*, vol. 47, no. 3, pp. 647–653, 1995.

[4] B. Bojko, A. Sułkowska, M. Maciazek, J. Równicka, F. Njau, and W. W. Sułkowski, "Changes of serum albumin affinity for aspirin induced by fatty acid," *International Journal of Biological Macromolecules*, vol. 42, no. 4, pp. 314–323, 2008.

[5] D. van der Spoel, E. Lindahl, B. Hess, G. Groenhof, A. E. Mark, and H. J. C. Berendsen, "GROMACS: fast, flexible, and free," *Journal of Computational Chemistry*, vol. 26, no. 16, pp. 1701–1718, 2005.

[6] W. F. van Gunsteren, S. R. Billeter, A. A. Eising et al., *Biomolecular Simulation: The GROMOS96 Manual and Userguide*, Hochschulverlag AG an der ETH Zürich, 1996.

[7] B. Hess, H. Bekker, H. J. C. Berendsen, and J. G. E. M. Fraaije, "LINCS: a linear constraint solver for molecular simulations," *Journal of Computational Chemistry*, vol. 18, no. 12, pp. 1463–1472, 1997.

[8] S. Miyamoto and P. A. Kollman, "Settle: an analytical version of the SHAKE and RATTLE algorithm for rigid water models," *Journal of Computational Chemistry*, vol. 13, pp. 952–962, 1992.

[9] W. Humphrey, A. Dalke, and K. Schulten, "VMD: Visual molecular dynamics," *Journal of Molecular Graphics*, vol. 14, no. 1, pp. 33–38, 1996.

[10] http://www.ks.uiuc.edu/Research/vmd/.

[11] http://plasma-gate.weizmann.ac.il/Grace/.

[12] N. Guex and M. C. Peitsch, "SWISS-MODEL and the Swiss-PdbViewer: an environment for comparative protein modeling," *Electrophoresis*, vol. 18, no. 15, pp. 2714–2723, 1997.

[13] H. J. C. Berendsen, J. R. Grigera, and T. P. Straatsma, "The missing term in effective pair potentials," *Journal of Physical Chemistry*, vol. 91, no. 24, pp. 6269–6271, 1987.

[14] S. Leekumjorn, Y. Wu, A. K. Sum, and C. Chan, "Experimental and computational studies investigating trehalose protection of HepG2 cells from palmitate-induced toxicity," *Biophysical Journal*, vol. 94, no. 7, pp. 2869–2883, 2008.

[15] D. P. Tieleman and H. J. C. Berendsen, "A molecular dynamics study of the pores formed by Escherichia coli OmpF porin in a fully hydrated palmitoyloleoylphosphatidylcholine bilayer," *Biophysical Journal*, vol. 74, no. 6, pp. 2786–2801, 1998.

[16] H. J. C. Berendsen, J. P. M. Postma, W. F. van Gunsteren, A. DiNola, and J. R. Haak, "Molecular dynamics with coupling to an external bath," *The Journal of Chemical Physics*, vol. 81, no. 8, pp. 3684–3690, 1984.

[17] A. A. Spector, "Fatty acid binding to plasma albumin," *Journal of Lipid Research*, vol. 16, no. 3, pp. 165–179, 1975.

[18] G. Hummer and A. Szabo, "Free energy reconstruction from nonequilibrium single-molecule pulling experiments," *Proceedings of the National Academy of Sciences of the United States of America*, vol. 98, no. 7, pp. 3658–3661, 2001.

[19] C. Jarzynski, "Equilibrium free-energy differences from nonequilibrium measurements: a master-equation approach," *Physical Review E*, vol. 56, no. 5, pp. 5018–5035, 1997.

[20] C. Jarzynski, "Nonequilibrium equality for free energy differences," *Physical Review Letters*, vol. 78, no. 14, pp. 2690–2693, 1997.

[21] S. Park, F. Khalili-Araghi, E. Tajkhorshid, and K. Schulten, "Free energy calculation from steered molecular dynamics simulations using Jarzynski's equality," *Journal of Chemical Physics*, vol. 119, no. 6, pp. 3559–3566, 2003.

[22] E. G. D. Cohen and M. David, "A note on the Jarzynski equality," *Journal of Statistical Mechanics*, vol. 2004, Article ID P07006, 17 pages, 2004.

[23] C. Jarzynski, "Nonequilibrium work theorem for a system strongly coupled to a thermal environment," *Journal of Statistical Mechanics*, vol. 2004, Article ID P09005, 2004.

17

Cytoskeletal Strains in Modeled Optohydrodynamically Stressed Healthy and Diseased Biological Cells

Sean S. Kohles,[1,2] **Yu Liang,**[3] **and Asit K. Saha**[3]

[1] *Regenerative Bioengineering Laboratory, Department of Biology, Science Research & Teaching Center (SRTC), Portland State University, P.O. Box 751, Portland, OR 97207, USA*
[2] *Department of Surgery, Oregon Health & Science University, Portland, OR 97239, USA*
[3] *Center for Allaying Health Disparities through Research and Education (CADRE) and Department of Mathematics & Computer Science, Central State University, Wilberforce, OH 45384, USA*

Correspondence should be addressed to Sean S. Kohles, ssk@kohlesbioengineering.com

Academic Editor: George Perry

Controlled external chemomechanical stimuli have been shown to influence cellular and tissue regeneration/degeneration, especially with regards to distinct disease sequelae or health maintenance. Recently, a unique three-dimensional stress state was mathematically derived to describe the experimental stresses applied to isolated living cells suspended in an optohydrodynamic trap (optical tweezers combined with microfluidics). These formulae were previously developed in two and three dimensions from the fundamental equations describing creeping flows past a suspended sphere. The objective of the current study is to determine the full-field cellular strain response due to the applied three-dimensional stress environment through a multiphysics computational simulation. In this investigation, the multiscale cytoskeletal structures are modeled as homogeneous, isotropic, and linearly elastic. The resulting computational biophysics can be directly compared with experimental strain measurements, other modeling interpretations of cellular mechanics including the liquid drop theory, and biokinetic models of biomolecule dynamics. The described multiphysics computational framework will facilitate more realistic cytoskeletal model interpretations, whose intracellular structures can be distinctly defined, including the cellular membrane substructures, nucleus, and organelles.

1. Introduction

The current research on human diseases primarily focuses on the molecular, microbiological, immunological, and pathological influences. The mechanical basis of disease is now often being explored to decipher any direct contributions toward the physiological response [1, 2]. In functionally loaded tissues such as cartilage and bone, cells (chondrocytes and osteocytes) experience multiaxial forces (hydrostatic, compressive, tensile, and shear), which play a significant role in modulating the biological function through maintenance of the phenotype and production of a neotissue [3]. Conversely, abnormal mechanical forces (either static or dynamic) can lead to altered cell behavior resulting in pathological matrix synthesis, increased catabolic activity (degradation), and ultimately osteoarthritis or osteoporosis (apoptosis) [4]. Our previous investigations have indicated

that chondrocytes and likely other cell types respond to their stress-strain environments in a temporal and spatial manner [5]. It has also just been shown that individual cellular mechanical properties may indicate the regenerative potential of mesenchymal stem cells [6]. Investigations of the biomechanics at the cellular level have also identified the biomarkers of disease. Cytoskeletal stiffness of metastatic cancer cells was reported as more than 70% softer than the benign cells that line the body cavity in patients with suspected lung, breast, and pancreatic cancer [7]. These approaches highlight the utility of single-cell biomechanics as a critical component of advancing microscale therapeutics.

Advancements in laser technology and microfluidics now allow the use of optical tweezers or traps and fluid mechanics to manipulate isolated single cells [8, 9]. Isolated loads can be applied experimentally to single cells in culture to quantify cellular and cytoskeletal biomechanics. One can then apply forces and displacements as small as pico-Newtons and nanometers, respectively [10–12]. The local microenvironment can therefore be precisely manipulated to facilitate biomechanical test sequences on individual biological cells and molecular structures.

In order to explore the connection between external mechanical stimulation and cellular regeneration or degeneration, we developed a three-dimensional, multiphysics computational model to fully characterize a unique micromechanical environment. The applied stress state within our custom-fabricated optical and hydrodynamic (optohydrodynamic) trap have been mathematically developed from the fundamental equations describing microfluidic creeping flows past a suspended sphere. The objective described in the following paper is to explore the full-field cellular strain response to a range of applied stresses and cellular moduli. The described computational framework will now allow us to develop more realistic cellular models, whose intracellular structures are distinctly identified. This approach is specifically focused on addressing our ongoing efforts in health disparity research [13].

2. Methods

2.1. Single Cell Biomechanics and Optohydrodynamic Trapping.
Living non-adhered, suspended osteoblasts, chondrocytes, fibroblasts, and myoblasts have recently been isolated and mechanically manipulated [12, 16]. Primary cultures of chondrogenic and osteogenic tissues were generated directly from rat long bones, while muscle cells were acquired from the mouse-derived myoblast C2C12 cell line (ATCC, CRL-1772, Manassas, VA, USA). All cells were tested at room temperature experiments (~20.5°C) in a flow media consisting of a physiological buffer resulting in a media viscosity of ~1 mPa s. This single cell biomechanical manipulation was made available by combining optical trapping with microfluidics to create the optohydrodynamic trap. This work was facilitated by an instrument, which integrates two laser-based techniques for the mechanical characterization of cellular and biomolecular structures [8, 12].

The optical tweezers or the trap component of the device utilizes an infrared laser (λ = 1,064 nm) to suspend micron-sized objects with nanometer position control and pico-Newton constraining forces. In the Mie regime, where objects are larger in dimension than the wavelength of the trapping laser (here biological cells), a ray optics description indicates the transfer of refracted light and the associated momentum into trapping forces (Figure 1). Micron-particle image velocimetry can be engaged by incorporating two frequency-doubled lasers (λ = 532 nm) aligned through the same optical path as the OT for full-field flow velocity characterization. However, the nanoparticles associated with velocity imaging have proven deleterious to cellular health, but provide useful experimental validation of flows around synthetic micron-sized particles.

The hydrodynamic component of this approach is facilitated through a microfluidic chip design configured in the form of a cross-junction channel (Figure 2). This geometry creates an extensional flow environment and a stagnation point at the channel's geometric center. Cells are positioned at the centroid with the optical trap and manipulated with microfluidics, thus creating the optohydrodynamic trap. The cell experiences a total drag force equal to zero, confirmed by integrating the stress tensor as defined by the normal (form drag) and shear (friction drag) stresses (Figure 3). This reflects the mechanical stability or the saddle-point nature of the optohydrodynamic trap. Previous studies describe chip fabrication in detail including the control of the gravity-driven flow initially associated with microfluidic manipulation [8].

2.2. Applied Fluid Stress Analysis.
The two- and three-dimensional stress states were previously developed as applied to the surface of a nonrotating spheroid cell of radius a, within the optohydrodynamic trap [2, 12, 14, 15]. Briefly, the full-field fluid velocity vector $\mathbf{u}$ was constructed from the constitutive equations describing a non-rotating sphere suspended in a general linear flow with viscosity μ and pressure distribution p [17]. In the polar-spherical components (r-θ-ϕ magnitudes and $\mathbf{e}_r$-$\mathbf{e}_\theta$-$\mathbf{e}_\phi$ vectors), the generalized flow field produces the individual velocity components:

$$\mathbf{u} = U\left(\frac{r}{a} - \frac{5}{2}\left(\frac{a}{r}\right)^2 + \frac{3}{2}\left(\frac{a}{r}\right)^4\right)\sin^2\phi \cos 2\theta\, \mathbf{e}_r$$

$$+ U\left(\frac{r}{a} - \left(\frac{a}{r}\right)^4\right)\sin\phi\cos\phi\cos 2\theta\, \mathbf{e}_\phi \qquad (1)$$

$$+ U\left(\frac{r}{a} - \left(\frac{a}{r}\right)^4\right)\sin\phi\cos 2\theta\, \mathbf{e}_\theta$$

including the pressure distribution

$$p = p_\infty - \frac{5\mu U}{a}\left(\frac{a}{r}\right)^3\sin^2\phi\cos 2\theta. \qquad (2)$$

The velocity gradients can be converted into the applied fluid stresses by applying the constitutive equation for an incompressible, Newtonian fluid [18]:

$$\mathbf{T} = -p\mathbf{I} + 2\mu\mathbf{E}, \qquad (3)$$

where $\mathbf{T}$ is the stress tensor and $\mathbf{I}$ is the identity matrix associated with the local isotropic (hydrostatic) pressure distribution p. The strain rate tensor $\mathbf{E}$ can be characterized by the flow velocity gradient tensor and its transpose:

$$\mathbf{E} = \left(\frac{1}{2}\right)\left[\nabla\mathbf{u} + \nabla\mathbf{u}^T\right]. \qquad (4)$$

By incorporating the velocity and pressure fields into the gradient analysis and then in turn into the constitutive equation, the fluidic stress tensor can be fully defined as

$$\mathbf{T}_{rr} = -p_\infty + \frac{\mu U}{a}\left(2 + 15\left(\frac{a}{r}\right)^3 - 12\left(\frac{a}{r}\right)^5\right)\sin^2\phi\cos 2\theta\, \mathbf{e}_r,$$

$$\mathbf{T}_{r\phi} = \frac{\mu U}{a}\left(2 - 5\left(\frac{a}{r}\right)^3 + 8\left(\frac{a}{r}\right)^5\right)\sin\phi\cos\phi\cos 2\theta\, \mathbf{e}_\phi,$$

$$\mathbf{T}_{r\theta} = -\frac{\mu U}{a}\left(2 - 5\left(\frac{a}{r}\right)^3 + 8\left(\frac{a}{r}\right)^5\right)\sin\phi\sin 2\theta\, \mathbf{e}_\theta.$$

$$(5)$$

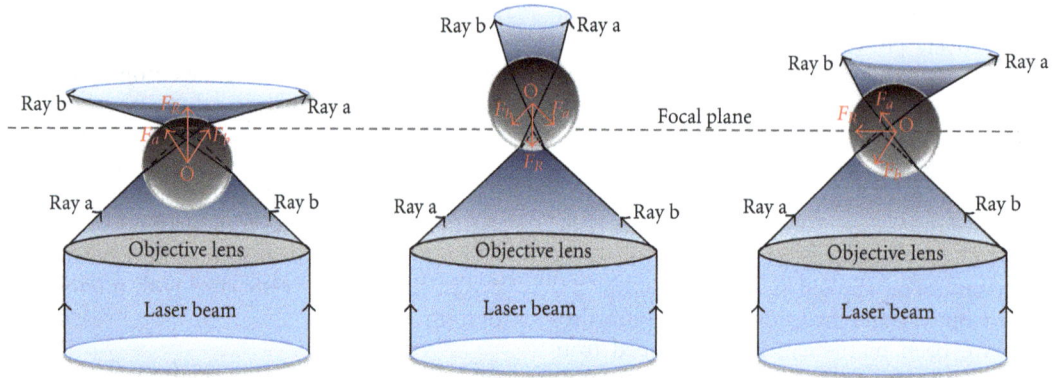

FIGURE 1: Qualitative illustration of the induced (F_a, F_b) and resultant (F_R) forces created from isolated refracted light rays (a, b) driving the centroid of the cell (O) back to the focal plane associated with the high aperture objective lens.

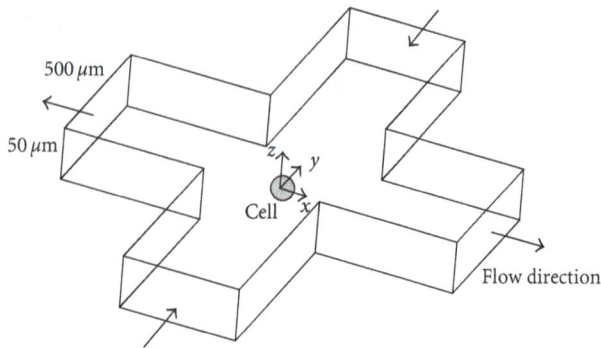

FIGURE 2: Schematic of the cross-junction microfluidic channel creating a fluid flow environment and the hydrodynamic multiaxial loading of an optically trapped biological cell [12, 14, 15].

Defining the stress tensor at the cellular surface, $r = a$, produces the volumetric fluidic stress state applied to the cell:

$$\mathbf{t}_r = \mathbf{T}\mathbf{e}_r = \sigma_{rr}\mathbf{e}_r + \tau_{r\phi}\mathbf{e}_\phi + \tau_{r\theta}\mathbf{e}_\theta, \qquad (6)$$

where the full-field stress state can then be defined in terms of distinct normal, σ_{rr}, and shear, $\tau_{r\theta}$ and $\tau_{r\theta}$, stress components, respectively, in polar spherical coordinates:

$$\sigma_{rr} = -p_\infty + \frac{5\mu U}{a}\sin^2\phi\cos 2\theta,$$

$$\tau_{r\phi} = \frac{5\mu U}{a}\sin\phi\cos\phi\cos 2\theta, \qquad (7)$$

$$\tau_{r\theta} = -\frac{5\mu U}{a}\sin\phi\sin 2\theta.$$

A three-dimensional presentation of the stresses was developed (MATLAB, MathWorks, Inc., Natick, MA) for demonstration of the site-specific nature of the stress distributions (Figure 4).

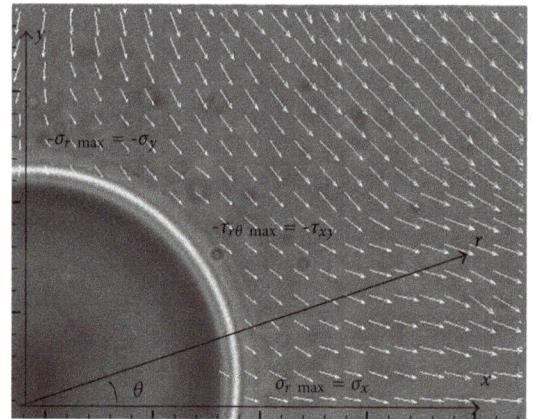

FIGURE 3: Quarter section image of an optically and hydrodynamically trapped glass bead with the surrounding fluid velocity vectors defined experimentally by the particle image velocimetry [16]. Analytically characterized flow velocity gradients are able to impart controlled normal and shear stresses onto trapped biological cells [12]. The relationship between the Cartesian and polar coordinates is indicated here within the x-y and r-θ planes, respectively.

2.3. Stress versus Strain Modeling. The six deviatoric stresses were combined as an effective stress value, σ_{eff} [19], as a means to model the three-dimensional stress state:

$$\sigma_{\text{eff}} = \frac{1}{\sqrt{2}}\left[\left(\sigma_x - \sigma_y\right)^2 + \left(\sigma_y - \sigma_z\right)^2 + \left(\sigma_z - \sigma_x\right)^2\right.$$
$$\left. + 6\left(\tau_{xy}^2 + \tau_{yz}^2 + \tau_{zx}^2\right)\right]^{1/2}. \qquad (8)$$

Here, the maximum polar coordinate derived stresses were converted to the Cartesian coordinate stresses such that the maximum normal stresses are located along the x-y-z axes and the maximum shear stresses are defined along the 45° orientation off-axis locations.

Volumetric strain, e, based on the strain invariants, I_i, was defined to encompass the full-field deformation response of the cell within the multiaxial fluidic loading environment and does not apply the small-displacement

(a)

(b)

(c)

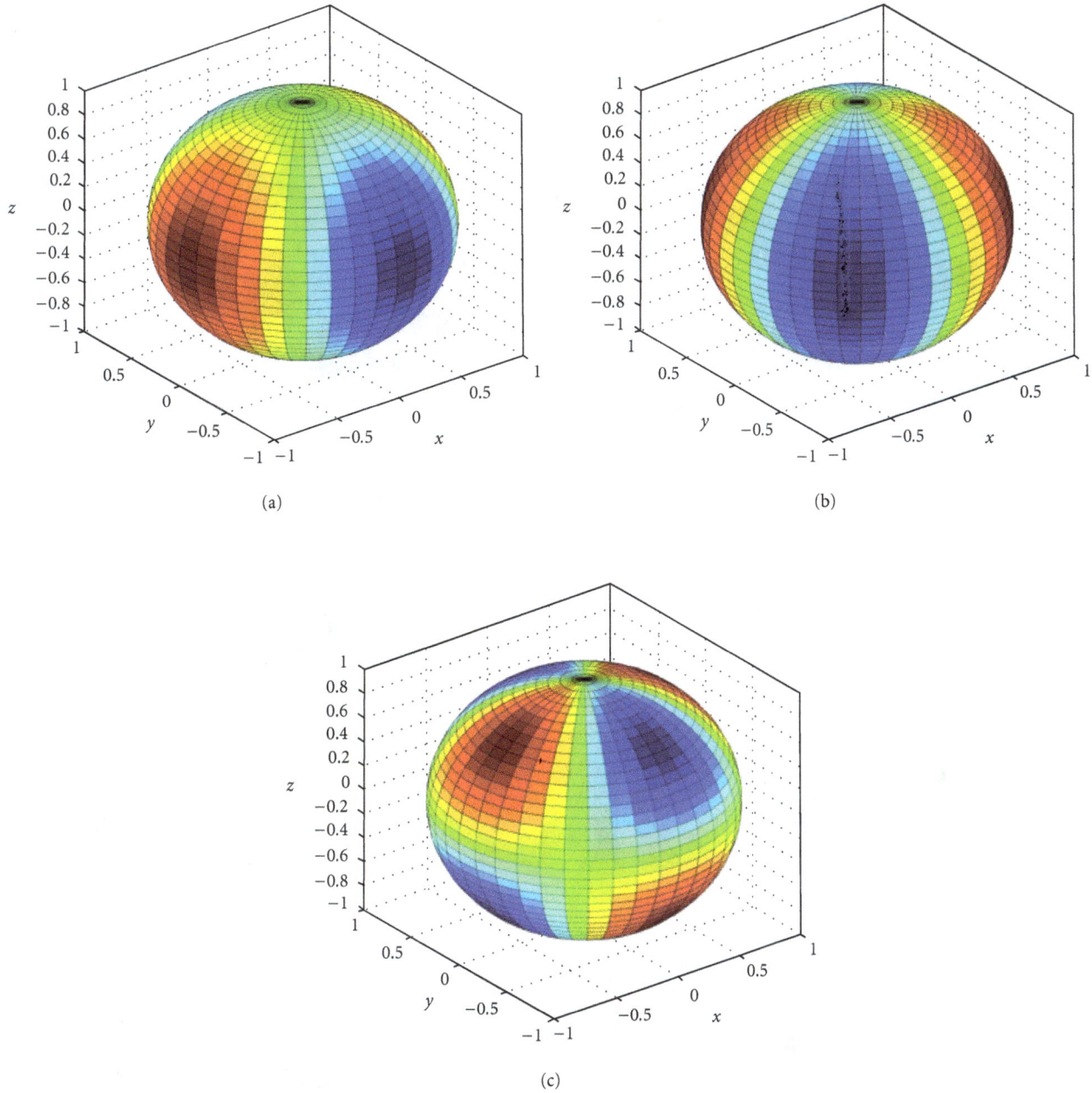

FIGURE 4: Three-dimensional representation of the applied (a) normal stresses, (b) x-y and r-θ shear stresses, and (c) y-z and r-ϕ shear stresses [14, 15] based on the original derived equations (7). Maximum positive stress levels are indicated in dark red while maximum negative stresses are in dark blue.

theory assumption. The Cartesian axes associated with the maximum normal strains were determined from a Mohr's circle analysis [12, 20]:

$$e = I_1 + I_2 + I_3 = \varepsilon_x + \varepsilon_y + \varepsilon_z + \varepsilon_x \varepsilon_y + \varepsilon_y \varepsilon_z + \varepsilon_z \varepsilon_x + \varepsilon_x \varepsilon_y \varepsilon_z. \tag{9}$$

The experimental approach described earlier is a planarwise measurement technique; thus the optical depth strain value, ε_z, can be determined through a transposition of the

planar loading scenario, again through a Mohr's circle analysis [21]:

$$\varepsilon_z = -\frac{\nu}{1-\nu}\left(\varepsilon_x + \varepsilon_y\right). \tag{10}$$

However, in this modeling presentation here, the full-field stress state and the corresponding strains are fully characterized.

The optohydrodynamic deviatoric stress state was applied to an isotropic, homogenous biological cell (20 μm in

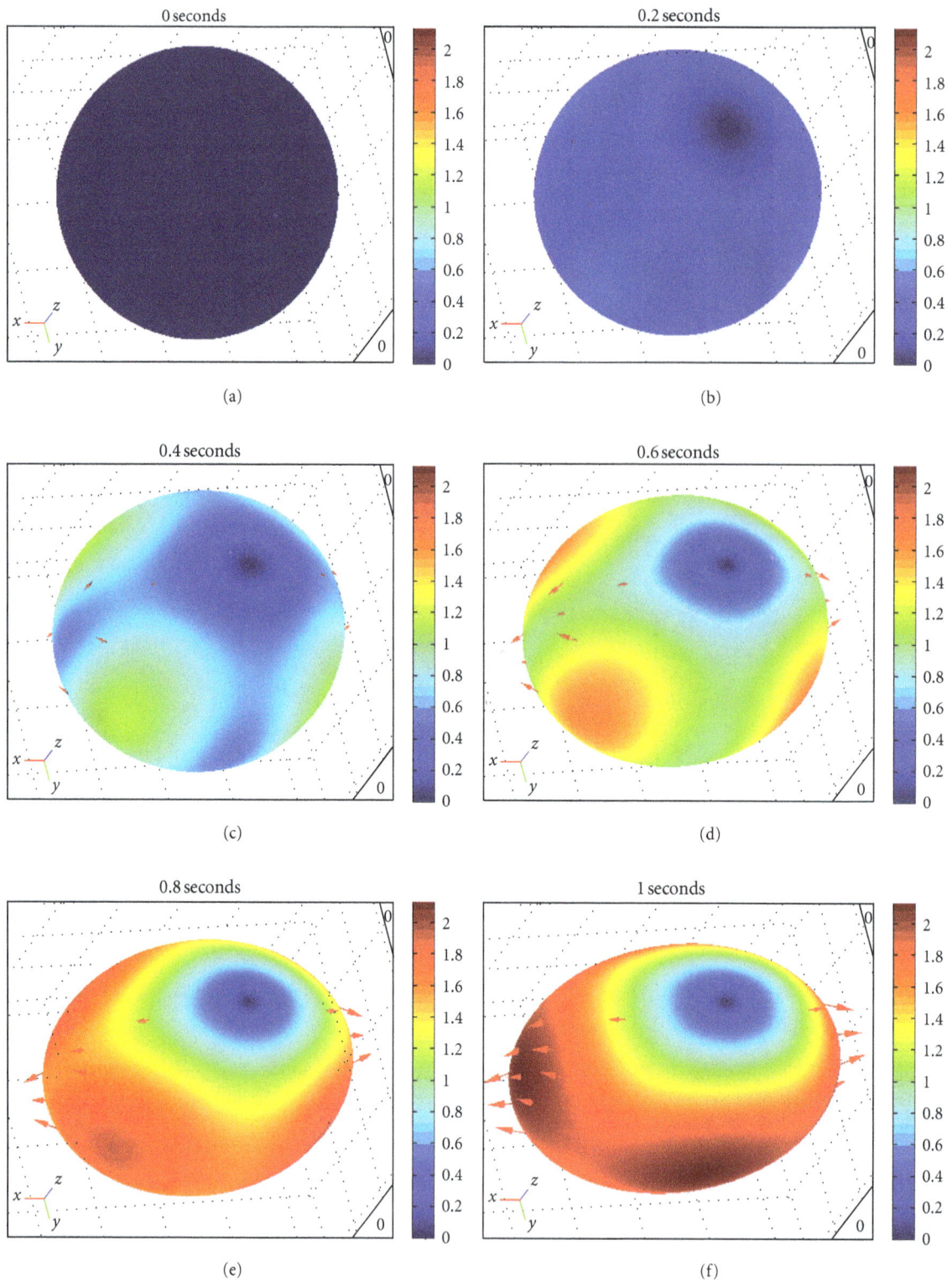

FIGURE 5: Step wise, time-sequenced volumetric cytoskeletal deformations due to applied microfluidic stresses created within an optohydrodynamic trap environment. Simulated deformations were modeled with a multiphysics computational software (COMSOL v4.0). Minimum deformation is indicated by dark blue while maximum deformation is in dark red. Units for the deformation scale are in micrometers (μm).

diameter) within a multiphysics computational environment (COMSOL v4.0, Palo Alto, CA, USA) in order to determine the individual principal strains and in turn the volumetric strains.

3. Results

In this work, we continue to examine the cellular biomechanics induced within a controlled micromechanical environment. Three-dimensional cellular strains were computationally modeled *in silico* with multiphysics software applying the analytically defined stress state (Figure 5). Volumetric stress and strain relationships indicated both the nonlinear response of the spherical cellular structure as well as the logarithmic increase in strain with subsequently softer cellular moduli (Figure 6). This biomechanical softening represents our hypothesized transition from healthy to diseased cellular states. This relationship is further explored when examining the direct relationship of strain with elastic moduli (Figure 7). The extreme strain response induced in diseased cells indicates their further vulnerability when passively or actively resisting applied stresses.

4. Discussion

We explored the full-field cellular strain response to a range of applied hydrodynamic stresses and cellular moduli, representing various degrees of functional loading and health/disease, respectively. The computational framework now allows us to develop more realistic cellular models with intracellular and membranous structures, distinct in spatial, elastic, and active transport characteristics [22].

The mechanical properties of a single cell are often formulated using either macroscopic or microscopic approaches [23]. Macroscopic approaches, as partially described in this work, homogenize every cellular component to produce an isotropic or anisotropic yet homogeneous continuum model so that the mechanical properties of cells can be formulated using temporal and spatially continuous partial differential equations [24]. Future efforts will incorporate a more microscopic approach, which regards the cell as a biocomposite material consisting of randomly or uniformly spaced anisotropic reinforcement cytoskeletal materials within an isotropic medium. The microscopic approach generally obtains the biomechanical response of a single cell by applying the mechanical boundary conditions at the individual fiber and matrix level, scaling "up" to the cellular level. Microscopic approaches often provide much more detail into the subtle interaction between the cytoskeletal fibers and matrix, which potentially leads to a more accurate model of the cellular behavior, such as characterizing the irreversible deformation of the cellular skeleton [25]. Unfortunately, refined microscopic models suffer from inhibitory computational and storage costs [26, 27].

When interpreting the potential geometric changes in cellular shape associated with mechanical loading, the cell may experience some interesting membrane transitions triggering unique biologic cues. Under suspension and hydrostatic loading, the cell's volume (V) can be defined as

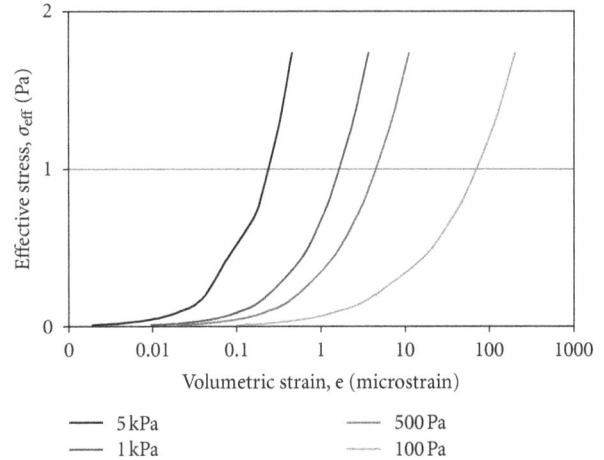

FIGURE 6: Graphical plot of the effective stresses versus volumetric strains (logarithmic scale) in modeled biological cells with varying linear elastic moduli representing the range in diseased to normal cells (low to high moduli, resp.). The modeled environment characterizes the fluid-induced stresses derived from flow velocities of $U = 10$ to $2{,}000 \, \mu m/s$ within the optohydrodynamic trap.

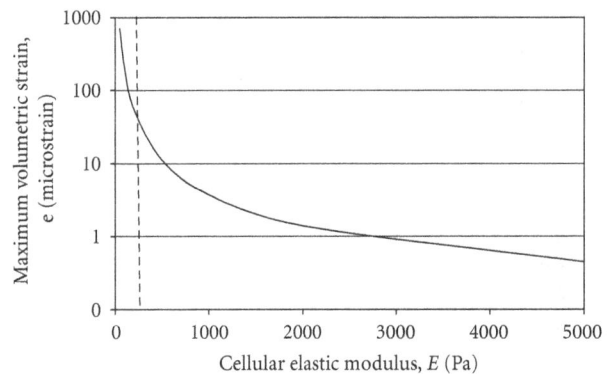

FIGURE 7: Graphical plot of the relationship between the cellular elastic moduli and the maximum volumetric strains created with a modeled flow velocity of $U = 2{,}000 \, \mu m/s$. The dashed line indicates the hypothetical distinction between the load response for soft and stiff cells, a biomechanical response that may exacerbate the degrading health and weakened mechanical state of diseased cells during functional loading.

a sphere, $V = 3/4\pi abc$, with radii $a = b = c$. As seen here during the hydrodynamic extensional loading state, the isotropic, homogenous cell model deformed into a scalene ellipsoid ($a \neq b \neq c$) with equal and opposite tensile and compressive strains in the horizontal plane. However, with the future inclusion of intracellular structures and organelles as well as nonlinear elastic properties assigned to the membrane and cytoskeleton, cellular models may also deform into either an oblate spheroid (formed when an ellipse is rotated about its minor axis, $a = b > c$) or a prolate spheroid (formed when an ellipse is rotated about its major axis, $a = b < c$). The resulting shape here replicated a scalene ellipsoid likely due to different maximum stresses applied in the three orthogonal axes. Ongoing experiments and modeling will continue to explore multiaxial single-cell biomechanics as well as the biologic triggers associated with geometric shape-shifting [28].

The described multiphysics computational framework will facilitate more realistic cytoskeletal model interpretations, whose intracellular structures can be distinctly defined, including the cellular membrane substructures, nucleus, and organelles. Future results will provide mathematical outcomes supporting the ongoing investigations in tissue and cellular engineering.

Acknowledgments

Support was provided by a National Institute of Biomedical Imaging and Bioengineering Academic Research Enhancement Award (Grant no. EB007077), a National Center on Minority Health and Health Disparities Exploratory Program Grant (Grant no. MD003350), creating the CSU Center for Allaying Health Disparities through Research and Education, and awards from the Collins Medical Trust.

References

[1] G. Y. H. Lee and C. T. Lim, "Biomechanics approaches to studying human diseases," *Trends in Biotechnology*, vol. 25, no. 3, pp. 111–118, 2007.

[2] Z. D. Wilson and S. S. Kohles, "Two-dimensional modeling of nanomechanical stresses-strains in healthy and diseased single-cells during microfluidic manipulation," *Journal of Nanotechnology in Engineering and Medicine*, vol. 1, no. 2, 2010.

[3] S. Grad, D. Eglin, M. Alini, and M. J. Stoddart, "Physical stimulation of chondrogenic cells in vitro: a review," *Clinical Orthopaedics and Related Research*, vol. 469, no. 10, pp. 2764–2772, 2011.

[4] M. G. Ehrlich, A. L. Armstrong, B. V. Treadwell, and H. J. Mankin, "The role of proteases in the pathogenesis of osteoarthritis," *Journal of Rheumatology*, vol. 14, pp. 30–32, 1987.

[5] S. S. Kohles, C. G. Wilson, and L. J. Bonassar, "A mechanical composite spheres analysis of engineered cartilage dynamics," *Journal of Biomechanical Engineering*, vol. 129, no. 4, pp. 473–480, 2007.

[6] R. D. González-Cruz, V. C. Fonseca, and E. M. Darling, "Cellular mechanical properties reflect the differentiation potential of adipose-derived mesenchymal stem cells," *Proceedings of the National Academy of Sciences of the United States of America*, vol. 109, no. 24, pp. E1523–E1529, 2012.

[7] S. E. Cross, Y. S. Jin, J. Rao, and J. K. Gimzewski, "Nanomechanical analysis of cells from cancer patients," *Nature Nanotechnology*, vol. 2, no. 12, pp. 780–783, 2007.

[8] N. Nève, J. K. Lingwood, J. Zimmerman, S. S. Kohles, and D. C. Tretheway, "The μPIVOT: an integrated particle image velocimeter and optical tweezers instrument for microenvironment investigations," *Measurement Science and Technology*, vol. 19, no. 9, Article ID 095403, 2008.

[9] L. M. Walker, Å. Holm, L. Cooling et al., "Mechanical manipulation of bone and cartilage cells with optical tweezers," *FEBS Letters*, vol. 459, no. 1, pp. 39–42, 1999.

[10] J. Guck, R. Ananthakrishnan, H. Mahmood, T. J. Moon, C. C. Cunningham, and J. Käs, "The optical stretcher: a novel laser tool to micromanipulate cells," *Biophysical Journal*, vol. 81, no. 2, pp. 767–784, 2001.

[11] G. Bao and S. Suresh, "Cell and molecular mechanics of biological materials," *Nature Materials*, vol. 2, no. 11, pp. 715–725, 2003.

[12] S. S. Kohles, N. Nève, J. D. Zimmerman, and D. C. Tretheway, "Mechanical stress analysis of microfluidic environments designed for isolated biological cell investigations," *Journal of Biomechanical Engineering*, vol. 131, no. 12, Article ID 121006, 10 pages, 2009.

[13] A. K. Saha, Y. Liang, and S. S. Kohles, "Biokinetic mechanisms linked with musculoskeletal health disparities: stochastic models applying Tikhonov's theorem to biomolecule Homeostasis," *Journal of Nanotechnology in Engineering and Medicine*, vol. 2, no. 2, Article ID 021004, 9 pages, 2011.

[14] S. S. Kohles, "Opto-hydrodynamic trapping for multiaxial single-cell biomechanics," in *Advances in Cell Mechanics, Advances in Materials and Mechanics (AMM)*, S. Li and B. Sun, Eds., pp. 237–255, Springer, New York, NY, USA, 2011.

[15] S. S. Kohles, Y. Liang, and A. K. Saha, "Volumetric stress-strain analysis of optohydrodynamically suspended biological cells," *Journal of Biomechanical Engineering*, vol. 133, no. 1, Article ID 011004, 6 pages, 2011.

[16] N. Neve, S. S. Kohles, S. R. Winn, and D. C. Tretheway, "Manipulation of suspended single cells by microfluidics and optical tweezers," *Cellular and Molecular Bioengineering*, vol. 3, no. 3, pp. 213–228, 2010.

[17] L. G. Leal, *Advanced Transport Phenomena: Fluid Mechanics and Convective Transport Processes*, Cambridge University Press, New York, NY, USA, 2007.

[18] T. C. Papanastasiou, G. C. Georgio, and A. N. Alexandrou, *Viscous Fluid Flow*, CRC Press, New York, NY, USA, 2000.

[19] R. D. Cook and W. C. Young, *Advanced Mechanics of Materials*, Macmillan, New York, NY, USA, 1985.

[20] H. L. Langhaar, *Energy Methods in Applied Mechanics*, John Wiley & Sons, New York, NY, USA, 1962.

[21] F. P. Beer, E. R. Johnston Jr., and J. T. DeWolf, *Mechanics of Materials*, McGraw-Hill, Boston, Mass, USA, 3rd edition, 2002.

[22] X. Zeng, S. Li, and S. S. Kohles, "Multiscale biomechanical modeling of stem cell-extracellular matrix interactions," in *Advances in Cell Mechanics, Advances in Materials and Mechanics (AMM)*, S. Li and B. Sun, Eds., pp. 27–54, Springer, New York, NY, USA, 2011.

[23] C. T. Lim, E. H. Zhou, and S. T. Quek, "Mechanical models for living cells—a review," *Journal of Biomechanics*, vol. 39, no. 2, pp. 195–216, 2006.

[24] G. Holzapfel, *Nonlinear Solid Mechanics*, Wiley, Chichester, UK, 2000.

[25] J. C. M. Lee and D. E. Discher, "Deformation-enhanced fluctuations in the red cell skeleton with theoretical relations to elasticity, connectivity, and spectrin unfolding," *Biophysical Journal*, vol. 81, no. 6, pp. 3178–3192, 2001.

[26] F. Guilak and V. C. Mow, "The mechanical environment of the chondrocyte: a biphasic finite element model of cell-matrix interactions in articular cartilage," *Journal of Biomechanics*, vol. 33, no. 12, pp. 1663–1673, 2000.

[27] W. Kim, D. C. Tretheway, and S. S. Kohles, "An inverse method for predicting tissue-level mechanics from cellular mechanical input," *Journal of Biomechanics*, vol. 42, no. 3, pp. 395–399, 2009.

[28] P. A. Janmey and C. A. McCulloch, "Cell mechanics: integrating cell responses to mechanical stimuli," *Annual Review of Biomedical Engineering*, vol. 9, pp. 1–34, 2007.

18

Redox Regulation of Calcium Signaling in Cancer Cells by Ascorbic Acid Involving the Mitochondrial Electron Transport Chain

Grigory G. Martinovich, Elena N. Golubeva, Irina V. Martinovich, and Sergey N. Cherenkevich

Department of Biophysics, Belarusian State University, Nezavisimosti Avenue 4, 220030 Minsk, Belarus

Correspondence should be addressed to Grigory G. Martinovich, martinovichgg@bsu.by

Academic Editor: Eaton Edward Lattman

Previously, we have reported that ascorbic acid regulates calcium signaling in human larynx carcinoma HEp-2 cells. To evaluate the precise mechanism of Ca^{2+} release by ascorbic acid, the effects of specific inhibitors of the electron transport chain components on mitochondrial reactive oxygen species (ROS) production and Ca^{2+} mobilization in HEp-2 cells were investigated. It was revealed that the mitochondrial complex III inhibitor (antimycin A) amplifies ascorbate-induced Ca^{2+} release from intracellular stores. The mitochondrial complex I inhibitor (rotenone) decreases Ca^{2+} release from intracellular stores in HEp-2 cells caused by ascorbic acid and antimycin A. In the presence of rotenone, antimycin A stimulates ROS production by mitochondria. Ascorbate-induced Ca^{2+} release in HEp-2 cells is shown to be unaffected by catalase. The results obtained suggest that Ca^{2+} release in HEp-2 cells caused by ascorbic acid is associated with induced mitochondrial ROS production. The data obtained are in line with the concept of redox signaling that explains oxidant action by compartmentalization of ROS production and oxidant targets.

1. Introduction

Redox processes involving transfer of electrons or hydrogen atoms are central processes of energy conversion in respiratory organisms. Recently, it has become apparent that numerous functionally significant biological processes proceed with participation of physical mechanisms ensuring intermolecular electron transfer. Electron transfer between low-molecular weight components of cytosol and intracellular proteins leads to the change of a functional state of both cellular proteins and cells as a whole [1, 2]. All biological systems contain redox elements that play an important role in transcriptional regulation, cell proliferation, apoptosis, hormonal signaling, and other fundamental cell functions [3]. Organization and coordination of the redox activity of these elements occur through redox circuits and depend on the intracellular concentration of redox-active molecules [4, 5]. Redox active molecules may cause both regulatory and

toxic effects depending on the value of cellular redox state parameters [5, 6]. However, little is known about mechanisms of regulation, structural organization, and interaction between electron-transport participants inside the cell and other signal and regulatory systems.

Recently new effects of such a redox-active molecule as ascorbic acid have been found. Beside numerous regulatory properties (hydroxylation of collagen, biosynthesis of carnitine and noradrenaline, etc.), selective cytotoxicity of high concentrations of ascorbic acid towards cancer cells has been described. Ascorbic acid in concentrations of 1–10 mM was shown to induce the death of prostate cancer cells, stomach cancer cells, and acute myeloid leukemia cells [7, 8]. In experiments *in vitro*, ascorbate cytotoxicity ($EC_{50} < 4$ mM) was observed in many types of cancer cell lines, whereas normal cells were resistant [9, 10]. Ascorbic acid treatment in high pharmacological concentrations significantly impeded tumor progression *in vivo* without toxicity to normal tissues [11, 12]. Thus, ascorbic acid at high doses possesses anticancer properties, but mechanisms of its selective cytotoxicity and targets of its action are still obscure.

It was shown that in the presence of transition metal ions and ascorbic acid, H_2O_2 is formed [13]. These data suggest that ascorbate cytotoxicity may be due to its ability to generate H_2O_2 [7, 12]. But this hypothesis does not appear to be compatible with facts. Recent studies have indicated that the intravenous injection of ascorbic acid in high (up to 8 mM) concentrations was not accompanied by the formation of H_2O_2 in blood [14]. It was also shown that even in the presence of transition metal ions and H_2O_2, ascorbate acted as an antioxidant that prevented lipid peroxidation in human plasma *in vitro* [15]. Previously, we have found that ascorbic acid regulates calcium signaling in human larynx carcinoma cells [16]. Ascorbate at concentrations in the range 3–10 mM activated cytosol pH value decrease and Ca^{2+} release from thapsigargin-sensitive intracellular Ca^{2+} stores. Although the ability of ascorbic acid to induce Ca^{2+} mobilization is shown, the precise mechanisms involved in Ca^{2+} release are not known. We proposed that mitochondria can be involved in ascorbate-induced calcium signaling. The aim of this study was to explore the participation of mitochondrial enzymes in regulation of calcium signaling in human larynx carcinoma HEp-2 cells by ascorbic acid.

2. Methods and Materials

2.1. Cell Culture and Reagents. Ascorbic acid was obtained from Himhrom Ltd. (Minsk, Belarus). Dulbecco's modified Eagle's medium (DMEM), fura-2-acetoxymethyl (AM) ester, 2′,7′-dichlorodihydrofluorescein diacetate (H_2DCF-DA), rotenone, antimycin A, catalase, and HEPES were purchased from Sigma-Aldrich (St. Louis, MO, USA). Human larynx carcinoma HEp-2 cells were purchased from the Republican Research and Practical Center for Epidemiology and Microbiology (Minsk, Belarus). The cells were cultured at 37°C under a humidified atmosphere with 5% CO_2 in DMEM supplemented with 10% fetal calf serum, 2 mM glutamine, and 80 mg/mL gentamicin. Rotenone (50–150 μM), antimycin A (5–40 μM), ascorbate (3–10 mM), and catalase (500 U/mL) were used.

2.2. Fluorescent Spectrofluorimetry. Measurements of the free calcium ions' cytosol concentration ($[Ca^{2+}]_{cyt}$) were performed as previously described [17]. HEp-2 cells were loaded with 2.5 μM fura-2 AM, washed, and mounted under continuous stirring in the chamber of the spectrofluorimeter (LSF 1211A, Minsk, Belarus). The standard recording medium (KRH) contained 131 mM NaCl, 5 mM KCl, 1.3 mM $CaCl_2$, 1.3 mM $MgSO_4$, 6 mM glucose, and 20 mM HEPES, pH 7.4 (NaOH).

Intracellular ROS generation was recorded using H_2DCF-DA, which is a nonpolar compound that is converted into a nonfluorescent polar derivative (H_2DCF) by cellular esterases after incorporation into cells. Membrane-impermeable H_2DCF is rapidly oxidized to highly fluorescent 2′,7′-dichlorofluorescein (DCF) in the

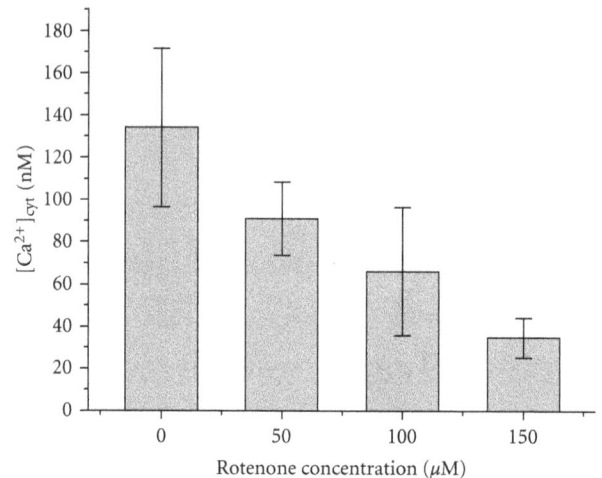

FIGURE 1: Influence of rotenone on ascorbate-induced increase of $[Ca^{2+}]_{cyt}$ in HEp-2 cells. Concentration of ascorbic acid in KRH— 5 mM. Number of cells in 1 mL—2.5×10^6.

presence of intracellular ROS [18]. After loading with 10 μM H_2DCF-DA for 30 min, HEp-2 cells were rinsed two times with KRH and DCF fluorescence intensity was measured using 488 nm excitation/530 nm emission settings. All experiments were carried out at 37°C.

2.3. Statistics. The data were expressed as means ± standard error of the mean (SEM). Statistical significances between means were assayed using Student's *t*-test. The values were taken as significantly different when $P < 0.05$.

3. Results and Discussion

In a previous study, ascorbic acid was found to cause a transient increase in $[Ca^{2+}]_{cyt}$ in human larynx carcinoma HEp-2 cells accompanied by depletion of intracellular thapsigargin-sensitive calcium stores [16]. Thapsigargin is known to induce Ca^{2+} release, from mitochondria and inositol-1,4,5-triphosphate-responsive Ca^{2+} stores. To examine involvement of mitochondria in ascorbate-induced Ca^{2+} release we used rotenone, mitochondrial complex I inhibitor. It was shown that treatment of cells with rotenone led to the decrease of ascorbate-induced Ca^{2+} release in HEp-2 cells (Figure 1), indicating that participation of mitochondrial enzymes in ascorbate induces Ca^{2+} release. There is no evidence that any specific receptors for ascorbate exist in mitochondria. To explain participation of mitochondria in ascorbate-induced Ca^{2+} release, we proposed a mechanism according to which an increase in ascorbate concentration could intensify ROS production by mitochondrial enzymes.

It is generally accepted that superoxide anion ($O_2^{\bullet-}$) is the primary free radical in mitochondria, which is formed as the result of electron "leak" from the electron transport chain elements to oxygen [19]. Ubisemiquinones generated in the respiratory chain were identified as possible donors of electrons for oxygen [20]. Here, it is reasonable to mention that ascorbate can reduce c-type cytochromes and b-type

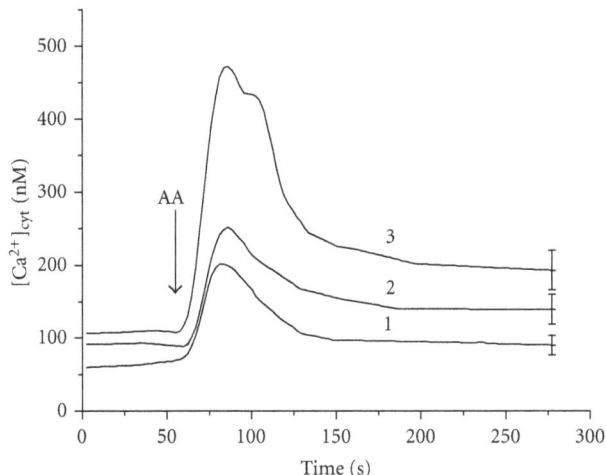

FIGURE 2: Influence of antimycin A on the ascorbate-induced Ca^{2+} release from HEp-2 cells mitochondria. Concentrations in KRH: ascorbic acid—5 mM; antimycin A (μM): 1—0, 2—10, 3—15. Number of cells in 1 mL—3×10^6. The arrow shows the instant of ascorbate addition.

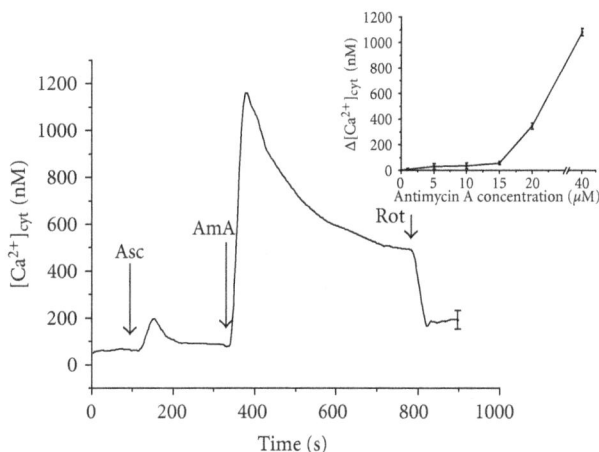

FIGURE 3: Influence of ascorbic acid and rotenone on the Ca^{2+} release from HEp-2 cells mitochondria under the antimycin A treatment. Concentrations in KRH: ascorbic acid—5 mM, antimycin A—10 μM, rotenone—50 μM. Number of cells in 1 mL—3×10^6. Arrows show the instants of antimycin A (AmA), ascorbic acid (Asc), and rotenone (Rot) addition. The inset shows dependence of increase in [Ca^{2+}]$_{cyt}$ on the antimycin A concentration.

cytochromes [21, 22]. The evidence, taken together, suggests that electron transfer from ascorbic acid to cytochrome c$_1$ or cytochrome c should decrease electron flow from ubiquinol to Rieske iron-sulfur cluster resulting in a rise of ubisemiquinone concentration and superoxide production. Dismutation of O$_2^{\bullet-}$ by the mitochondrial matrix Mn-superoxide dismutase leads to the formation of H$_2$O$_2$ [20]. H$_2$O$_2$, in turn, can oxidize thiol groups of targets and regulate Ca^{2+} release [23, 24]. To exclude possible generation of H$_2$O$_2$ by ascorbate in the extracellular solution catalase was used. Ascorbate-induced Ca^{2+} release from mitochondria was shown to be unaffected by catalase in an concentration of 500 U/mL (not shown). Thus, the important point in our model is that ROS production by ascorbic acid proceeds nearby specific redox sensors.

According to the proposed model, the increase of ROS generation in mitochondria is thought to induce Ca^{2+} release. The impact of this mechanism was assessed by using mitochondrial complex inhibitors antimycin A and rotenone. The increase in [Ca^{2+}]$_{cyt}$ in HEp-2 cells was detected after the application of mitochondrial complex III inhibitor antimycin A in concentrations above 20 μM (Figure 3, inset). On the other hand, the increase in the amplitude and duration of ascorbate-induced Ca^{2+} release was observed even at lower concentrations of antimycin A (10 μM) used in the study (Figure 2). Moreover, after the addition of ascorbate to HEp-2 cell suspension, antimycin A in this concentration induced a significant increase in [Ca^{2+}]$_{cyt}$. Subsequent addition of mitochondrial complex I inhibitor rotenone resulted in the decrease of [Ca^{2+}]$_{cyt}$ in cancer HEp-2 cells (Figure 3). These findings support our suggestion that ascorbate-induced Ca^{2+} release from HEp-2 cells mitochondria proceeds with the participation of ROS produced in the electron transport chain.

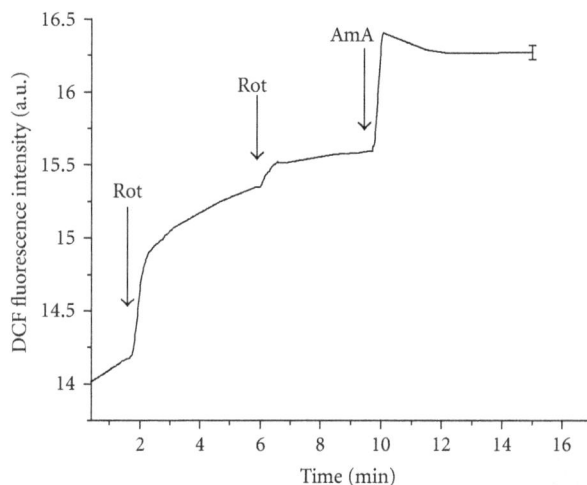

FIGURE 4: DCF fluorescence intensity in HEp-2 cells. Concentrations in KRH: rotenone—50 μM, antimycin A—10 μM. Number of cells in 1 mL—3×10^6. Arrows show the instants of antimycin A (AmA) and rotenone (Rot) addition.

The major sites of superoxide formation within the mitochondrial respiratory chain are linked to NADH: ubiquinone oxidoreductase (complex I) and ubiquinol: cytochrome c oxidoreductase (complex III) [19]. In our experiments both rotenone and antimycin A were ascertained to enhance ROS production in HEp-2 cells. Superoxide production by complex I is supposed to occur during the reverse electron transport (RET) from ubiquinol to NAD$^+$ and during the forward electron transport (FET) from NADH to ubiquinone, the former being faster than the latter. Recent observations have led to the conclusion that rotenone enhances ROS formation during the FET and inhibits it

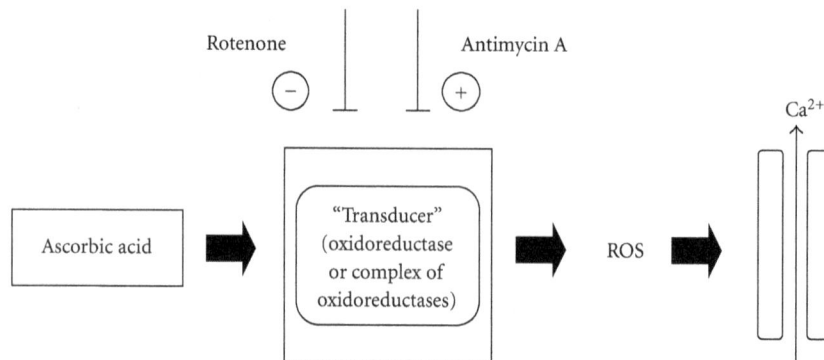

FIGURE 5: A schematic representation of the mechanism of ascorbate-induced redox regulation of Ca^{2+} release. Signal transduction form redox-active molecule to target occurs with additional participants—oxidoreductases (in our case—mitochondrial enzymes). Mitochondria-derived ROS induce Ca^{2+} release. Signal decoding in cells depends on the transducer activity of additional participant. Antimycin A may increase the ROS production and Ca^{2+} response in cells, while rotenone decreases it.

during the RET [19]. Moreover, the RET-induced ROS production is regulated by $\Delta\psi$. Generation of $O_2^{\cdot-}$ and H_2O_2 by the electron transport chain is amplified by an increase in $\Delta\psi$ [25]. The inhibition of electron transport by antimycin A in complex III resulted in ROS formation even after rotenone treatment (Figure 4). These results indicate that ROS formation by mitochondria in HEp-2 cells under physiological conditions is not connected with the RET in complex I. The RET is observed under conditions of high $\Delta\psi$ [19]. Previously we have shown that ascorbate at high concentrations induces the decrease in intracellular pH value that can lead to the increase in $\Delta\psi$ [16]. In such conditions, rotenone inhibits ROS production in mitochondria. Therefore, rotenone decreased ascorbate-induced Ca^{2+} release (Figure 1) and antimycin A-induced Ca^{2+} release (Figure 3). The results obtained indicate that antimycin A activates ROS production by complex III leading to the increase in ascorbate-induced calcium response of HEp-2 cells. Rotenone, in turn, decreases the rise in $[Ca^{2+}]_{cyt}$ caused by the action of ascorbate and antimycin A as it blocks electron transfer in complex I of the mitochondrial respiratory chain and inhibits the H_2O_2 production in complex III. Decrease in the intracellular pH value enhances oxidative processes in cells [26]; therefore after the ascorbate treatment, the increase in $[Ca^{2+}]_{cyt}$ caused by antimycin A occurred even at low concentrations of the inhibitor (Figure 3). Taken together, our results suggest that ascorbic acid can regulate Ca^{2+} release in HEp-2 cells by locally induced ROS production.

The data of this preliminary study point out that complex III may be a possible player of ROS production in mitochondria under ascorbate treatment. It is important to emphasize that multiple sources of ROS generation have been identified in mitochondria [27, 28]. Recently, proapoptotic protein p66[Shc] has been shown to localize in the mitochondrial intermembrane space and redox cycle with cytochrome c to produce H_2O_2 that induces permeability transition [29]. In conditions of rise of ascorbic acid concentration, activation of p66[Shc] pathways may in part explain enhancement of ROS generation. Further investigation will improve our

understanding of mechanisms by which mitochondria are stimulated to produce oxidants and what the proximate targets of such oxidants are.

On the other hand, the results obtained indicate a new possible way of redox regulation of Ca^{2+} signaling (Figure 5). Calcium signaling occurs when the cell is stimulated to release Ca^{2+} from intracellular stores. According to our data, the antioxidant ascorbic acid can regulate Ca^{2+} release by mitochondria-derived ROS. Ascorbate-induced changes of intracellular redox state can be transduced in increase of ROS concentration by components of electron transport chain. Thus in the proposed mechanism of regulation in addition to redox-active molecules and their targets, an additional participant of signal transduction—oxidoreductases (components of electron transport chain)—appears. The important implication of mitochondrial enzymes in redox regulation is that in this case signal decoding in cells depends also on the "transducer" activity in different types of cells. Thus the redox modulation of Ca^{2+} sparks occurs in species- and tissue-specific fashion. Some agents (antimycin A) may increase the ROS production and Ca^{2+} response in cells, while others (rotenone) decrease it (see Figure 5). Local ROS production modifies the amplitude of Ca^{2+} sparks that are sensed and decoded into defined cell actions by a broad variety of cellular effectors.

In general, results of our research are in agreement with the concept of redox signaling that explains oxidant action by compartmentalization of ROS production and oxidant targets [30]. Our observations lead to the conclusion that the key condition of ascorbate cytotoxicity is ROS generation by mitochondria. In these conditions, Ca^{2+} release proceeds as a result of local effects of mitochondrial oxidants.

Earlier was supposed that selectivity of ascorbate effect on cancer cells was mediated by an altered acid-base balance in tumor tissues [16]. Changes in the activity of electron transport chain components observed in many cancer cells including carcinoma cells [31] may also promote cytotoxic action of ascorbate towards cancer cells. Ascorbic acid, capable of regulating both acid-base and redox states of cancer cells, may serve as a prototype for the development

Redox Regulation of Calcium Signaling in Cancer Cells by Ascorbic Acid Involving the Mitochondrial Electron
Transport Chain

161

of new anticancer agents with the mechanism of binary regulatory action. Therefore, a detailed understanding of the mechanism of ascorbate-induced ROS production could aid in the development of new anticancer strategies.

Abbreviations

$\Delta \psi$:   Transmembrane electric potential difference
DCF: 2′,7′-Dichlorofluorescein
H_2DCF: 2′,7′-Dichlorodihydrofluorescein
H_2DCF-DA: 2′,7′-Dichlorodihydrofluorescein diacetate
FET: Forward electron transport
KRH: Krebs-Ringer-HEPES buffer
RET: Reverse electron transport
ROS: Reactive oxygen species.

Acknowledgment

This work was partially supported by a funding from the Belarusian Republican Foundation for Fundamental Research (B11-027).

References

[1] F. Rusnak and T. Reiter, "Sensing electrons: protein phosphatase redox regulation," *Trends in Biochemical Sciences*, vol. 25, no. 11, pp. 527–529, 2000.

[2] G. M. Ullmann and E. W. Knapp, "Electrostatic models for computing protonation and redox equilibria in proteins," *European Biophysics Journal*, vol. 28, no. 7, pp. 533–551, 1999.

[3] D. P. Jones, "Redox sensing: orthogonal control in cell cycle and apoptosis signalling," *Journal of Internal Medicine*, vol. 268, no. 5, pp. 432–448, 2010.

[4] D. P. Jones, "Radical-free biology of oxidative stress," *American Journal of Physiology*, vol. 295, no. 4, pp. C849–C868, 2008.

[5] G. G. Martinovich, I. V. Martinovich, and S. N. Cherenkevich, "Redox regulation of cellular processes: a biophysical model and experiment," *Biophysics*, vol. 56, no. 3, pp. 444–451, 2011.

[6] G. G. Martinovich, I. V. Martinovich, S. N. Cherenkevich, and H. Sauer, "Redox buffer capacity of the cell: theoretical and experimental approach," *Cell Biochemistry and Biophysics*, vol. 58, no. 2, pp. 75–83, 2010.

[7] S. Park, S. S. Han, C. H. Park et al., "L-Ascorbic acid induces apoptosis in acute myeloid leukemia cells via hydrogen peroxide-mediated mechanisms," *International Journal of Biochemistry and Cell Biology*, vol. 36, no. 11, pp. 2180–2195, 2004.

[8] Y. X. Sun, Q. S. Zheng, G. Li, D. A. Guo, and Z. R. Wang, "Mechanism of ascorbic acid-induced reversion against malignant phenotype in human gastric cancer cells," *Biomedical and Environmental Sciences*, vol. 19, no. 5, pp. 385–391, 2006.

[9] Q. Chen, M. G. Espey, M. C. Krishna et al., "Pharamacologic ascorbic acid concentrations selectively kill cancer cells: action as a pro-drug to deliver hydrogen peroxide to tissue," *Proceedings of the National Academy of Sciences of the United States of America*, vol. 102, no. 38, pp. 13604–13609, 2005.

[10] Q. Chen, M. G. Espey, A. Y. Sun et al., "Pharmacologic doses of ascorbate act as a prooxidant and decrease growth of aggressive tumor xenografts in mice," *Proceedings of the National Academy of Sciences of the United States of America*, vol. 105, no. 32, pp. 11105–11109, 2008.

[11] S. Ohno, Y. Ohno, N. Suzuki, G. I. Soma, and M. Inoue, "High-dose vitamin C (ascorbic acid) therapy in the treatment of patients with advanced cancer," *Anticancer Research*, vol. 29, no. 3, pp. 809–815, 2009.

[12] H. B. Pollard, M. A. Levine, O. Eidelman, and M. Pollard, "Pharmacological ascorbic acid suppresses syngeneic tumor growth and metastases in hormone-refractory prostate cancer," *In Vivo*, vol. 24, no. 3, pp. 249–255, 2010.

[13] G. R. Buettner and B. A. Jurkiewicz, "Catalytic metals, ascorbate and free radicals: combinations to avoid," *Radiation Research*, vol. 145, no. 5, pp. 532–541, 1996.

[14] Q. Chen, M. G. Espey, A. Y. Sun et al., "Ascorbate in pharmacologic concentrations selectively generates ascorbate radical and hydrogen peroxide in extracellular fluid in vivo," *Proceedings of the National Academy of Sciences of the United States of America*, vol. 104, no. 21, pp. 8749–8754, 2007.

[15] J. Suh, B. Z. Zhu, and B. Frei, "Ascorbate does not act as a pro-oxidant towards lipids and proteins in human plasma exposed to redox-active transition metal ions and hydrogen peroxide," *Free Radical Biology and Medicine*, vol. 34, no. 10, pp. 1306–1314, 2003.

[16] G. G. Martinovich, I. V. Martinovich, and S. N. Cherenkevich, "Effects of ascorbic acid on calcium signaling in tumor cells," *Bulletin of Experimental Biology and Medicine*, vol. 147, no. 4, pp. 469–472, 2009.

[17] R. A. Hirst, C. Harrison, K. Hirota, and D. G. Lambert, "Measurement of $[Ca^{2+}]$i in whole cell suspensions using fura-2," *Methods in Molecular Biology*, vol. 312, pp. 37–45, 2006.

[18] X. Chen, Z. Zhong, Z. Xu, L. Chen, and Y. Wang, "2′,7′-Dichlorodihydrofluorescein as a fluorescent probe for reactive oxygen species measurement: forty years of application and controversy," *Free Radical Research*, vol. 44, no. 6, pp. 587–604, 2010.

[19] G. Lenaz and M. L. Genova, "Structure and organization of mitochondrial respiratory complexes: a new understanding of an old subject," *Antioxidants and Redox Signaling*, vol. 12, no. 8, pp. 961–1008, 2010.

[20] S. Raha and B. H. Robinson, "Mitochondria, oxygen free radicals, disease and ageing," *Trends in Biochemical Sciences*, vol. 25, no. 10, pp. 502–508, 2000.

[21] A. I. Al-Ayash and M. T. Wilson, "The mechanism of reduction of single-site redox proteins by ascorbic acid," *Biochemical Journal*, vol. 177, no. 2, pp. 641–648, 1979.

[22] D. Njus, M. Wigle, P. M. Kelley, B. H. Kipp, and H. B. Schlegel, "Mechanism of ascorbic acid oxidation by cytochrome b561," *Biochemistry*, vol. 40, no. 39, pp. 11905–11911, 2001.

[23] A. P. Halestrap, K. Y. Woodfield, and C. P. Connern, "Oxidative stress, thiol reagents, and membrane potential modulate the mitochondrial permeability transition by affecting nucleotide binding to the adenine nucleotide translocase," *The Journal of Biological Chemistry*, vol. 272, no. 6, pp. 3346–3354, 1997.

[24] I. N. Pessah, K. H. Kim, and W. Feng, "Redox sensing properties of the ryanodine receptor complex," *Frontiers in Bioscience*, vol. 7, pp. a72–a79, 2002.

[25] A. A. Starkov and G. Fiskum, "Regulation of brain mitochondrial H_2O_2 production by membrane potential and NAD(P)H redox state," *Journal of Neurochemistry*, vol. 86, no. 5, pp. 1101–1107, 2003.

[26] G. G. Martinovich, I. V. Martinovich, E. N. Golubeva, and S. N. Cherenkevich, "Role of hydrogen ions in the regulation of the redox state of erythrocytes," *Biofizika*, vol. 54, no. 5, pp. 846–851, 2009.

[27] A. Y. Andreyev, Y. E. Kushnareva, and A. A. Starkov, "Mito-chondrial metabolism of reactive oxygen species," *Biochemistry*, vol. 70, no. 2, pp. 200–214, 2005.

[28] A. A. Starkov, "The role of mitochondria in reactive oxygen species metabolism and signaling," *Annals of the New York Academy of Sciences*, vol. 1147, pp. 37–52, 2008.

[29] M. Giorgio, E. Migliaccio, F. Orsini et al., "Electron transfer between cytochrome c and p66Shc generates reactive oxygen species that trigger mitochondrial apoptosis," *Cell*, vol. 122, no. 2, pp. 221–233, 2005.

[30] L. S. Terada, "Specificity in reactive oxidant signaling: think globally, act locally," *Journal of Cell Biology*, vol. 174, no. 5, pp. 615–623, 2006.

[31] H. Simonnet, N. Alazard, K. Pfeiffer et al., "Low mitochondrial respiratory chain content correlates with tumor aggressiveness in renal cell carcinoma," *Carcinogenesis*, vol. 23, no. 5, pp. 759–768, 2002.

19

Examinations of tRNA Range of Motion Using Simulations of Cryo-EM Microscopy and X-Ray Data

Thomas R. Caulfield,[1] Batsal Devkota,[2] and Geoffrey C. Rollins[1]

[1] *School of Chemistry & Biochemistry, Georgia Institute of Technology, 901 Atlantic Avenue, Atlanta, GA 30332-0230, USA*
[2] *School of Biology, Georgia Institute of Technology, 901 Atlantic Avenue, Atlanta, GA 30332-0230, USA*

Correspondence should be addressed to Thomas R. Caulfield, thomas.caulfield@chemistry.gatech.edu

Academic Editor: Eaton Edward Lattman

We examined tRNA flexibility using a combination of steered and unbiased molecular dynamics simulations. Using Maxwell's demon algorithm, molecular dynamics was used to steer X-ray structure data toward that from an alternative state obtained from cryogenic-electron microscopy density maps. Thus, we were able to fit X-ray structures of tRNA onto cryogenic-electron microscopy density maps for hybrid states of tRNA. Additionally, we employed both Maxwell's demon molecular dynamics simulations and unbiased simulation methods to identify possible ribosome-tRNA contact areas where the ribosome may discriminate tRNAs during translation. Herein, we collected >500 ns of simulation data to assess the global range of motion for tRNAs. Biased simulations can be used to steer between known conformational stop points, while unbiased simulations allow for a general testing of conformational space previously unexplored. The unbiased molecular dynamics data describes the global conformational changes of tRNA on a sub-microsecond time scale for comparison with steered data. Additionally, the unbiased molecular dynamics data was used to identify putative contacts between tRNA and the ribosome during the accommodation step of translation. We found that the primary contact regions were H71 and H92 of the 50S subunit and ribosomal proteins L14 and L16.

1. Introduction

tRNA is a key component for protein synthesis in the cell. tRNA delivers amino acids to the ribosome, where they are incorporated to the growing nascent polypeptide chains. The characteristic L-shaped tertiary structure of tRNA is intimately related to its function, and it has intrigued investigators for decades [1, 2].

With the first high-resolution crystal structure for tRNA [1, 2], it was suggested that the molecule may possess a flexible hinge between the D-stem and the anticodon stem (Figure 1). Figure S6 (see Figure in Supplementary Material available online at doi: 10.1155/2011/219515) shows a diagrammed version of tRNA for illustration. The interarm consists of the hinge formed between the acceptor stem and the anticodon stem (Supplementary Figure S6). Figure 1(a) schematic shows a two-dimensional layout for the RNA nucleotides that account for the acceptor stem, anticodon

stem, D-stem, and T-stem. Early light scattering experiments were interpreted in terms of bending motions between the two arms of the L-shaped tRNA that were thought to facilitate functional flexibility [3]. tRNA binds to aminoacyl tRNA synthetases, elongation factors, as well as different sites on the ribosome.

The first concrete experimental evidence of tRNA flexibility came with determination of the tRNA^{Asp} crystal structure for which an interarm bend angle of approximately 89° was observed, larger than that observed in tRNA^{Phe} [1, 4–9]. Recently, a larger conformational difference was demonstrated using cryo-electron microscopy (cryo-EM) density maps for the kirromycin-stalled state of the ribosome, which has two tRNAs bound at the A and P sites of the small subunit [10–15]. In the cryo-EM-derived model, the incoming tRNA is shown bound in a preaccommodation state called "A/T" (Figures 1(b) and 1(e)). The cryo-EM-derived model shows a notable kink between the D-stem

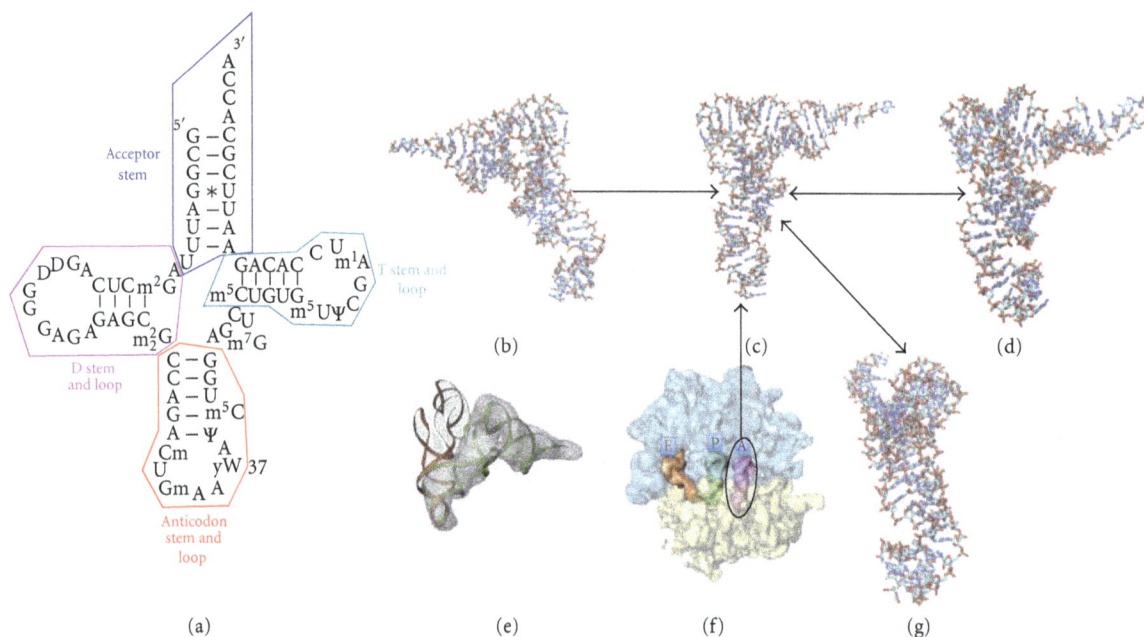

FIGURE 1: tRNA Conformations. Both kinked structures for tRNA^Phe (PDB code: 1EHZ) and the hybrid state for cryo-EM densities of tRNA are shown. The different orientations of tRNA depict the unique differences caused from kinking or twisting (A/T and P/E states, resp.). The arrows depict the transition from A/T state to A-site (A/A state) to P-site and P/E hybrid state. (a) Secondary structure for Phe-tRNA^Phe (1EHZ) [33]. (b) Structure of kinked A/T-tRNA, as it enters the ribosome (pre-accommodation) [13]. (c) A-site tRNA after accommodation occurs (post accommodation). (d) Hybrid P/E-tRNA following peptidyl transfer. (e) Comparison of A-site tRNA and kinked A/T-tRNA from modeling with cryo-EM density. (f) Position of tRNA in A-site indicated in cryo-EM density. (g) Conformation sampled from MD demonstrating an experimentally unidentified twisted conformation for tRNA^Phe.

and the anticodon stem, where it was postulated that this might indicate that a loaded spring-like action is possible that would drive tRNA through accommodation to the A/A state [10, 13, 15].

tRNA flexibility has been a motivating topic for early molecular modeling studies on tRNA [16–18]. Prior modeling studies of tRNA have examined B-factor and H-bonding correlation with X-ray data [17–19]. More recently, large-scale tRNA motions through the ribosome were studied using biased molecular dynamics simulations [20, 21]. Another recent study focused on the structure and dynamics of a fragment of tRNA^Ala [22].

MD simulations have been successfully characterized as structural dynamics, base pairing, specific hydration, cation binding, and electrostatic potential distributions, including mutagenesis [20, 23–30]. Sponer et al. studied the structural dynamics, elasticity, and deformability of helix 44 (H44) from the small subunit 16S rRNA, finding that H44 has an intrinsic ability to deform on the nanosecond timescale [31]. They also observed "bulge-induced switches" in nucleic bases 1466C, 1431C, and 1467C that formed at a cost of approximately 5–7 kcal/mol [28, 32].

In the present study, we report global range of motion for tRNA derived from molecular dynamics simulations. The simulation method was a novel application first reporting the conjunction of X-ray data driven toward that of the shape given from the cryo-EM density map (Biophysical Society Meeting 2006 and 2007) [34]. We assembled a database of specific similar tRNA structures from the PDB with common

stems. Using the database, insights into the modes global motion of tRNA were examined, which we then compared with our simulations. We collected two kinds of simulation data. The first type was derived used a biasing method called Maxwell's demon molecular dynamics (MdMD), which is used to flexibly fit a high-resolution crystal structure of tRNA toward that of a cryo-EM density maps for other conformations of tRNA. The second sort of simulations was an unbiased, long-running simulation of aa-tRNA. Some of the unbiased simulations included a tethered or restrained anticodon, which mimics the hydrogen bonding of the anticodon to mRNA during translation. Post simulation, we docked these tethered tRNA trajectories into a static model of the ribosome to identify putative contact regions between tRNA and the ribosome.

2. Methods

2.1. MD Protocols. Multiple simulations of aa-tRNAs were completed using NAMD version 2.62 with the CHARMM27 force field, and then the simulation was conducted in duplicate using NAMD version 2.62 with the AMBER force field [35–39]. Using duplicate simulations under different force fields allowed us to account for differences between force fields and test for differences on flexibility. A typical tRNA simulation was comprised of a water box containing between 60,000 and 100,000 TIP3P water molecules that were added around the tRNA to provide a depth of 12 to

18 Å from the edge of the molecule [40]. Starting structures for the tRNA consisted of tRNAs in the literature [9, 13, 16, 23, 33], and cryo-EM densities were given directly from our collaborator (J. Frank, personal communication). The cryo-EM densities used for simulation are obtained from isolated *E. coli* ribosome microscopy data of tRNA/RNA/ribosome complexes.

Several tRNA simulations were run only using neutralizing Na^+ ions. These were initially placed using the Xleap module of AMBER9 at the positions of the lowest electrostatic potential. In one case, we neutralized with 76 Na^+ counterions to the tRNA. In another case, we neutralized with counterions and then created a solvent with 150 mM Na^+ Cl^- to recreate physiological strength. We observed similar tRNA flexibility in both cases.

The AMBER force field parameters for the naturally occurring modified nucleosides of RNA were obtained from the web-server at the lab of SantaLucia and from the Sanbonmatsu lab (personal communication) [41]. We used the following van der Waals parameters for Na^+: radius 1.868 Å and well depth of 0.00277 kcal/mol [42].

Simulations were carried out using the particle mesh Ewald technique (PME) with repeating boundary conditions in a box approximately 70–105 Å^3 with 9 Å nonbonded cutoff [43], using SHAKE with a 2 fs timestep [44]. Pre-equilibration was started with 10,000 steps of minimization followed by 250 ps of heating under MD, with the atomic positions of tRNA fixed. Then, two cycles of minimization (1000 steps each) and heating (200 ps) were carried out with restraints of 10 and 5 kcal/(mol·Å^2) applied to all tRNA atoms. Then, 1000-steps of minimization were performed with solute restraints reduced by 1 kcal/(mol·Å^2). Then, 200 ps of unrestrained MD were carried out, and the system was slowly heated from 1 to 310 K. The production MD runs were carried out with constant pressure boundary conditions (relaxation time of 1.0 ps). A constant temperature of 300 K was maintained using the Berendsen weak-coupling algorithm with a time constant of 1.0 ps. SHAKE constraints were applied to all hydrogen atoms to eliminate X-H vibrations, which yielded a longer simulation time step (2 fs). Our method of equilibration and production simulation is similar to protocol in the literature [30, 31].

Translational and rotational center-of-mass motions (COMs) were initially removed. Periodically, simulations were interrupted to have the COM removed again by a subtraction of velocities to account for the "flying ice-cube" effect [45]. Following the simulation, the individual frames were superposed back to the origin, to remove rotation and translation effects.

2.2. MdMD Protocols. In the MdMD algorithm, we run a short "sprint" of MD (typically ranging from 50 fs to 5 ps) [34]. We then compute the value of a global progress variable. If the MD sprint has moved the system toward the goal value of the progress variable, we save the state of the system and carry out another MD sprint. If the system has moved in the wrong direction with respective to its goal, we reset the system to the last archived state, reset the velocities, and

repeat the MD sprint. In this way, we only retain the MD steps that move the system toward its goal. We repeat this cycle of MD sprints until the system reaches its goal. In this study, we defined the global progress variable to be the cross-correlation between the electron density of the simulation structure and the experimental density from the cryo-EM density map. We rejected MD sprints that decreased the cross-correlation.

2.3. Cryo-EM Fitting Using MdMD. This application of MdMD allows the user to alter the shape of X-ray crystallographic structures to match the cryogenic-electron microscopic data, which may present an alternative conformation of the structure. In doing so, the cryo-EM density can drive the MD toward an unknown conformation. By automating this process, human biases and errors are minimized for the model making process. The first step is conversion of the MD all-atom structure into a representative low-resolution cryo-EM density, and the second step is the comparison of the MD-generated density to the target cryo-EM density. If the correlation between the MD density and the target density increases, then the iteration is saved and the process continues; otherwise the structure returns to the prior state and another MD sprint is carried out. We verified the correlation of the densities using both SPIDER and NMFF [46–48]. Our MdMD method of flexible cryo-EM fitting, as described herein, is cited within the Flexible Fitting program (FFMD) [34, 49]. In the FFMD program [49], a gradient is applied to the potential based upon the cross-correlation coefficient (CC), whereas in the MdMD-type cryoEM fitting algorithm there are external entropic forces applied in selection.

The CC represents the numerical correlation between the electron density of the target cryo-EM density map and the simulated density map from the current MD structure. The CC is defined as follows:

$$C.C. = \frac{\sum_{ijk} \rho^{exp}(i,j,k) \rho^{sim}(i,j,k)}{\sqrt{\sum_{ijk} \rho^{exp}(i,j,k)^2 \sum_{ijk} \rho^{sim}(i,j,k)^2}}, \quad (1)$$

where ρ^{Exp} and ρ^{Sim} are the experimental and simulated densities for the voxels (i,j,k). The simulated density map is generated from the MD structure, using a Gaussian function on every atom, and then integrating for each of these atom in all of the given voxels, as determined by its set of atomic coordinates (x_n, y_n, z_n), where $\rho^{Sim}(i,j,k) = \Sigma \int V_{ijk} dxdydz$ for N atoms, and (i,j,k) being a given voxel of $g(x,y,z,x_n,y_n,z_n)$. Using the Gaussian, $g(x,y,z,x_n,y_n,z_n) = \exp[-3/2\sigma^2((x-x_n)^2 + (y-y_n)^2 + (z-z_n)^2)]$, with sigma as a parameter of the resolution used. We use a 2σ cutoff for the resolution in our simulated maps in order to approximate the experimental resolution. This method of calculating a density is similar to that used to generate electron densities by methods in the literature; however we also tested the program SPIDER to generate simulated electron densities [46, 47]. SPIDER only differed slightly in the amount of CPU time to calculate a correlation and a slight drop in overall efficiency.

3. Results

3.1. Database of Crystal Structures. We compiled a set of tRNA crystal structures from structural databases to quantify any tRNA flexibility from experimental data. The Nucleic Acid Database and the Protein Data Bank RCSB [50, 51] contain over 80 structures of tRNAs alone or in complexes with other molecules at a resolution of 3.3 Å or better. For the study presented we focused on a subset that were in good alignment with tRNA^{Phe}. Thus, thirty-four of the crystal structures are tRNA-protein complexes. The remaining six are free, unbound tRNA structures. Most of the complexes are aminoacyl-tRNA synthetases (32 structures), but there is also one cocrystal of tRNA^{Glu} with its amidotransferase, and another between tRNA^{fMet} and its formyltransferase (Figure 2(b)). Of the isolated tRNA structures, six are unique and of the highest available resolution, while four are distinct: Asp from *E. coli* and *S. cerevisae*, Lys from *B. taurus*, Phe from *S. cerevisae* and *T. thermophilus*, and fMet from *E. coli* and *S. cerevisae* (Figure 2(b)) [3]. tRNAs for fMet and 16 other amino acids are represented (Figures 2(a) and 2(b)). Supplementary Table S1 summarizes the crystal structures that we examined. Figure 2 demonstrates the conformational variability among the full set of structures, and Figure 2(c) shows the variability in just phenylalanine-tRNA^{Phe}.

We used the interarm angle between the anticodon stem and the acceptor stem as a measure of flexibility. There were eight species of tRNA for which multiple X-ray crystal structures were available: Asp, Glu, Gln, Met, Phe, Trp, Tyr, and Val. All eight of these had similar ranges of interarm angles, somewhere between 70° and 100°, which suggests that interarm flexibility is an intrinsic feature of the topology of tRNA and does not depend strongly on the amino acid bound to the acceptor stem (Supplementary Table S1).

In addition, we computed the RMSDs of the crystal structures, relative to 1EHZ tRNA^{Phe}, and found a range from 0.5 Å to approximately 6 Å (Supplementary Table S1) for superposition of the main stems. Consequently, on a residue-by-residue basis, the greatest regions of flexibility were the 3′-CCA end of tRNA and the anticodon stem-loop (ASL) (Figure 3(a), Supplementary Movies S1, S2, S5).

3.2. Fitting to Cryo-EM Maps via MdMD. Cryo-EM reconstructions of tRNA bound to the 70S *E. coli* ribosome at different steps in the translation-elongation cycle were examined [14, 52]. From the reconstructions, it is possible to identify subtle conformational changes in tRNA (Figure 1(e)). For example, at a resolution of 7.8 Å a cryo-EM density map can reveal the helical structure of tRNA for the hybrid "A/T" state, which is deformed at the anticodon stem arm through interactions with the D-stem loop residues 26, 44, and 45 of phenylalanine-tRNA^{Phe} (Figure 1(e)).

We examined four distinct tRNA conformations derived from cryo-EM density maps. These tRNA conformations included the A, P, A/T, and P/E states. Using Maxwell's demon Molecular Dynamics (MdMD), the starting structure of the native tRNA structures (1EHZ) is driven into conformations that match the cryo-EM data based upon a cross-correlation calculation between a theoretical density for the

modeled structure and the experimental cryo-EM density. MdMD is derived from existing methods in the literature for biasing simulations [53]. However, in addition to having an entropic penalty for directional control, there is an adaptive component that dictates the amount of sampling time per interval of MD [34]. A brief explanation of MdMD consists of the following iterative steps: (1) a short interval of MD simulation (MD sprint) is completed, (2) the MD is paused and a "check" of some user defined variable is assessed, which determines a numerical value for the *progress* of the variable, and (3) the algorithm either accepts or rejects the result of the progress variable, which is based on the current value of the progress variable plus several previous archived states. The adaptive component will accept an MD sprint and increase or decrease the amount of sampling time for the next MD sprint based on the new value for the progress variable. Likewise, the sampling size may shrink if the success criteria opposite the goal.

The steps are archived into a single trajectory based upon the outcome of the progress variable. During the MD sprint the variable is free to fluctuate into space that would otherwise be forbidden by the Maxwell demon.

In the case of cryoEM-fitting with MdMD, our progress variable is the cross-correlation between the electron density of the simulation structure and the experimental density from cryo-EM, where we rejected MD sprints that decrease the cross-correlation.

Using MdMD, we obtained pathway transitions from the crystal structure state (1EHZ) to the cryo-EM states A, P, A/T, and P/E using less than 10 nanoseconds of molecular dynamic simulation (Supplementary Movies S4, S6). From these simulations, the ensembles of structures are consistent with the density distributions from the cryo-EM density maps. Figure 4 shows the quality of fit for the four structures. Supplementary Table S2 identifies the cryo-EM density maps of each structure and gives the RMSD of each structure relative to the crystal structure of yeast tRNA^{Phe} (PDB code: 1EHZ) [33]. All-atom root mean-squared deviation (RMSD) analyses between the crystal structure and regions from our molecular dynamics model are <2 Å at these superposed regions, indicating that the fine structure was maintained. Using MdMD, the A-site conformation can be forced rapidly to the kinked conformation required for the A/T state. The kinked structure generated using MdMD is within 1.0 Å RMSD of the original manually modeled structure of the A/T tRNA (Supplementary Movie S6) [13].

The cross correlations between the fit structures and the maps give a good indication that the fit structures are not deformed. Specifically, the pathways between the fit structures reveal the transitional motion, dynamically, between modes of tRNA: from A/T to A, from A to P, and from P to P/E, respectively, exploring stochastically reversible excursions along a tRNA pathway between experimentally verified states [54].

3.3. Testing tRNA Motion Using Unbiased Molecular Dynamics. We carried out two sets of unbiased MD simulations on tRNA: (1) We ran ~0.5 μs of simulations starting from the

(a)

(b)

(c)

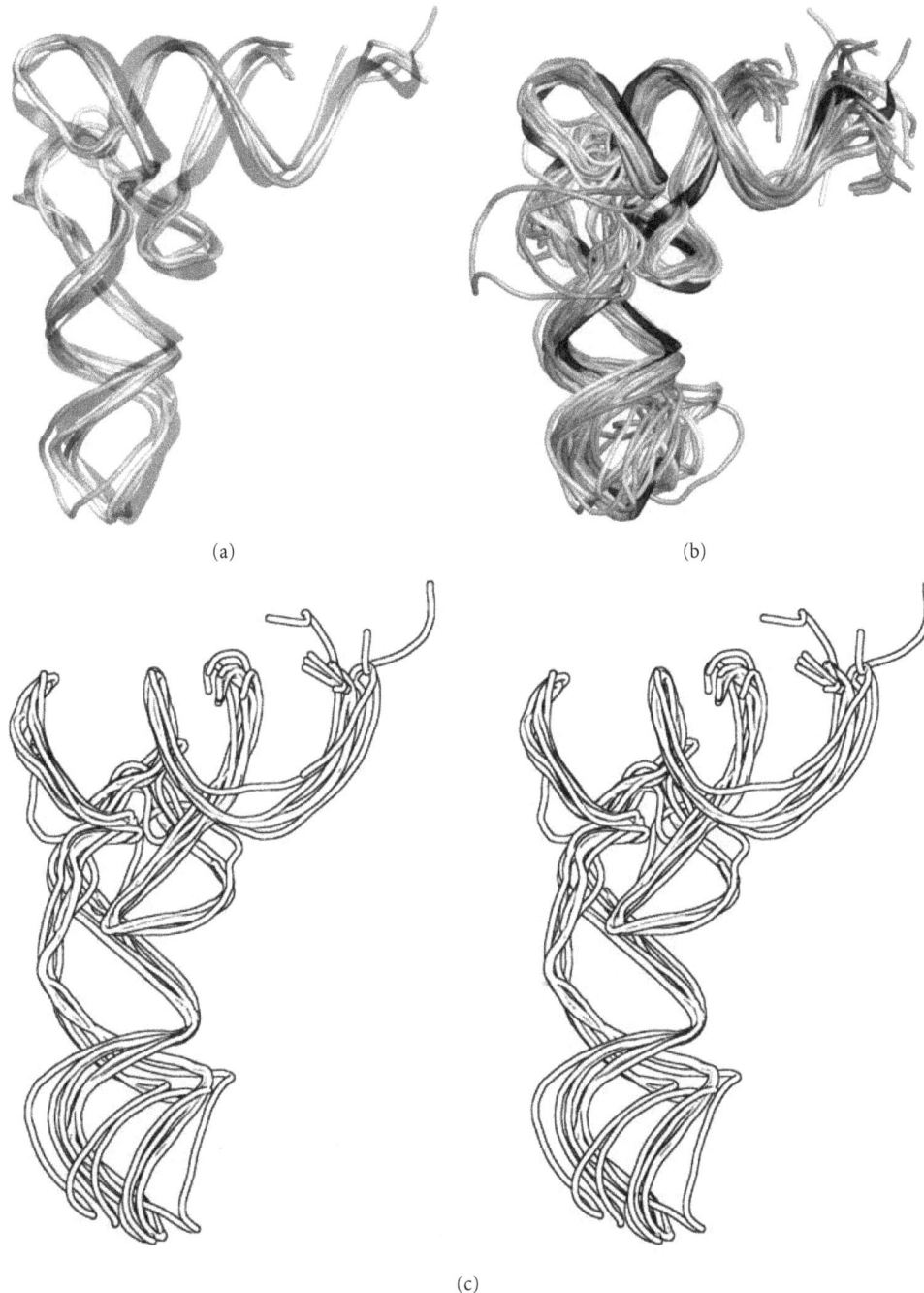

FIGURE 2: tRNA Catalog. (a) Six crystal structures of unbound Phe-tRNAPhe. The ribbon in the middle is from 1EHZ. (b) Crystal structures (41) of tRNA free and in complexes, including synthetases and ribosomal (Table S1) aligned at the acceptor stems and the anticodon stems. The ribbon in the middle is from 1EHZ. (c) Phe-tRNAPhe crystal structures (6) aligned at the acceptor stem (shown in divergent stereo), thus illustrating the range of motion at the anticodon stem loops, and notice the closely aligned acceptor stem regions.

crystal structure of yeast tRNA^{Phe} (1EHZ) to examine the inherent flexibility of tRNA. (2) We asked whether tRNA in the twisted A/T conformation would spontaneously move toward the native crystal structure for yeast tRNA^{Phe} using unbiased MD [13]. In this case, the anticodon stem must untwist to find the native state. We found that A/T tRNA does so quite rapidly (Supplementary Movie S1). The unkinking occurs below the hinge, toward the anticodon.

We repeated this simulation with a tethered anticodon to mimic the hydrogen bonding with mRNA. We imposed restraints on anticodon nucleotides 34, 35, and 36 (1 to 10 kcal/(mol∗Å^2) per hydrogen bonding atom). One would expect that this tether would result in dampening of tRNA motion, because the D-loop, T-loop, and acceptor-stem must rotate in order to relieve the twist in the anticodon stem. This simulates the effect of being bound to the ribosome

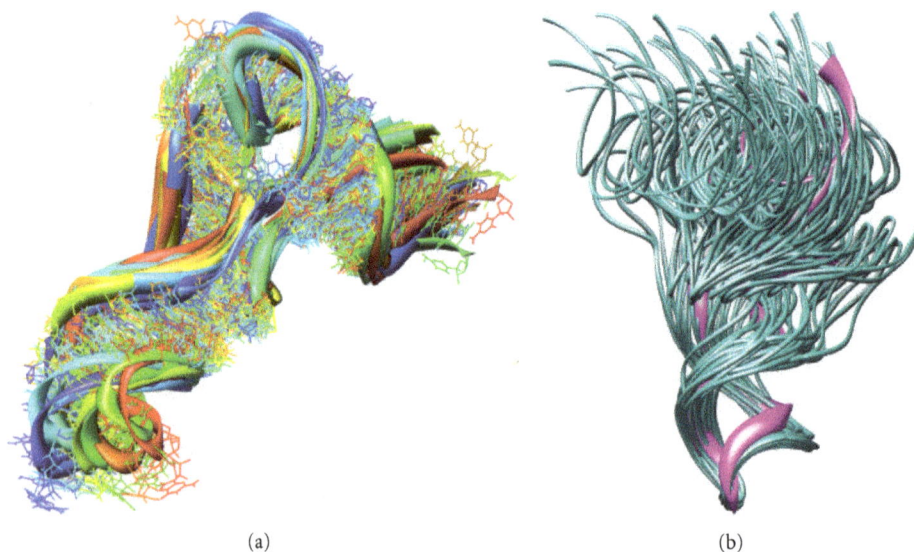

(a)   (b)

FIGURE 3: Jittergram of kinked A/T-tRNA from molecular dynamics simulations. (a) Using multiple unbiased molecular dynamics for kinked A/T-tRNA, we generated a set of structures. These are superposed based on all P-atoms found in the backbone of the stem regions. This orientation optimally shows the alignment of the stems and the variation at the acceptor stem loop and anticodon stem loop. The different tRNA colors help distinguish the different conformations. (b) Illustrated in cyan as a series of snapshots is a simulation for kinked A/T-tRNA using free molecular dynamics with tethering constraints at the anticodon atoms (H-bond atoms of nucleotides 34, 35, 36), which was run for >10 ns. We used tethering (harmonic constraints) at the anticodon atoms to mimic base pairing at the codon. This data was used for the contact map (Figure 5). This orientation best shows the wide range of motion that the tRNA covers (over 70 Å from beginning to end for the 3'CCA end).

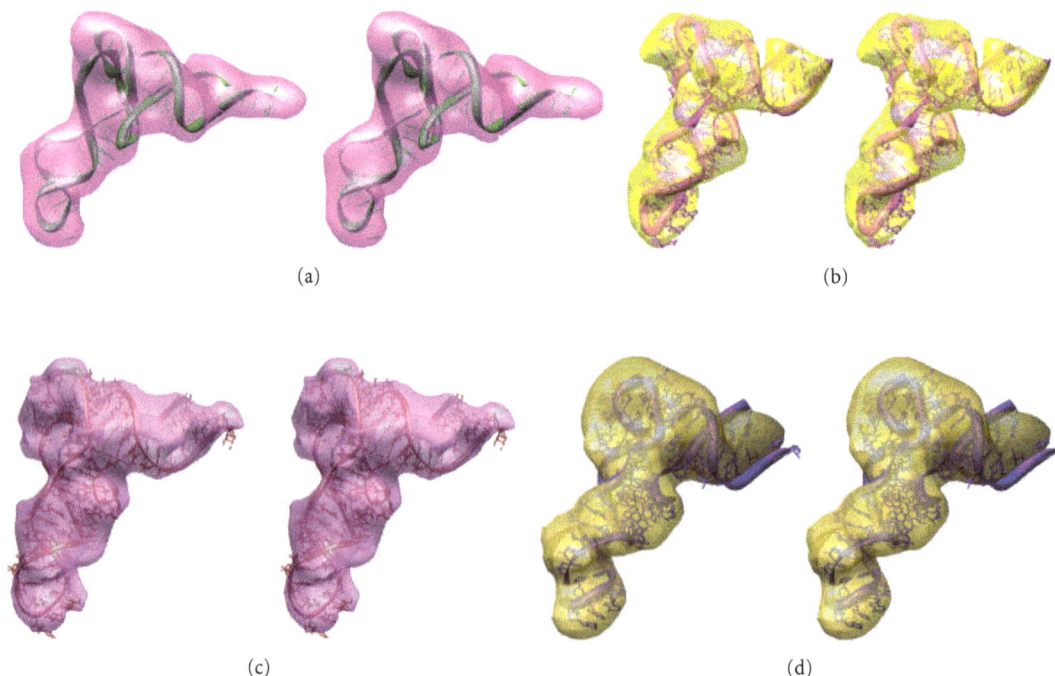

(a)   (b)

(c)   (d)

FIGURE 4: Cryogenic-electron microscopy fitting using Maxwell's demon Molecular Dynamics for tRNAs A/T, A, P, and P/E is shown. All tRNA atomic models are shown in ribbons/CPK, while densities are as solid. (a) A/T-site structure for tRNA fit to density, cross-correlation coefficient (CCC) of 95% between the structures theoretical density and the cryo-EM experimental density (Divergent stereo). (b) A-site structure for tRNA fit to density with a CCC of 83% (Divergent stereo). (c) P-site structure for tRNA with a CCC of 86% (Divergent stereo). (d) P/E-site structure with a CCC >90% (Divergent stereo).

(a)

(b)

(c)   (d)   (e)

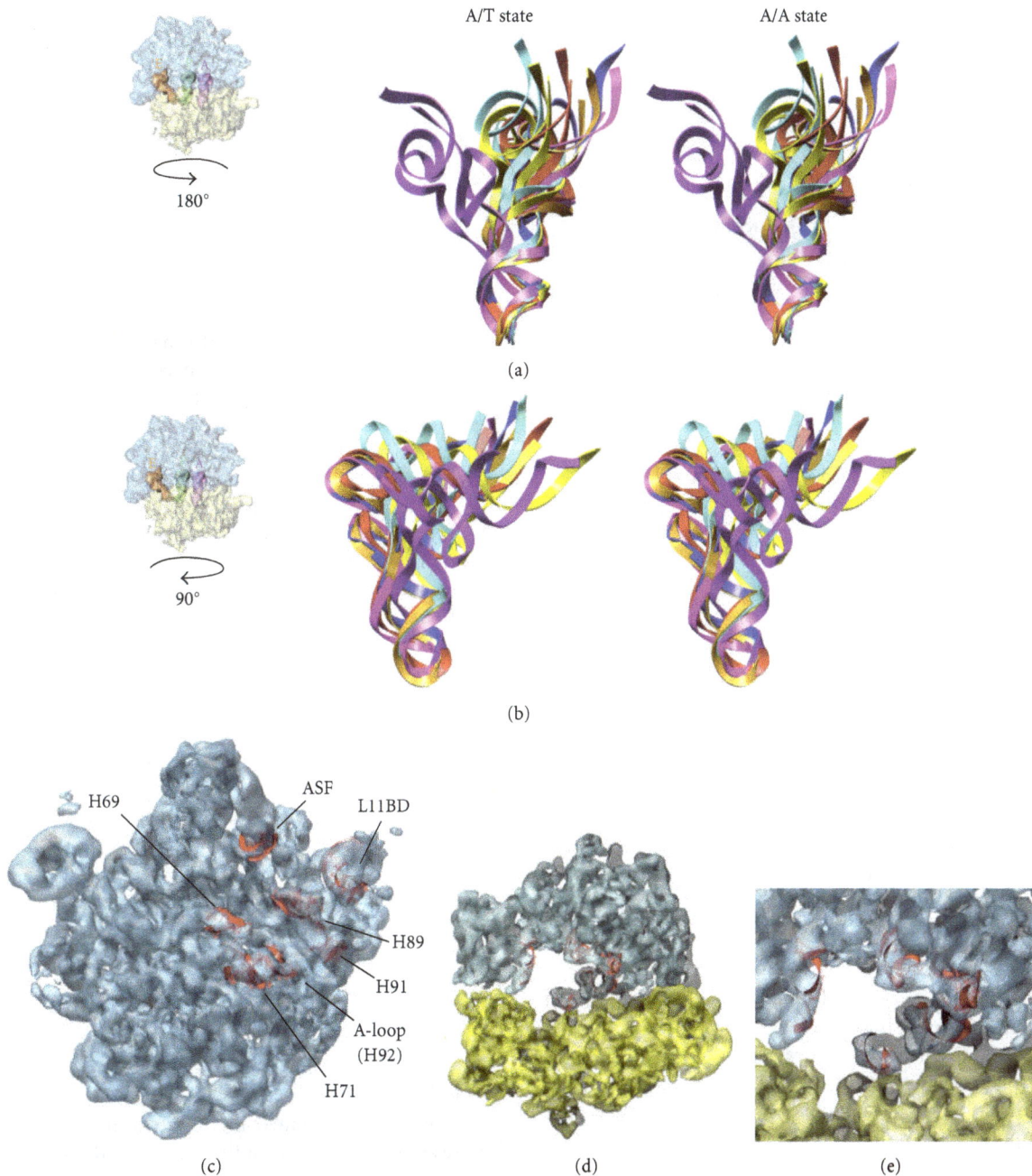

FIGURE 5: Various snapshots from multiple simulations of A/T-tRNA and native tRNA are shown, where the time sampled over is >200 ns (Supplementary Movie S1). (a) A/T-site tRNA (shown in purple) moves toward the A-site tRNA (red) which is shown in reverse view of the ribosome density, to illustrate the >70 Å motion covered by the 3′-CCA acceptor end. The tRNA shows an oscillation about the A-site tRNA structure (red) (Divergent stereo). (b) *Side view* of Figure 5(a) (Divergent stereo). (c) Contact map for tRNA trajectory superposed with the ribosome. (d) The side view panel on right shows the accommodation corridor for the A/T-tRNA in the ribosome. The A/T-tRNA has to make a snug fit because of the narrow space. (e) The lower right panel is zoomed in closer to illustrate the interaction areas.

(Supplementary Movie S2). This situation is similar to the simulation of Sanbonmatsu et al., except that there is no ribosome present to hinder swinging, and no biasing forces were used [20, 54].

From our unbiased MD simulations, we found that tRNA[Phe] unkinked in the 2 ns following equilibration, when the anticodon atoms were unrestrained. However, when the

anticodon was restrained, unkinking occurred between 6 and 10 ns. We also found that simulations of the A/T to the A/A state provide an approximate range for tRNA swinging-type motion that agrees with the structural data in our database of crystal structures.

Our results suggest that tRNA deforms quite readily at the anticodon stem from an initial free tRNA state.

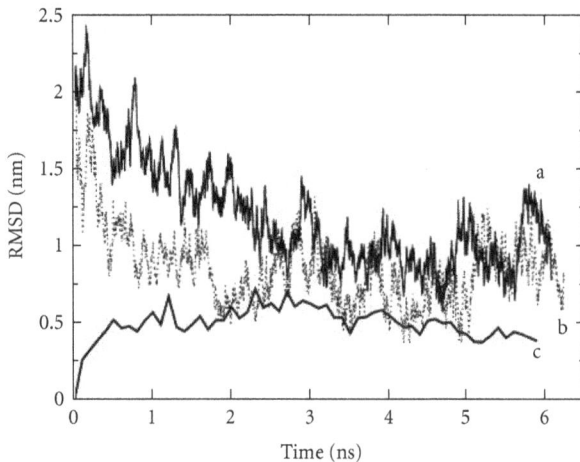

FIGURE 6: RMSD graph for three types of tRNA simulations presented in this study. This graph represents the root mean square deviation between the simulation of tRNA^{Phe} with that of the crystal structure for tRNA^{Phe} (1EHZ) [33]. As time progresses the initial tRNA^{Phe} relaxes under the force field and the biased and unbiased states of tRNA progress toward the crystal form. (a) "Line a" represents the constrained anticodon MD simulation of the A/T state kinked model, which converges to the A-site ensemble after 4 ns. (b) "Line b" represents the free MD simulation of the A/T state kinked model that converges much quicker to the A-site ensemble of structures (2 ns). (c) "Line c" represents the free molecular dynamics simulation for the A-site crystal structure model, which moves from the starting structure to an average ensemble around 4 Å RMSD from the original as it relaxes under MD.

We found that (1) the kinked A/T structure's trajectory converged toward the crystal structure of tRNA^{Phe} (Figure 6, Supplementary Figure S2), and (2) tRNA^{Phe} (1EHZ) relaxed under unbiased MD to an equilibrated state [13, 33]. This relaxed structure for tRNA^{Phe} sampled regions of conformational space that were similar to those sampled by the kinked A/T structure. The trajectory of the kinked A/T and the trajectory of the native tRNA^{Phe} structure overlapped in conformational space around 1.4 ns, 2.7 ns, and after 5 ns (Supplementary Figure S2). The RMSD matrix plot identifies several well-populated regions of conformational space between the A/T and A/A states (Figure S2). Also, we were able to rapidly drive the crystal structure of tRNA^{Phe} (1EHZ) to form the kinked A/T structure with MdMD (~8 ns) (Supplementary Movie S3, Figure S2). Both force fields were successful in achieving unkinking, but the helical parameters of base pairing were better modeled with the amber force field [36, 37, 39].

Figures 3(a) and 3(b) and Figures 5(a) and 5(b) show the range-of-motion accessible to tRNA during multiple 100+ ns simulations. In Figure 3(a), we aligned the trajectories of structures at their center-of-mass. The anticodon stem loop region fluctuated 7–10 Å, while the 3′-CCA end of the acceptor stem fluctuated 13–15 Å (Movies S1, S2). Figure 3(b) shows a progressive pathway for tRNA with a constrained anticodon during simulations of unkinking from the A/T state to the A/A state. The aminoacyl-3′-CCA

tip of the acceptor stem moved approximately 70 Å from start to finish. Figures 5(a) and 5(b) show several snapshots from unbiased simulations. They appear to fluctuate around a structure that resembles the A-site structure of tRNA. The A-site structure also resembles the crystal structure of tRNA^{Phe} (1EHZ). The native tRNA and A-site tRNA exhibited an average RMSD of 4 Å among all the tRNAs and 2.2 Å for the Phenylalanine-tRNA^{Phe} structure. The A/T structure is shown for reference in purple.

The RMSD values from multiple simulations of kinked and unkinked tRNA, as well as native tRNA, show that these different conformations often converge within 10 ns, indicating that the favored conformation in all cases occurs in a region of conformational space similar to the relaxed native structure (Figure 6). For all simulations addressed here, RMSD convergence occurs during the initial 10 ns and then again periodically. Here, convergence is defined as an RMSD below 3 Å. Longer sampling (>100 ns) showed that twisted conformations similar to the hybrid states like P/E or A/T are accessible from the native form of tRNA (1EHZ). Our long trajectories also sampled many of the conformations found in our set of tRNA crystal structures.

3.4. A Map of Ribosomal Contact Points from Simulation Data Overlap. We docked the unbiased MD simulation of tethered tRNA into the ribosome in order to study contacts between tRNA and the 70 S *E. coli* ribosome. We compiled the results of our simulations (0.5 μs in total) into a contact map (Figures 5(c) and 5(d)). Our definition of a contact is based on C-alpha-to-RNA base distances and base-to-base distances within a 4 Å cutoff. The map shows that tRNA interacts with ribosomal proteins L14 and L16 as well as H71 and H92 of the 50S subunit during accommodation. These areas may act to filter out less energetically kinked tRNAs during accommodation. If near-cognate tRNAs are less kinked than cognate tRNAs, this process might filter them out. There is significant agreement between our proposed contact map and the transition pathways observed in the literature [20, 21, 54].

Recent data presented in the literature demonstrates a low-resolution structure for the 80S P-site tRNA that suggests a state in which the anticodon stem-loop is significantly bent [55]. This occurs during programmed ribosomal frame shifting, when the ribosome is subverted into a frame shift by maintaining the 3′-CCA acceptor end bound, while the anticodon takes a −1 step, thus placing the message into a new reading frame [55]. We observed similar deformations during our simulations (Movie S6). Additionally, during the formation of the kinked A/T state, we observed deformations of the D-loop relative to the acceptor stem (Movies S1, S5).

4. Conclusions

Our simulations of tRNA support a complex flexible hinge motion between the anticodon stem and acceptor stem with a characteristic relaxation time on the nanosecond timescale.

The hinge is centered on nucleotides 26, 44, and 45, and additionally there are complex stem deformations visible in the dynamics movies. In simulations where the anticodon was tethered, which mimic hydrogen bonding to the mRNA codon during translation, we found that the correlation time for rotational diffusion of tRNA about the fixed codon-anticodon duplex was approximately 10 ns. We found that tethering the anticodon bases in space resulted in diffusion being driven by an unkinking force, which was able to swing the entire tRNA 3′-terminus along a trajectory mimicking tRNA accommodation in the ribosome during translation. As the A/T-tRNA moved away from the kinked state, it progressed into a range of conformations similar to those found in other catalogued crystal structures.

Our results confirm that tRNA has a stochastic molecular spring-like motion from the biasing method [52, 54], and from the unbiased simulations a nonlinear Brownian type motion from the A/T-structure toward the A-site state model, which matches the free tRNA^{Phe} crystal structure (1EHZ). Moreover, all tRNAs tested possess an intrinsic ability to use flexibility to achieve a kinked state. Hybrid states of tRNA might be utilized by the ribosome to load the spring as a result of translocation. Thus, the ribosome may be inducing tRNA into the next position, twisting it into the hybrid state. Following this loading process, tRNA would untwist via a stochastic spring-like motion.

Here, we report that a Maxwell demon-type algorithm that acts externally to the potential one may utilize experimental cryo-EM densities to "drive" molecular dynamics simulations into preferred conformations. This result is promising for future work to obtain transition pathways. We found that our biased MdMD method can sample multiple hybrid conformations of tRNA that have been experimentally observed from cryo-EM and X-ray crystallography, thus bridging between two experimental methods.

Finally, we constructed a putative ribosomal contact map from the microsecond dynamics of tRNA bound to the ribosome. We observed contacts with H71 and H92 of the 50S subunit and ribosomal proteins L14 and L16. It is possible that these areas of the ribosome act to filter near-cognate tRNA in a test of anticodon base pairing during accommodation.

Acknowledgments

This work was supported by a GAANN and Sam Nunn-MacArthur Foundation doctoral fellowship to T. R. Caulfield., and by the College of Computing Computer Resources Allocation at the Georgia Institute of Technology, and by the Bluegene/L computing facility at the UAB Center for Computational and Structural Biology at the University of Alabama at Birmingham. The authors thank M. Wolf of the College of Computing at Georgia Institute of Technology for resource administration, M. Hanby, M. Perez, and J. Segrest at U.A.B. for being computational liaisons for the Bluegene/L, and Drs. S. C. Harvey and A. Petrov for insightful contributions to computer simulations, ideation, and collaboration.

References

[1] J. D. Robertus, J. E. Ladner, J. T. Finch et al., "Structure of yeast phenylalanine tRNA at 3 Å resolution," Nature, vol. 250, no. 5467, pp. 546–551, 1974.

[2] J. D. Robertus, J. E. Ladner, J. T. Finch et al., "Correlation between three-dimensional structure and chemical reactivity of transfer RNA," Nucleic Acids Research, vol. 1, no. 7, pp. 927–932, 1974.

[3] T. Olson, M. J. Fournier, K. H. Langley, and N. C. Ford, "Detection of a major conformational change in transfer ribonucleic acid by laser light scattering," Journal of Molecular Biology, vol. 102, no. 2, pp. 193–203, 1976.

[4] P. Dumas, J. P. Ebel, R. Giege, D. Moras, J. C. Thierry, and E. Westhof, "Crystal structure of yeast tRNA(Asp): atomic coordinates," Biochimie, vol. 67, no. 6, pp. 597–606, 1985.

[5] D. Moras, A. C. Dock, P. Dumas et al., "The structure of yeast tRNA(Asp). A model for tRNA interacting with messenger RNA," Journal of Biomolecular Structure and Dynamics, vol. 3, no. 3, pp. 479–493, 1985.

[6] M. W. Friederich, F. U. Gast, E. Vacano, and P. J. Hagermant, "Determination of the angle between the anticodon and aminoacyl acceptor stems of yeast phenylalanyl tRNA in solution," Proceedings of the National Academy of Sciences of the United States of America, vol. 92, no. 11, pp. 4803–4807, 1995.

[7] M. W. Friederich, E. Vacano, and P. J. Hagerman, "Global flexibility of tertiary structure in RNA: yeast tRNA as a model system," Proceedings of the National Academy of Sciences of the United States of America, vol. 95, no. 7, pp. 3572–3577, 1998.

[8] E. Vacano and P. J. Hagerman, "Analysis of birefringence decay profiles for nucleic acid helices possessing bends: the τ-ratio approach," Biophysical Journal, vol. 73, no. 1, pp. 306–317, 1997.

[9] M. M. Yusupov, G. Yusupova, A. Baucom et al., "Crystal structure of the ribosome at 5.5 Å resolution," Science, vol. 292, no. 5518, pp. 883–896, 2001.

[10] J. C. Schuette, F. V. Murphy, A. C. Kelley et al., "GTPase activation of elongation factor EF-Tu by the ribosome during decoding," The EMBO Journal, vol. 28, no. 6, pp. 755–765, 2009.

[11] H. Stark, M. V. Rodnina, H. J. Wieden, F. Zemlin, W. Wintermeyer, and M. van Heel, "Ribosome interactions of aminoacyl-tRNA and elongation factor Tu in the codon-recognition complex," Nature Structural Biology, vol. 9, no. 11, pp. 849–854, 2002.

[12] M. Valle, R. Gillet, S. Kaur, A. Henne, V. Ramakrishnan, and J. Frank, "Visualizing tmRNA entry into a stalled ribosome," Science, vol. 300, no. 5616, pp. 127–130, 2003.

[13] M. Valle, A. Zavialov, W. Li et al., "Incorporation of aminoacyl-tRNA into the ribosome as seen by cryo-electron microscopy," Nature Structural Biology, vol. 10, no. 11, pp. 899–906, 2003.

[14] M. Valle, A. Zavialov, J. Sengupta, U. Rawat, M. Ehrenberg, and J. Frank, "Locking and unlocking of ribosomal motions," Cell, vol. 114, no. 1, pp. 123–134, 2003.

[15] E. Villa, J. Sengupta, L. G. Trabuco et al., "Ribosome-induced changes in elongation factor Tu conformation control GTP hydrolysis," Proceedings of the National Academy of Sciences of the United States of America, vol. 106, no. 4, pp. 1063–1068, 2009.

[16] S. C. Harvey and J. A. McCammon, "Intramolecular flexibility in phenylalanine transfer RNA," Nature, vol. 294, no. 5838, pp. 286–287, 1981.

[17] S. C. Harvey, M. Prabhakaran, B. Mao, and J. A. McCammon, "Phenylalanine transfer RNA: molecular dynamics simulation," *Science*, vol. 223, no. 4641, pp. 1189–1191, 1984.

[18] M. Prabhakaran, S. C. Harvey, B. Mao, and J. A. McCammon, "Molecular dynamics of phenylalanine transfer RNA," *Journal of Biomolecular Structure and Dynamics*, vol. 1, no. 2, pp. 357–369, 1983.

[19] P. Auffinger and E. Westhof, "Singly and bifurcated hydrogen-bonded base-pairs in tRNA anticodon hairpins and ribozymes," *Journal of Molecular Biology*, vol. 292, no. 3, pp. 467–483, 1999.

[20] K. Y. Sanbonmatsu, S. Joseph, and C. S. Tung, "Simulating movement of tRNA into the ribosome during decoding," *Proceedings of the National Academy of Sciences of the United States of America*, vol. 102, no. 44, pp. 15854–15859, 2005.

[21] K. Y. Sanbonmatsu and C. S. Tung, "High performance computing in biology: Multimillion atom simulations of nanoscale systems," *Journal of Structural Biology*, vol. 157, no. 3, pp. 470–480, 2006.

[22] M. Nina and T. Simonson, "Molecular dynamics of the tRNA acceptor stem: comparison between continuum reaction field and particle-mesh Ewald electrostatic treatments," *Journal of Physical Chemistry B*, vol. 106, no. 14, pp. 3696–3705, 2002.

[23] P. Auffinger, L. Bielecki, and E. Westhof, "Symmetric K and Mg ion-binding sites in the 5 S rRNA loop e inferred from molecular dynamics simulations," *Journal of Molecular Biology*, vol. 335, no. 2, pp. 555–571, 2004.

[24] P. Auffinger, L. Bielecki, and E. Westhof, "The Mg binding sites of the 5S rRNA loop E motif as investigated by molecular dynamics simulations," *Chemistry and Biology*, vol. 10, no. 6, pp. 551–561, 2003.

[25] V. Cojocaru, R. Klement, and T. M. Jovin, "Loss of G-A base pairs is insufficient for achieving a large opening of U4 snRNA K-turn motif," *Nucleic Acids Research*, vol. 33, no. 10, pp. 3435–3446, 2005.

[26] W. Li, B. Ma, and B. A. Shapiro, "Binding interactions between the core central domain of 16S rRNA and the ribosomal protein S15 determined by molecular dynamics simulations," *Nucleic Acids Research*, vol. 31, no. 2, pp. 629–638, 2003.

[27] K. Y. Sanbonmatsu and S. Joseph, "Understanding discrimination by the ribosome: stability testing and groove measurement of codon-anticodon pairs," *Journal of Molecular Biology*, vol. 328, no. 1, pp. 33–47, 2003.

[28] K. Réblová, N. Špačková, R. Štefl et al., "Non-Watson-Crick basepairing and hydration in RNA motifs: molecular dynamics of 5S rRNA loop E," *Biophysical Journal*, vol. 84, no. 6, pp. 3564–3582, 2003.

[29] J. Sarzynska, T. Kulinski, and L. Nilsson, "Conformational dynamics of a 5S rRNA hairpin domain containing loop D and a single nucleotide bulge," *Biophysical Journal*, vol. 79, no. 3, pp. 1213–1227, 2000.

[30] N. Špačková and J. Šponer, "Molecular dynamics simulations of sarcin-ricin rRNA motif," *Nucleic Acids Research*, vol. 34, no. 2, pp. 697–708, 2006.

[31] K. Réblová, F. Lankaš, F. Rázga, M. V. Krasovska, J. Koča, and J. Šponer, "Structure, dynamics, and elasticity of free 16S rRNA helix 44 studied by molecular dynamics simulations," *Biopolymers*, vol. 82, no. 5, pp. 504–520, 2006.

[32] A. C. Vaiana and K. Y. Sanbonmatsu, "Stochastic gating and drug-ribosome interactions," *Journal of Molecular Biology*, vol. 386, no. 3, pp. 648–661, 2009.

[33] H. Shi and P. B. Moore, "The crystal structure of yeast phenylalanine tRNA at 1.93 A resolution: a classic structure revisited," *RNA*, vol. 6, pp. 1091–1105, 2000.

[34] T. R. Caulfield and S. C. Harvey, "Conformational fitting of atomic models to cryogenic-electron microscopy maps using Maxwell's demon molecular dynamics," *Biophysical Journal*, pp. 368A–368A, 2007.

[35] B. R. Brooks, R. E. B. Olafson, D. J. States, S. Swaminathan, and M. Karplus, "CHARMM: a program for macromolecular energy, minimization, and dynamics calculations," *Journal of Computational Chemistry*, vol. 4, no. 2, pp. 187–217, 1983.

[36] W. D. Cornell, P. Cieplak, C. I. Bayly et al., "A second generation force field for the simulation of proteins, nucleic acids, and organic molecules," *Journal of the American Chemical Society*, vol. 117, no. 19, pp. 5179–5197, 1995.

[37] A. Pérez, I. Marchán, D. Svozil et al., "Refinement of the AMBER force field for nucleic acids: improving the description of α/γ conformers," *Biophysical Journal*, vol. 92, no. 11, pp. 3817–3829, 2007.

[38] L. Kalé, R. Skeel, M. Bhandarkar et al., "NAMD2: greater scalability for parallel molecular dynamics," *Journal of Computational Physics*, vol. 151, no. 1, pp. 283–312, 1999.

[39] D. A. Pearlman, D. A. Case, J. W. Caldwell et al., "AMBER, a package of computer programs for applying molecular mechanics, normal mode analysis, molecular dynamics and free energy calculations to simulate the structural and energetic properties of molecules," *Computer Physics Communications*, vol. 91, no. 1–3, pp. 1–41, 1995.

[40] W. L. Jorgensen, J. Chandrasekhar, J. D. Madura, R. W. Impey, and M. L. Klein, "Comparison of simple potential functions for simulating liquid water," *The Journal of Chemical Physics*, vol. 79, no. 2, pp. 926–935, 1983.

[41] R. Aduri, B. T. Psciuk, P. Saro, H. Taniga, H. B. Schlegel, and J. SantaLucia Jr., "AMBER force field parameters for the naturally occurring modified nucleosides in RNA," *Journal of Chemical Theory and Computation*, vol. 3, no. 4, pp. 1464–1475, 2007.

[42] W. S. Ross and C. C. Hardin, "Ion-induced stabilization of the G-DNA quadruplex: free energy perturbation studies," *Journal of the American Chemical Society*, vol. 116, no. 14, pp. 6070–6080, 1994.

[43] T. Darden, D. York, and L. Pedersen, "Particle mesh Ewald: an N·log(N) method for Ewald sums in large systems," *The Journal of Chemical Physics*, vol. 98, no. 12, pp. 10089–10092, 1993.

[44] J. P. Ryckaert, G. Ciccotti, and H. J. C. Berendsen, "Numerical integration of the cartesian equations of motion of a system with constraints: molecular dynamics of n-alkanes," *Journal of Computational Physics*, vol. 23, no. 3, pp. 327–341, 1977.

[45] S. C. Harvey, R. K. Tan, and T. E. Cheatham III, "The flying ice cube: velocity rescaling in molecular dynamics leads to violation of energy equipartition," *Journal of Computational Chemistry*, vol. 19, no. 7, pp. 726–740, 1998.

[46] J. Frank, M. Radermacher, P. Penczek et al., "SPIDER and WEB: processing and visualization of images in 3D electron microscopy and related fields," *Journal of Structural Biology*, vol. 116, no. 1, pp. 190–199, 1996.

[47] F. Tama, O. Miyashita, and C. L. Brooks III, "Flexible multi-scale fitting of atomic structures into low-resolution electron density maps with elastic network normal mode analysis," *Journal of Molecular Biology*, vol. 337, no. 4, pp. 985–999, 2004.

[48] F. Tama, O. Miyashita, and C. L. Brooks III, "Normal mode based flexible fitting of high-resolution structure into low-resolution experimental data from cryo-EM," *Journal of Structural Biology*, vol. 147, no. 3, pp. 315–326, 2004.

[49] M. Orzechowski and F. Tama, "Flexible fitting of high-resolution X-ray structures into cryoelectron microscopy

maps using biased molecular dynamics simulations," *Biophysical Journal*, vol. 95, no. 12, pp. 5692–5705, 2008.

[50] H. M. Berman, W. K. Olson, D. L. Beveridge et al., "The nucleic acid database. A comprehensive relational database of three- dimensional structures of nucleic acids," *Biophysical Journal*, vol. 63, no. 3, pp. 751–759, 1992.

[51] P. E. Bourne, K. J. Addess, W. F. Bluhm et al., "The distribution and query systems of the RCSB Protein Data Bank," *Nucleic Acids Research*, vol. 32, pp. D223–D225, 2004.

[52] J. Frank, J. Sengupta, H. Gao et al., "The role of tRNA as a molecular spring in decoding, accommodation, and peptidyl transfer," *FEBS Letters*, vol. 579, no. 4, pp. 959–962, 2005.

[53] S. C. Harvey and H. A. Gabb, "Conformational transitions using molecular dynamics with minimum biasing," *Biopolymers*, vol. 33, no. 8, pp. 1167–1172, 1993.

[54] P. C. Whitford, P. Geggier, R. B. Altman, S. C. Blanchard, J. N. Onuchic, and K. Y. Sanbonmatsu, "Accommodation of aminoacyl-tRNA into the ribosome involves reversible excursions along multiple pathways," *RNA*, vol. 16, no. 6, pp. 1196–1204, 2010.

[55] O. Namy, S. J. Moran, D. I. Stuart, R. J. C. Gilbert, and I. Brierley, "A mechanical explanation of RNA pseudoknot function in programmed ribosomal frameshifting," *Nature*, vol. 441, no. 7090, pp. 244–247, 2006.

20

F-Ratio Test and Hypothesis Weighting: A Methodology to Optimize Feature Vector Size

R. M. Dünki[1] and M. Dressel[2, 3, 4]

[1] Physics Institute, CAP, University of Zürich, CH 8057 Zürich, Switzerland
[2] Research Department, Cantonal Psychiatric Hospital, CH 8462 Rheinau, Switzerland
[3] Department of Psychology, University of Konstanz, 78457 Konstanz, Germany
[4] Verhaltenstherapie, Post-Straße 3, 79098 Freiburg, Germany

Correspondence should be addressed to R. M. Dünki, rmd@physik.uzh.ch

Academic Editor: Serdar Kuyucak

Reducing a feature vector to an optimized dimensionality is a common problem in biomedical signal analysis. This analysis retrieves the characteristics of the time series and its associated measures with an adequate methodology followed by an appropriate statistical assessment of these measures (e.g., spectral power or fractal dimension). As a step towards such a statistical assessment, we present a data resampling approach. The techniques allow estimating $\sigma^2(F)$, that is, the variance of an F-value from variance analysis. Three test statistics are derived from the so-called F-ratio $\sigma^2(F)/F^2$. A Bayesian formalism assigns weights to hypotheses and their corresponding measures considered (hypothesis weighting). This leads to complete, partial, or noninclusion of these measures into an optimized feature vector. We thus distinguished the EEG of healthy probands from the EEG of patients diagnosed as schizophrenic. A reliable discriminance performance of 81% based on Taken's χ, α-, and δ-power was found.

1. Introduction

The reduction of a feature vector to an optimized dimensionality is a common problem in the context of signal analysis. Consider for example, the assessment of the dynamics of biomedical/biophysical signals (e.g., EEG time series). These may be assessed with either linear (mainly: power spectral) and/or nonlinear (mainly: fractal dimension) analysis methods [1–5]. Each of the methods used for analysis of the time series extracts one or several measures out of a signal like peak frequency, band power, correlation dimension, K-entropy, and so forth. Some, but not necessarily all of these measures are supposed to exhibit state-specific information connected to the underlying biological/physiological process. Let us denote a collection of these measures a feature vector. An appropriately weighted collection of these information, specific measures may span an optimal feature vector in the sense that the states may be best separated.

The temporal variation of these signals often has to be regarded as being almost stationary over limited segments only and not as being stationary in a strict sense, a property which is sometimes denoted as "quasistationarity". This suggests regarding a specific outcome as being randomly drawn from a distribution of outcomes around a state-specific mean. Hence any inference made on such outcomes must be based on statistics relating the effect of interest to that stochastic variation even when regarding a single individual. If a comparative study is conducted, one has to select samples of probands, and this again introduces sources of random variations into analysis. The problem to solve is hence twofold. Efforts must be made (1) to retrieve effects out of the random variations for the different measures and (2) to reduce the set of all measures to the set of those which allow for a reliable state identification.

A widespread statistical method used to attack the first type of problem is known as analysis of variance. Given the ith measurement of a biophysical/biomedical signal, the perhaps most simple variance analytic model for this signal reads as

$$\text{signal}_{ji} = \alpha_j + \text{error}_i, \tag{1}$$

where i denotes the ith measurement of the signal which was obtained under experimental condition j. The so-called effect (or treatment) term α_j may be a fixed or a random effect and either continuous or discrete (cf. below). With regard to model (1), the analysis of variance infers the extent to which the estimates of the squared differences among the effects α_j rise above the squared error. Testing the significance of the effect then depends upon whether the levels α_j are regarded as fixed or random, whereby the null hypothesis is normally formulated as having equal levels.

A typical situation for this problem is when a study is based on a sample of probands. The probands must be viewed as a random sample drawn out of the reservoir of all possible individuals.

If no correction is made, the analysis result applies specifically to the sample at the end. This is in most cases not the effect hunted for because one searches results applicable also to those (normally vast majority of) humans who were not included in the study, for example, reliable discriminant functions. The classical approach in variance analysis splits the effect term into two parts, fixed and random, and also enriches the error term with an estimate of the random part.

As an alternative to this classical approach, one may consider the family of the so-called F-ratio tests which are based on randomly splitting and recollecting the sample. One hereby chooses repeatedly random subsets of the original data to gain an estimate of the variance of F, namely, $\sigma^2(F)$, and inspects the ratios $\sigma^2(F)/F^2$ or variants therefrom [6]. Here F denotes the quantity obtained from a F-test (cf. Section 2.1). Such resampling methods have proven capabilities to enhance statistical inference on parameter estimates which are not available otherwise. The most popular examples of such methods are known as Jackknife or Bootstrap. F-ratio test statistics have indicated to (a) better retrieve fixed effects by fading away the random parts and (b) allow for an incremental test, that is, testing the effect of the inclusion of additional variables into an existing feature vector. The latter property makes them especially interesting when one tries to reduce the dimension of a feature vector to an optimal size. The different combinations with additional variables included lead to different probabilities under the hypotheses of interest which, in turn, allow for a weighted inclusion of these measures into an optimal feature vector. One may thus perform an adaptive model selection.

A traditional way of model selection would be to perform analysis on all combination of features under interest and then to make a decision with the help of some information criterion (AIC, BIC, etc.). These try to select the optimal combination by weighting the number of measures in the model against residual error. This kind of selection leads to an inclusion of a measure with weight of either one or zero, however, and may neglect knowledge gained from incremental tests as those mentioned above. This pecularity motivated us to search for alternatives. Weighting information of different sources to an optimal degree is frequently conducted via Bayes' theorem. The Bayesian view will be adapted to derive weights different from zero and one for the construction of feature vectors, that is, to allow for partial inclusion. We note that reduced inclusion is also an important property of the so-called shrinkage or penalized regression methods [7].

The rest of the paper is organized as follows. We first recapitulate the derivation of three different F-ratio test statistics and outline the computational scheme to construct the corresponding confidence intervals by means of Monte Carlo simulations. A comparison to the outcome of the traditional method is made. We then show the inclusion of the outcome of these multivariate statistical methods into a selection scheme following a Bayesian heuristic by weighting hypotheses. This allows for reliably constructing weights for the measures. These weights are the basis for constructing reliable feature vectors suitable for further analysis, for example, discriminance procedures.

We demonstrate our approach on the reanalysis of an earlier study and address the problem of state specificity: psychosis versus nonpsychosis as expressed in the EEG. It is shown that an optimal combination of the so-called relative unfolding (or Taken's) χ and two power spectral estimates (α, δ) will allow for a correct classification of at least 81% of the probands, even in absence of active mental tasks.

2. Recapitulation of the F-Ratio Test

2.1. Recapitulation of ANOVA/MANOVA. The usage of analysis of variance is the traditional approach to distinguish systematic effects from noise. The methods of analysis of variance (ANOVA/MANOVA) try to decompose the variance of a population of outcomes (e.g., the results of EEG assessments obtained under different well-defined conditions) into two parts, namely, the treatment effect and the error effect. We adopt the notation of Bortz [8] and denote the treatment effect as h^2 and the error effect as e^2. The treatment effect h^2 explains how much of the total sum of squares may be due to a systematic effect of the different conditions (treatments). The second part, e^2, is an estimator of the remaining sum of squares due to other random or noise effects. In the light of (1), the term "error" affects both, e^2 and h^2, whereas α affects h^2 only [8]. The important question is: to what extent the treatment effect significantly rises above the level of a possible error effect. The quantity entering this test is (univariate case)

$$c = \frac{h^2}{e^2}. \tag{2}$$

As stated above, h^2 denotes the sum of squares due to treatment and e^2 the sum of squares due to error. If the influence of the treatment is zero, h^2 also reflects only the error influence. Hence the test may be formulated as an F-test, that is, to test whether a calculated value of F might have occurred by chance or if the value deviates significantly from an outcome by chance. This might be done classically by comparing the evaluated value of F with the values in a table displaying F-value probabilities or get it from an appropriate statistical software package.

The F-value is given as

$$F = \frac{cg}{g} \cdot \frac{df_e}{df_h} = \frac{h^2}{e^2} \cdot \frac{df_e}{df_h} := \frac{\sigma_h^2}{\sigma_e^2}, \tag{3}$$

where g is some appropriate weight (without having an effect in the univariate case, however), and df_e and df_h are the corresponding degrees of freedom, respectively. The univariate case (ANOVA) tests the influence of one or more treatment effects upon the outcome of a single variable, for example, how the nonlinear correlation-dimension estimate b_0 [9] is affected by group, mental situation, and proband (cf. Section 4).

The possible existence of an overall effect must be tested not only on b_0 but also simultaneously on all evaluated measures, however. So the appropriate test is not a sequence of ANOVA tests but a multivariate approach (MANOVA). This is because the outcome of the variables might be statistically dependent to some degree, and thus the simultaneous effect is different from the set of the effects of the individual variables. Hence, (3) must be converted to the multivariate case. The quantities h^2 and e^2 turn into their corresponding matrices H and E [8]. The *F-test* depends now on the eigenvalues of the matrix HE^{-1} which is analogous to (3), but the single weight g splits up into the weights g_i, and these may be different for different axes$_i$. The most common of such F-values are

$$F_H = \frac{\sum_i^s c_i/(g)}{\sum_i^s 1/(g)} \cdot \frac{df_e}{df_h} \qquad (4)$$

(i.e., $g_i = 1/g \ \forall i$),

$$F_P = \frac{\sum_i^s c_i/(1+c_i)}{s - \sum_i^s c_i/(1+c_i)} \cdot \frac{df_e}{df_h} \qquad (5)$$

(i.e., $g_i = 1/(1+c_i)$), or

$$F_R = \frac{c_1/(1+c_1)}{1 - c_1/(1+c_1)} \cdot \frac{df_e}{df_h} \qquad (6)$$

(i.e., $g_1 = 1/(1+c_1)$; $g_i = 0 \ \forall i \geq 2$), where c_i is the ith (ordered by value) eigenvalue of the matrix HE^{-1}, and $s = \text{rank}(HE^{-1})$. Equation (4) is known as Hotelling's (generalized) T^2, [10], (5) as Pillais' trace [11], and (6) as Roy's largest root [12]. For a sufficiently large number of observations, F_H, F_R, and F_P become equivalent and, in the $s = 1$ case, they become identical. As in the univariate case, testing for significance of an effect is done by evaluating the probability that a calculated F-value might occur by chance. The software packages that perform MANOVA do normally return this probability together with further properties on the sum of squares involved in H and E.

2.2. Outline of the Problem Separating Fixed and Random Effects. To motivate the derivation of our algorithm, we consider the influence of a randomly chosen sample of persons out of a population, whereby other effects might also be present, but fixed. The effect term h^2 may then be decomposed into

$$h^2 = (\Delta a)^2 + (\Delta pa)^2 + (\Delta e)^2, \qquad (7)$$

where $(\Delta a)^2$ denotes here the influence of fixed conditions, $(\Delta pa)^2$ the effect of the (randomly chosen) persons, and

$(\Delta e)^2$ the influence of the random error effects [8]. (We note that the quantities $(\Delta a)^2$ and $(\Delta pa)^2$ are sometimes also called treatment effects in a biomedical context). Under the null hypothesis of having no fixed effect, $(\Delta a)^2$ is assumed to be zero. The same holds—in principle—for $(\Delta pa)^2$. Generally, if an observable stems from a subpopulation drawn from a larger set, the corresponding effect may itself become random. This is normally the case when regarding person as condition (one will never be able to assess all humans). Hence, $(\Delta pa)^2$ is zero only within the bounds of statistical deviations. The classical approach to solve this problem within the ANOVA/MANOVA framework is a modification of the F-test. The error term is hereby enhanced from e^2 to $(e^2 + (\Delta pa)^2)$, and the effect is tested through $h^2/(e^2 + (\Delta pa)^2)$ instead of (2). The obvious disadvantage is the requirement of a higher level of the effect $(\Delta a)^2$ which has to rise significantly above the "noise-"term $(e^2 + (\Delta pa)^2)$ as compared to the pure noise level due to e^2.

So an attempt to test $(h^2 - (\Delta pa)^2)/e^2$ seems more favorable. But this might lead to a negative variance estimate, and it is not clear what effective degrees of freedom would have to be assigned to such a variance estimate.

2.3. Derivation of the F-Ratio Test Statistics. To overcome this situation, we propose a statistic estimating the influence of the population with the help of a resampling technique. This statistic is based on the decreasing sample-to-sample variation when a fixed term is present as compared to the influence of purely random effects.

Following [6], we rely (a) upon the classical error propagation rule and (b) upon the variance's variance. The error propagation rule is given as [13]

$$\sigma^2(g(x)) = \left(\frac{\partial g}{\partial x}\right)^2 \sigma^2(x) + \text{h.o.t.}, \qquad (8)$$

where g is a smooth function, x a random variable, and h.o.t denote higher order terms. As usual in error propagation considerations, this formula neglects correlational and higher order effects. We mention further that neglecting variations around absolute means the variance of an empirical variance estimate may be written as [14]

$$\hat{\sigma}^2(\sigma^2) = \frac{2\hat{\sigma}^4}{df}. \qquad (9)$$

We denote the variance with $\hat{\sigma}^2$ and the empirical variance estimate with σ^2. This conforms to (3).

As our last step (c), we decompose $\hat{\sigma}^2(h^2)$, the variance of the effect term

$$\hat{\sigma}^2(h^2) = \hat{\sigma}^2\left((\Delta pa)^2\right) + \hat{\sigma}^2\left((\Delta e)^2\right). \qquad (10)$$

We assumed here all error terms to be uncorrelated to the rest. Essential here is the fact that the fixed effect does not contribute to the variation of h^2 and accordingly does not enter into the variance $\hat{\sigma}^2(h^2)$. With (9), (8), and (7), we may write the variance of the F-value defined in (3) as

$$\sigma^2(F) = F^2\left[\frac{\sigma^2(h^2)}{h^4} + \frac{\sigma^2(e^2)}{e^4}\right] + \text{h.o.t.} \qquad (11)$$

Using (8), this turns into

$$\sigma^2(F) = 4F^2 \left[\frac{\nu}{2df_k} + \frac{1}{2df_{ek}} \right], \qquad (12)$$

where df_k denotes the degrees of freedom of the effect considered, df_{ek} the corresponding error degrees of freedom, and ν is the ratio

$$\nu = \frac{\left((\Delta pa)^2 + (\Delta e)^2 \right)}{h^2}. \qquad (13)$$

We note that in the case of a pure random effect, ν becomes 1 and significant deviations towards a lower value point to a nonnegligible fixed effect. Equation (12) obviously suggests using the statistic $\sigma^2(F)/F^2$ to test for $\nu < 1$. According to (12), the expectation value of this statistic is—under the null hypothesis $\nu = 1$—given by $1/2df_k + 1/2df_{ek}$. To gain an estimate for $\sigma^2(F)$, one may randomly resample, m times, a subset encompassing an equal number of probands from the original sample and, each time, find the F-value corresponding to the particular subset. So the method becomes a variant of the so-called delete-d jackknife [15]. It has been shown that the following quantity estimates $\sigma^2(F)$ up to a factor [16, 17]

$$\sigma^2(F) = \frac{1}{m-1} \sum \left(F_j - \langle F \rangle \right)^2, \qquad (14)$$

where $E(\sigma^2(F)) = \hat{\sigma}^2(F)$. The number of random splittings conducted is denoted as m, the average $\langle F \rangle$ is defined as

$$\langle F \rangle = \frac{1}{m} \sum F_j, \qquad (15)$$

and F_j denotes the found F-value obtained from the jth of the m runs. The above mentioned factor depends on #probands and selected #probands per random sample [15]. (We abbreviate here "number of" with the symbol #.) This is important, because p, the probability of a person to appear in a particular random sample, increases with the ratio #probands per random sample/#probands per sample. In case of a small sample size, this may impose an additional restriction of the variance $\sigma^2(F)$ [6].

The cumulative distribution of the ratios $\sigma^2(F)/\langle F \rangle^2$ will hence depend on the parameters (df_k, df_{ek}, #random splittings, #probands, #probands per random sample). The #random splittings, m, hereby influences the cumulative distribution because higher values for m lead to a narrower deviation around $\hat{\sigma}^2(F)$. A deviation from a random result may be found by estimating the probability that a ratio $\sigma^2(F)/\langle F \rangle^2$ is by chance as small or smaller than the experimentally found estimate. If this probability is too low, the null hypothesis is rejected. We will come back to this point in the following section.

These ideas may be extended to the multivariate case [6]. We note that the error effects may again be assumed to be uncorrelated. Therefore the off-diagonal elements of E are

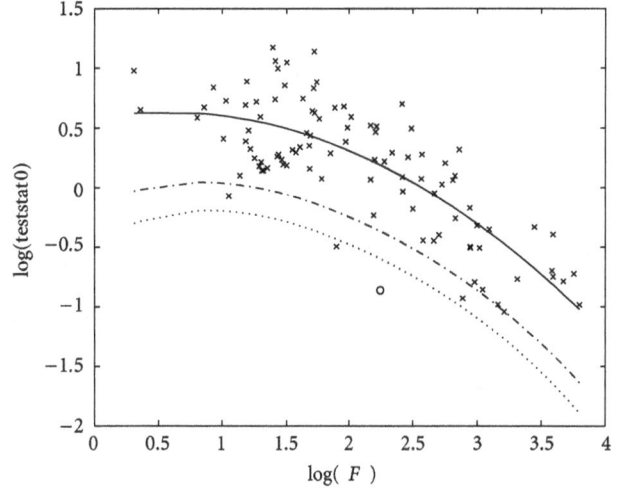

FIGURE 1: Outcome of an artificially generated signal with fixed effect (o) for our test statistics (testat0 (16) versus $\langle F \rangle$ (15), logarithmic scale) compared to outcomes of the corresponding random effects (x). The deviation from the expected value (solid line) of the latter is highly significant and below the 5% level (dash-dotted line) and even the 1% level (dotted line). The classical method according to Section 2.1 revealed the (insignificant) 13.95% level only. The proposed method recognizes the nonrandom effect correctly in this example while the classical approach does not.

random with an expectation value of zero. Furthermore, the trace of the matrix HE^{-1} remains unchanged when the basis is changed such that the eigenvectors build the new basis. Hence the diagonal terms of HE^{-1} are expected to represent, on the average, the individual F-values, and the trace is the sum over the individual F_i's. In case of a fixed effect with only two states ($s = 1$) and n random variables, this leads to a multivariate F with value $1/n \sum_{i=1}^{n} F_i$. To test the null hypothesis H_0 of having random effects only, we may again use the independence of $\sigma^2(F_i)$ and find testat0, our first test statistic,

$$\text{teststat0} = \frac{\sum \sigma^2(F_i)}{4(\sum F_i)^2}, \qquad (16)$$

whose distribution is a function of (df_k, df_{ek}, n, #random splittings, #probands, #probands per random sample). If random effects for the treatment term exist, things become a bit more complicated. In that case, the contributions of the individual $\sigma^2(F_i)$ may be unequal, and—in extremis—the sum may be dominated by one single term. A way to account for this effect is to consider df_{eff}, the effective degrees of freedom. The effective degrees of freedom are defined as $df_{eff} = (\sum \sigma_i^2)/(\sum(\sigma_i^2 / df_i))$ (cf. [8], Chapter 8). This quantity is minimized if one term is clearly dominant and maximized when there are equal contributions.

As stated above, if an empirical value of testat0 appears too low, one may conclude that there is a systematic nonrandom deviation in at least one variable between the treatment groups under consideration (see Figure 1).

In the case of a true multivariate statistic type, one has to replace the univariate individual F-values by the eigenvalues of HE^{-1} and modify testat0 into

$$\text{teststat1} = \frac{\sum_{i=1}^{s} \sigma^2 \left(1/g_i \sum_{j=1}^{n} k_i^j F_j \right)}{4 \left(\sum_{i=1}^{s} 1/g_i \sum_{j=1}^{n} k_i^j F_j \right)^2}, \qquad (17)$$

where $k_i^j F_j$ is the contribution of the individual univariate F-value F_j to the ith eigenvalue of (HE^{-1}) adjusted with the degrees of freedom, namely, $c_i\, df_e/ df_h$. This statistic depends on $(df_h, df_e, n, \#\text{simulations}, \#\text{probands}, \#\text{probands per random sample}, \text{stattype}, df_{\text{eff}})$. If stattype, the statistics type, is Hotelling's statistics, this obviously becomes equivalent to the $s = 1$ case because $g_i = \text{const.}$ and $F = \sum_{i=1}^{s} c_i\, df_e/df_h$ (cf. Section 2.1). In absence of a between-variable effect, one will have

$$\text{testat1} = \frac{\sigma^2(F_{\text{multi}})}{4 F_{\text{multi}}^2}. \qquad (18)$$

This suggests two normalized versions of our test statistic in the following way:

$$\text{teststat1}_R = \frac{\sum_{i=1}^{s} \sigma^2 \left(1/g_i \sum_{j=1}^{n} k_i^j F_j \right)}{4 \left(\sum_{i=1}^{s} 1/g_i \sum_{j=1}^{n} k_i^j F_j \right)^2} \Bigg/ \frac{\sigma^2(F_{\text{multi}})}{4 F_{\text{multi}}^2}. \qquad (19)$$

The expectation value under the null hypothesis (i.e., having no multivariate effect) is 1, and the cumulative distribution depends on $(df_h, df_e, n, \#\text{simulations}, \#\text{probands}, \#\text{probands per random sample}, \text{stattype})$. Significant deviations from 1 indicate that at least one variable shows a fixed effect or that a between-variable effect exists.

As a last step, we extend (19) to an incremental test statistic. In the case of having already knowledge on certain measures displaying a multivariate effect, one may wish to test for the influence of an additional measure. We therefore modify the test statistic testat1$_R$ into

$$\text{teststat1}_M = \frac{k^2 \sigma^2(F_c) + \sigma^2(F_{\text{add}})}{4(kF_c + F_{\text{add}})^2} \Bigg/ \frac{\sigma^2(F_{\text{multi}})}{4 F_{\text{multi}}^2}, \qquad (20)$$

where k is the number of those measures already showing a multivariate effect, and F_c is the F-value found with these measures. Our assumption of an existing effect implies $F_c > 1$, because $E(F_c) > E(F_{\text{random}})$ and $\sigma^2(F_c) \leq \sigma^2(F_{\text{random}})$. Hence testat1$_M$ tests the null hypothesis $(F_c > 1, \nu = \nu(F_c))$, that is, the additional variable has no influence. The cumulative distribution function then depends on $(df_h, df_e, n, \#\text{simulations}, \#\text{probands}, \#\text{probands per random sample}, F_c, \sigma^2(F_c), df_{\text{eff}}, \text{stattype})$ because $E(F_c) > E(F_{\text{random}})$ and $\sigma^2(F_c) \leq \sigma^2(F_{\text{random}})$. Because $\sigma^2(F_c)$ is assumed to be unequal to $\sigma^2(F_{\text{add}})$, we must again consider the so-called effective degrees of freedom df_{eff} of the pooled variances.

The assumptions entering this incremental test are the same as in testat1$_R$. The null hypothesis states that the additional measure contributes its univariate F-value F_{add} to the trace while F_{add} is built up from nonfixed effects only. If the testat1$_M$ becomes unexpectedly high, this may

be regarded as indicating an additional systematic effect due to the inclusion of this measure. If the statistic type is Hotelling's statistic, this becomes again equivalent to the $s = 1$ case.

These statistics are useful answering questions like the following: "are there measures providing significantly to the treatment term?" and, if so, "which ones may be identified?" and "to what extent do they provide to the effect?" The knowledge of such measures and its contribution to the treatment effect allows one, for example, to select them and collect them with appropriate weights into a feature vector useable for discriminance or predictive purposes.

2.4. The Computational Scheme to Determine Confidence Intervals for the F-Ratio Test Statistics and Comparison with the Classical Approach . The quantity of interest, namely, the distribution of the ratios $\sigma^2(F)/F^2$, must be evaluated numerically, and the dependence of the ratios from the number of random splittings and the number of persons involved calls for a calculation of the confidence intervals for each case. Generating the distribution of the F-ratios appropriately and, therefrom, the desired confidence interval is our method of choice to overcome this problem. This algorithm is basically a Monte Carlo technique generating L outcomes and their F-ratios. This leads to a population of L random deviates of the ratio $\sigma^2(F)/\langle F \rangle^2$ according to the appropriate null hypothesis (remember Figure 1). We note that both the F-value obtained for the whole sample as well as $\langle F \rangle$ (15) provide an estimate for F and calculating $\sigma^2(F)$ and $\langle F \rangle^2$ is done within the same procedure, so we prefer $\sigma^2(F)/\langle F \rangle^2$. From the population of the L ratios, one may derive a quantile and the associated probability P, for example, by building a histogram or ordering the population by rank and selecting the $P \cdot L$th value. This value estimates the quantile above which F-ratios occur by chance with probability P.

2.4.1. General Scheme. The general scheme of our algorithm is stated in more detail as follows [6].

(1) Restate the model through a separation of the desired factor. The multivariate model describing our null hypotheses may be derived from (1) and may be formulated as

$$\text{Signal}_{ijk} = \alpha_{i(j)} + \beta_j + \text{error}_{ijk}, \qquad (21)$$

where Signal_{ij} denotes the (uni- or multivariate) measured quantities, β_j the random factor considered (e.g., different clinical groups), α_i and the other factor(s), which may implicitly depend on the random factor.

(2) Determine/select the constants $k, L, m, \#n, p, \text{stattype}$ (if necessary) such that L is the number of deviates desired to estimate the quantile with acceptable accuracy, m is the number of random splittings needed for each deviate, $\#n$ the levels of the factor β (typically the number of persons involved, i.e., #probands), p the relative number of levels (or persons, i.e.,

#probands per random sample/#probands) entering one splitting, k the number of levels of α_i, and stattype is again the multivariate statistic type. The values k, m, #n, p, stattype must conform to the setting with which the original data was analyzed.

(3) Perform the Monte Carlo loop. This encompasses the following steps.

   (a) Generate a sequence of #n times k random numbers to mimic the random errors in (21). The amplitude must be chosen to match the value found for e^2 in the original analysis.

   (b) Generate another random #n-sequence to mimic the influence of the random factor. The amplitude must be chosen to match the null hypothesis. The random treatment effect assumed, $(\Delta pa)^2$, should be chosen such that $\langle F \rangle$ matches the found univariate outcome.

   (c) Add the different contributions to the simulated signal.

   (d) Build m random splittings and analyze it by the same procedures as the original sample was analyzed. Typically m is chosen to lie between 12 and 50. From the m splittings, build $\sigma^2(F)$, $\langle F \rangle^2$ (14), and (15), and the ratio $\sigma^2(F)/\langle F \rangle^2$. The analysis is normally done by means of a statistical software package estimating an appropriate F-value. This is sufficient for testat0. In the case of testat1, also build $\langle F_{\mathrm{multi}} \rangle^2$, $\sigma^2(F_{\mathrm{multi}})$, and the ratios $\sigma^2(F_{\mathrm{multi}}) \langle F_{\mathrm{multi}} \rangle^2$ and $(\sigma^2(F)/\langle F \rangle^2)/(\sigma^2(F_{\mathrm{multi}})/\langle F_{\mathrm{multi}} \rangle^2)$. These are necessary for the different variants of testat1 (18)–(20).

   (e) Repeat steps (a) to (d) L times and gain therefrom empirically the quantile(s) of interest. As stated above, this may be done by means of a histogram or a rank ordered sequence obtained from the L F-ratios $\sigma^2(F)/\langle F \rangle^2$ and $(\sigma^2(F)/\langle F \rangle^2)/(\sigma^2(F_{\mathrm{multi}})/\langle F_{\mathrm{multi}} \rangle^2)$. Depending on the probability P associated with the quantile and the desired accuracy, L will typically be on the order of $10^2, \ldots, 10^5$.

The statistic testat1$_M$ (20) requires some attention with respect to (a) simulation and (b) effective degrees of freedom. This is because we estimate $\sigma^2(F_c)$, where F_c is expected to be larger than one due to the already recognized fixed or common effect and, therefore, $\sigma^2(F_c) < \sigma^2_{\mathrm{random}}$.

F_c is carried over from the result obtained without the measure under consideration, so we test the additional measure under the constraints that the known effect equals F_c (or $F_{\mathrm{total}} = F_{\mathrm{sample\_total}}$). In the case that the measures contributing to F_c are expected to carry fixed effects, the model must also be adjusted with a fixed effect, such that the expected values $E(\sigma^2(F))$ and $E(\sigma^2(F_c))$ match the corresponding values of the original sample. The quantiles must be derived at the point where df_{eff} matches df_{eff} of the original sample. This may be done by repeating step (e) thus collecting a

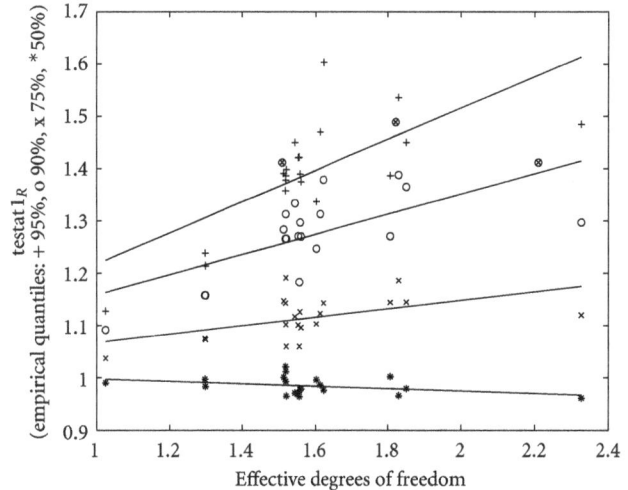

FIGURE 2: Variation of quantiles of the test statistics with the effective degrees of freedom df_{eff}. 50% (∗); 75% (x); 90% (o); 95% (+) for a variety of simulations and their corresponding functional fit. The ⊗ denotes the results presented in Table 1. These are from left to right χ-δ, $\chi - b0$, χ-δ-α.

population of empirical quantiles belonging to the same probability P and building a functional dependence quantile versus df_{eff} (cf. Figure 2, where dependencies quantile$_P$ = a_P + $b_P \cdot df_{\mathrm{eff}}$ were fitted). The alternative is waiting until L results with approximately equal effective degrees of freedom emerged by chance.

2.4.2. Particular Settings. The reconstruction of the model (21) is performed by generating streams of two types of uncorrelated random numbers from a normal distribution. The first type will mimic the error and has simulation parameters $(0, \sigma^2_e)$, that is, the estimated squared mean of the error$_{ij}$ of the original sample. The second type has simulation parameters $(0, \sigma^2_p)$, that is, the average squared effect due to the probands. Both quantities may be read out from the output of the classical ANOVA/MANOVA analysis (cf. Section 2.1) of the original sample. In this respect, the expected outcome of the simulation with the classical approach will correspond to the result obtained with the original sample, if the parameters k and #n also correspond to the original sample and the null hypothesis H_0: "no presence of a fixed effect due to person group" is true.

Our clinical sample consists of 30 persons from two clinical groups evaluated at four mental states ([18], see also Section 4.2). So we have $k = 4$ and #$n = 30$. Because the mental states have shown fixed effects in previous studies [18, 19], the simulated signals were offset by four fixed different levels. The amount of the offset values is not relevant, however, because the offset is fixed and the F-ratio test is set up to test for differences between the two groups. The offsets were introduced only to mimic better the original data. Hence a simulated person has four outcomes built by one choosing four times the same random deviate from $(0, \sigma^2_p)$ plus four times a different random deviate from $(0, \sigma^2_e)$ enriched with the state-specific offset. The first 15 simulated

TABLE 1: Outcomes of F-ratio test statistics with a considerable significance level for EEG feature vectors.

Test statistic used	Feature vector (measures)	F_{multi}	Ratio	Test statistic value	df_{eff}	Significance level
teststat1$_R$ (19)	χ, δ-power	6.168	0.233	1.412	1.507	>0.95
teststat1$_R$ (19)	χ, b0	10.393	0.145	1.489	1.822	>0.95
teststat1$_R$ (19)	χ, δ-power, α-power	6.890	0.158	1.416	2.21	>0.90
teststat1$_M$ (20)	χ, δ-power, α-power	6.890	0.158	1.192	2.21	$\simeq$ 0.90

persons were labeled as group 1 and the last 15 labeled as group 2. The F-ratio tests were conducted with $m = 30$ and $p = 2/3$, if not stated otherwise. A Monte Carlo loop was normally evaluated with $L = 100$ for each stattype. Hence getting results for each of the stattypes testat0, testat1$_R$, and testat1$_M$ requires three different runs of the Monte Carlo loop. Roy's largest root (6) was used as the classical method, if not stated otherwise.

The F-ratio test statistic obviously requires more numerical efforts than the classical approach. So one could ask if its usage might be worth these efforts. We therefore tested the sensitivity of the F-ratio tests to the presence of fixed effects of person categories, that is, we tested for H_0 in case when H_0 is false. A comparison of runs on 250 different artificial data sets was made. We evaluated for each data set the probability that a test outcome as high or higher may occur by chance. This was done for both the classical test and the F-ratio test (applying a nonparametric method). Then we built for each set ΔP the difference between the probability according to the classical and the probability according to the F-ratio test. The resulting 250 values of ΔP were then sampled into a histogram. In case of equivalence of the two methods, one would expect a symmetric distribution around zero. Our data (Figure 3) show a significant deviation from a symmetric distribution towards the F-ratio test ($\chi^2 = 5.6$, $P = 0.02$). The F-ratio test seems to be more sensitive to the presence of a fixed effect than the classical approach, thus a higher tendency to reject H$_0$ in the case when the test should reject it.

This seems not to be too surprising, however, because the deviations from the expected value of the quantity $\sigma^2(F)/\langle F \rangle^2$ occur in 4th power instead of the 2nd power as in the classical view. A further advantage of the F-ratio is its applicability to nonnormally distributed data because random number generation for nonnormal data bears no additional difficulties.

Having established this as a method for an incremental inclusion of measures, we will now turn to the problem of using this knowledge to construct optimized feature vectors.

3. Hypothesis Weighting

Consider the outcomes of the tests above of, say, three measures which occur with different significance levels. We

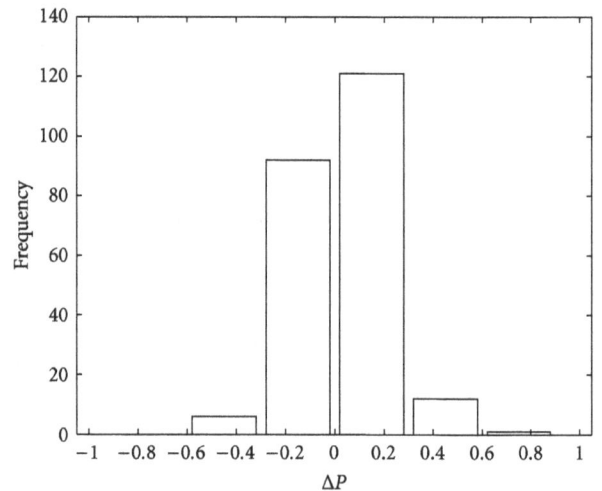

FIGURE 3: Comparison of the F-ratio test with the classical approach for 250 data samples. The probability of the spontaneous occurrence of the corresponding outcome is on the average smaller than with the classical approach. This is shown by the asymmetric distribution of ΔP, the differences between the two probabilities.

make the assumption that from these measures (or variables) the one with the least significance carries also the least information, while the others bear more information in accordance to their significance level. The problem with what weight they should enter into a feature vector is regarded from a Bayesian view. Bayes formula allows one to express a conditional probability $P[A_i \mid B]$ with the conditional probabilities $P[B \mid A_j]$ through

$$P[A_i \mid B] = \frac{P[A_i]P[B \mid A_i]}{\sum_j P[A_j]P[B \mid A_j]}. \qquad (22)$$

This may be used to express the probability of a hypothesis H_i to be correct by means of the probabilities of the outcomes corresponding to the different hypotheses tested for. Consider two hypotheses H_0 and H_1 concerning the quality of the measures/variables. We would like to weight the hypotheses H_0 (measures display no difference between groups) and H_1 (measures display a difference between groups). The probability $P(H_i)$, namely, H_i being correct, appears as a natural weight for this hypothesis. Let b denote the empirical outcome of an F-ratio test as obtained with the Monte Carlo technique above. Let B denote the set of possible outcomes which deviate at least as much as the quantile belonging to the significance level π. If b exceeds this quantile it is also an element of B. The set B then allows for weighting hypotheses by means of (22).

We may set the a priori probabilities $P[H_0] = 1 - P[H_1] = c = 0.5$, because we have no a priori preference neither for the hypothesis H_0 nor an alternative H_1. We may further assume the probability $P[B \mid H_1] = c_2$. The quantity $P[B \mid H_0] := \pi$ is our present knowledge, namely, the probability assigned to find an outcome b within B, given H_0, for example, $\pi = 0.05$, $\pi = 0.1$, and so forth.

The probability of "$H_0 =$ true" given the set B may be written as (22)

$$P[H_0 \mid B] = \frac{c\pi}{c\pi + c_2(1 - c)} \qquad (23)$$

and, similarly,

$$P[H_1 \mid B] = \frac{c_2(1 - c)}{c\pi + c_2(1 - c)}. \qquad (24)$$

In general, we find the quantities $p[H_1^i \mid B]$ and may formally assign an "expected hypothesis" through the weighted mean

$$\overline{H} = \frac{\sum H_1^i p\left[H_1^i \mid B\right]}{\sum p\left[H_1^i \mid B\right]}. \qquad (25)$$

The formulation of an "expected alternative hypothesis" seems somewhat purely formal at this stage. However, if each hypothesis is intrinsically connected to a specific feature vector f_i, this approach returns the expected feature vector $\overline{f}$ given the observation B, however,

$$\overline{f} = \frac{\sum f_i p[f_i \mid B]}{\sum p[f_i \mid B]}, \qquad (26)$$

because each feature vector f_i is spanned by its specific collection of measures

$$f_i = \{A, B, C, \ldots\}_i. \qquad (27)$$

From the weights of the hypotheses one immediately also gets the weights of the measures. In the context of EEG time series analysis, the measures $A, B, C, \ldots$ denote quantities like correlation dimension, peak frequency, spectral band power, and so forth.

A simple weighting follows for the case of two possible alternative hypotheses. The likelihood ratio $P[H_1|B]/P[H_0|B]$ then gives the weight with which the alternative is preferable to H_0 when the weight of H_0 is set to 1. It is expressed as

$$\frac{c_2(1 - c)/(c\pi + c_2(1 - c))}{c\pi/(c\pi + c_2(1 - c))} = \frac{c_2(1 - c)}{c\pi}. \qquad (28)$$

Now consider two alternatives H_1^1, H_1^2 and $P[B^1 \mid H_0] = \pi_1$, $P[B^2 \mid H_0] = \pi_2$, and $P[B \mid H_1^i] = c_2$ for all H_1^i (i.e., no

preference for any alternative). Their likelihood ratio may be expressed through the ratio of their likelihood ratios against the null hypothesis [6]

$$\frac{c_2(1 - c)/c\pi_1}{c_2(1 - c)/c\pi_2} = \frac{\pi_2}{\pi_1}. \qquad (29)$$

This may be regarded as the weight with which the second alternative should enter when the weight of the first alternative is set to 1. If in addition H_1^1 is a subset of the H_1^2, that is, the variables assigned to H_1^1 are a subset of the variables assigned to H_1^2, this weighting applies to that part of H_1^1 which is not common to H_1^2.

We have to note that the formulation of c_2 is correct only when each probability π_i is small. If this is not the case, some correction might be required [6].

The application to the problem optimizing a feature vector is straightforward. The ith feature vector is regarded as the ith combination of measures corresponding to the ith hypotheses. To find the weights with which the variables enter the feature vector, we assume assigning the weight 1 to that combination of measures with the highest significance level. Taking into account the implicit dependence of c_2 as stated above, the subsequent variables will enter with weights according to (26). If a probability (thus weight) falls close to zero, it may be set to zero which results in dropping that particular feature vector and its corresponding measures. This reduces the dimension of the optimal feature vector.

4. Application to the Problem Discriminating EEG States

4.1. Motivation of the Problem and Results of Earlier EEG Analysis. As an application, we choose the problem of distinguishing the EEG of the two proband groups taken from a neuropsychologically oriented study [19] by their EEG. This choice was motivated by the following: it is well known that schizophrenic patients show abnormalities compared to healthy controls when the so-called evoked potentials are studied [20–22]. This may point to a threshold regulation problem in the activation of the neural network in schizophrenics [23], and there might be differences in the metabolism of the frontal cortex [24, 25]. Therefore one may expect differences in the spontaneous EEG. Such differences were indeed reported repeatedly, for example, [26–28] using linear (FFT) or nonlinear (correlation dimension) analysis.

An earlier study conducted with our proband samples (cf. below) revealed a significant difference between the two samples but only for a specific mental task [18]. While the EEG of the controls showed a drastic decrease in dimensionality, the EEG of the patients did not exhibit any peculiarity. Other studies, however, pointed to the existence of a difference in the "eyes-closed quiet" state [2, 9]. The degree to which this difference is visible in the "eyes-closed quiet" state, that is, in absence of external activation, however, is not yet established and was examined with the method proposed here.

4.2. Proband Sample and EEG Analysis. The neuropsychologically oriented EEG study consisted of two groups, namely, 15 acute hospitalized subjects diagnosed as schizophrenic and 15 controls in a healthy state. EEG measurements were repeated for four different mental tasks [19]. A trained clinical staff member ranked each patient's symptoms on a psychiatric rating scale, and the psychopharmaceuticals were noted. Both groups were exposed to the same mental tasks, while three 30-second segments of EEG were recorded [19]. We focus here mainly on the so-called "eyes-closed quiet" mental situation. The EEG were recorded according to the international 10–20 standard, which allows for the so-called parallel embedding scheme [2].

Our nonlinear EEG analysis follows a biparametric dimensional technique. In contrast to standard methods, this technique also considers attractor unfolding, and the outcomes provide several nonlinear measures, namely, the asymptotic correlation dimension (b_0), the so-called unfolding dimension m^*, and the relative unfolding (or Taken's) χ [9]. In addition, EEG analysis with conventional FFT techniques [29] was performed. This provided measures like α- or δ-power, that is, the spectral power from the so-called α (8–12 Hz) and δ (1–5 Hz) frequency band. A complete description of the proband samples, conditions, and technical settings is given elsewhere [18, 19]. With our experimental setup, the model consists of four fixed conditions (i.e., the four mental tasks) and two groups with 15 persons (i.e., patients and controls). According to our hypothesis, the influence of the group is in the focus of interest. Those persons building the two groups must be suspected to provide a sample-specific (or random) effect to the discriminant capacities between the groups (cf. Section 2), however, and demand for the application of our scheme. In each group, 10 from the 15 persons where chosen for the simulation, that is, at the point $p = 2/3$.

4.3. Results. The findings listed in the Section 4.1 led us to hypothesize differences in the absence of stimulated activation or medication. Therefore we applied our method to the EEG outcomes to the "eyes-closed quiet" situation. The results obtained with the different test statistics of this setting are shown in Table 1.

From here one sees that the relative unfolding χ seems to play the role of a major indicator, because χ occurs in all combinations of Table 1. This result is in agreement with findings from an earlier study [2] and with previous results from our sample [18, 19]. The δ power seems to be the best spectral measure because it appears in two combinations. An effect on the δ band is also in agreement with older findings in the literature [30].

This let us expect a reliable discrimination between the two states, schizophrenic versus healthy, by means of the EEG outcomes, if a combination of measures is appropriately selected. Among the triple combinations, only $f_i =(\chi, \delta$-power, α-power) seems to carry information. The combination (χ, δ-power, b_0) did not show any remarkable effect. So the effect on δ-power and b_0 seems somewhat opposite,

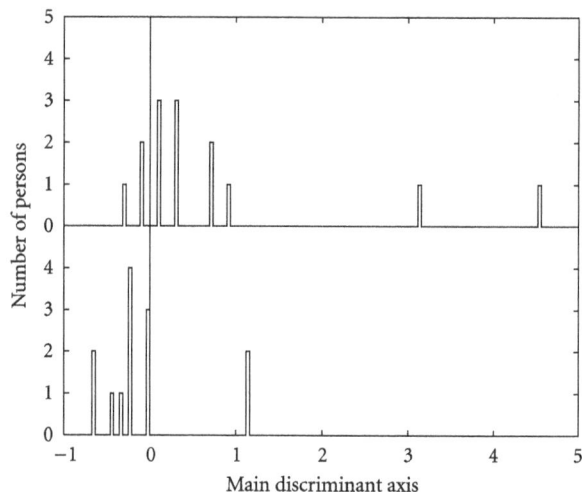

FIGURE 4: Discriminant analysis of EEG outcomes with weighted feature vector (eyes closed at rest). The number of persons is shown above the value on the main axis of the discriminant function where they appear. Upper: control group; lower: patient group (redisplayed from [6]).

and this combination was dropped. To discriminate between the two groups, it seems therefore reasonable to select the variables χ, δ-, and α power. The information obtained with these outcomes is used to build an appropriate feature vector.

Following Section 3 to find weights for feature vector components, we assume the 95% interval as significant and assign the weight 1. This conforms to π_1 and H_1^1: χ and δ-power. Applying our considerations to the 90% solution ($\pi_2 = 0.1$, H_1^2: χ, δ-power, α-power) reveals the weight 0.48. Hence, the variables χ, δ enter with weight 1.00 into the feature vector, while the variable α enters with weight 0.48 only. A discriminant analysis with this weighted feature vector reveals a correct classification with more than 81%. The result is displayed in Figure 4, where the outcome on the main axis of the discriminant function (essentially a rotation of the coordinate system [8], Chapter 18) is shown. The discriminant analysis could not be done on all 15 persons of each group. Due to failure to EEG-record quality requirements [19], one person of the control group and two persons of the patient group could not be evaluated, unfortunately.

We note that our *F*-ratio test statistics with its ability to perform multivariate and incremental testing on fixed effects allowed for this weighting of feature vectors. Furthermore, we may regard this result as reliable because this variable weighting has been done based on the emergence of fixed effects, therefore not optimizing across random (or sample-specific) discriminant capacities.

5. Discussion

We proposed and derived a computational scheme which is based on a random splitting method and which allows separating fixed and random effects in multivariate variance analysis. This approach seems to be advantageous in two respects. The classical method is implemented only for the univariate problem in most standard statistical software packages. So the decomposition of the effect matrix H into a fixed and a random effect requires additional matrix algebra programming efforts anyway. This may turn out to be a more difficult numerical problem than the generation of streams of random numbers.

Secondly, the normality assumptions inherent to the classical test also remain true for the multivariate test, namely, normally distributed random deviations around the effect levels. If this is not true, the statistics to be used do not follow an F-distribution and may be unknown, thus preventing a classical significance test.

In contrast, our method requires testing against quantiles derived from simulated outcomes. Thus the calculations can be done completely analogously when it seems more appropriate to use a distribution other than the normal distribution. Because our test statistic is based on relative ratios rather than absolute ratios, one might expect that an effect due to a particular distribution in the denominator will have a related effect in the numerator which could make our test statistic more robust.

Our tests for partial inclusion followed a Bayesian weighting of hypothesis. This leads to an optimized feature vector. This feature vector comprises those measures relevant to the fixed effect being tested for. This exceeds the classical model selection because each measure enters with an appropriate weight between one and zero rather than in an all or none fashion.

Another advantage of this approach is the simultaneous inclusion of linear and nonlinear measures. We note that the interpretation of the latter must be done with caution. It has been recognized for a long time that these measures are affected by noise and estimation errors when they are used for EEG analysis which then may circumvent their interpretation as chaos indicators (cf. e.g., [9, 31, 32] and the references concerning this matter therein). Despite this fact, these measures proved the ability to display individual properties of the EEG not seen with linear measures (cf. e.g., [2, 3]), and this is confirmed here.

As was shown with our EEG data, the above mentioned properties of our methods allowed for a clear distinction (>81%) between the two proband groups, controls versus schizophrenic patients, in a resting state with eyes closed. Earlier results stating that δ and χ seem to differentiate between the two groups are confirmed, but such a clear result has not yet been found in previous studies.

References

[1] H. H. Stassen, "The octave approach to EEG analysis," *Methods of Information in Medicine*, vol. 30, no. 4, pp. 304–310, 1991.

[2] R. M. Dünki and G. B. Schmid, "Unfolding dimension and the search for functional markers in the human electroencephalogram," *Physical Review E*, vol. 57, pp. 2115–2122, 1998.

[3] R. M. Dünki, G. B. Schmid, and H. H. Stassen, "Intraindividual specificity and stability of human EEG: comparing a linear vs a nonlinear approach," *Methods of Information in Medicine*, vol. 39, no. 1, pp. 78–82, 2000.

[4] P. Bob, J. Chladek, M. Susta, K. Glaslova, F. Jagla, and M. Kukleta, "Neural chaos and schizophrenia," *General Physiology and Biophysics*, vol. 26, no. 4, pp. 298–305, 2007.

[5] A. Khodayari-Rostamabad, G. M. Hasey, D. J. MacCrimmon, J. P. Reilly, and H. Bruin, "A pilot study to determine whether machine learning methodologies using pre-treatment electroencephalography can predict the symptomatic response to clozapine therapy," *Clinical Neurophysiology*, vol. 121, no. 12, pp. 1998–2006, 2010.

[6] R. M. Dünki and M. Dressel, "Statistics of biophysical signal characteristics and state specificity of the human EEG," *Physica A*, vol. 370, no. 2, pp. 632–650, 2006.

[7] T. Hastie, R. J. Tibshirani, and J. Friedman, *The Elements of Statistical Learning*, Springer, New York, NY, USA, 2001.

[8] J. Bortz, *Statistik: Für Sozialwissenschaftler*, Springer, Berlin, Germany, 1989.

[9] G. B. Schmid and R. M. Dünki, "Indications of nonlinearity, intraindividual specificity and stability of human EEG: the unfolding dimension," *Physica D*, vol. 93, no. 3-4, pp. 165–190, 1996.

[10] H. Hotelling, "A generalized t test and measure of multivariate dispersion," in *Proceedings of the 2nd Berkely Symposium on Mathematical Statistics and Probability*, J. Neyman, Ed., pp. 23–41, University of California Press, Berkeley, Calif, USA, July 1950.

[11] K. C. S. Pillai, "Some new test criteria in multivariate analysis," *The Annals of Mathematical Statistics*, vol. 26, no. 1, pp. 117–121, 1955.

[12] S. N. Roy, "On a heuristic method of test construction and its use in multivariate statistics," *The Annals of Mathematical Statistics*, vol. 24, no. 2, pp. 220–238, 1953.

[13] M. Kendall and A. Stuart, *The Advanced Theory of Statistics*, vol. 1, chapter 10, Griffin & Co., 1977.

[14] M. Fisz, *Wahrscheinlichkeitstheorie und Mathematische Statistik*, VEB Deutscher Verlag der Wissenschaften, Berlin, Germany, 11th edition, 1989.

[15] B. Efron and R. J. Tibshirani, *An Introduction to the Bootstrap*, Chapman & Hall, London, UK, 1993.

[16] J. Shao, "On resampling methods for variance and bias estimation in linear models," *The Annals of Statistics*, vol. 16, pp. 986–1006, 1988.

[17] J. Shao, "Consistency of jackknife variance estimators," *Statistics*, vol. 1, pp. 49–57, 1991.

[18] M. Dressel, B. Ambühl-Braun, R. M. Dünki, P. F. Meier, and T. Elbert, "Nonlinear dynamic in the EEG of schizophrenic patients and its variation with mental tasks," in *Workshop on Chaos in Brain?* K. Lehnertz, J. Arnold, P. Grassberger, and C. Elger, Eds., pp. 348–352, World Scientific, Singapore, 2000.

[19] M. Dressel, *Nichtlineare Dynamik im EEG schizophrener Patienten und deren Veränderungen in Abhängigkeit von mentalen Aufgaben*, Ph.D. thesis, Faculty of Social Sciences, University of Konstanz, 1999, http://deposit.ddb.de/cgi-bin/dokserv?idn=981087329&dok_var=d1&dok_ext=pdf&filename=981087329.pdf.

[20] W. S. Pritchard, "Cognitive event related potential correlates of schizophrenia," *Psychological Bulletin*, vol. 100, pp. 43–66, 1986.

[21] R. Cohen, "Event-related potentials and cognitive dysfunction in schizophrenia," in *Search in the Causes of Schizophrenia*, H. Haefner and W. F. Gatter, Eds., vol. 2, pp. 342–360, Springer, Berlin, Germany, 1990.

[22] D. Friedmann, "Endogenous scalp-recorded brain potentials in schizophrenia: a methodological review," in *Handbook of*

Schizophrenia, S. R. Steinhauer, J. H. Gruzelier, and J. Zubon, Eds., pp. 91–129, Elsevier Scence, New York, NY, USA, 1991.

[23] T. Elbert and B. Rockstroh, "Threshold regulation: a key to the understanding of the combined dynamics of EEG and event-related potentials," *Journal of Psychophysiology*, vol. 1, no. 4, pp. 317–333, 1987.

[24] S. W. Lewis, R. A. Ford, G. M. Syed, A. M. Reveley, and B. K. Toone, "A controlled study of 99mTc-HMPAO single-photo emission imaging in chronic schizophrenia," *Psychological Medicine*, vol. 22, no. 1, pp. 27–35, 1992.

[25] D. A. Yurgelun-Todd, C. Waternaux, B. M. Cohen, S. Gruber, C. English, and P. Renshaw, "Functional magnetic resonance imaging of schizophrenic patients and comparison subjects during word production," *American Journal of Psychiatry*, vol. 153, no. 2, pp. 200–205, 1996.

[26] M. Koukkou-Lehmann, *Hirnmechanismen Normalen und Schizophrenen Denkens: Eine Synthese von Theorien und Daten*, Springer, Berlin, Germany, 1987.

[27] M. Koukkou, D. Lehmann, J. Wackermann, I. Dvořák, and B. Henggeler, "The dimensional complexity of EEG brain mechanisms in untreated schizophrenia," *Biological Psychiatry*, vol. 33, no. 6, pp. 397–407, 1993.

[28] T. Elbert, "Slow cortical potentials reflect the regulation of cortical excitability," in *Slow Potential Changes of the Human Brain*, W. C. McCallum, Ed., pp. 235–251, Plenum Press, New York, NY, USA, 1990.

[29] P. D. Welch, "The use of fast fourier transform for the estimation of power spectra: a method based on time averaging over short modified periodograms," *IEEE Transactions on Audio and Electroacoustics*, vol. 15, no. 2, pp. 70–73, 1967.

[30] G. Winterer and W. M. Hermann, "Ueber das Elektroenzephalogramm in der Psychiatrie: eine kritische Bewertung," *Zeitschrift für Elektroenzephalographie, Elektromyographie*, vol. 26, pp. 19–37, 1995.

[31] R. M. Dünki, "The estimation of the Kolmogorov entropy from a time series and its limitations when performed on EEG," *Bulletin of Mathematical Biology*, vol. 53, no. 5, pp. 665–678, 1991.

[32] K. Lehnertz, J. Arnold, P. Grassberger, and C. Elger, *Workshop on Chaos in Brain? (1999 Bonn, Germany)*, World Scientific, Singapore, 2000.

Permissions

The contributors of this book come from diverse backgrounds, making this book a truly international effort. This book will bring forth new frontiers with its revolutionizing research information and detailed analysis of the nascent developments around the world.

We would like to thank all the contributing authors for lending their expertise to make the book truly unique. They have played a crucial role in the development of this book. Without their invaluable contributions this book wouldn't have been possible. They have made vital efforts to compile up to date information on the varied aspects of this subject to make this book a valuable addition to the collection of many professionals and students.

This book was conceptualized with the vision of imparting up-to-date information and advanced data in this field. To ensure the same, a matchless editorial board was set up. Every individual on the board went through rigorous rounds of assessment to prove their worth. After which they invested a large part of their time researching and compiling the most relevant data for our readers. Conferences and sessions were held from time to time between the editorial board and the contributing authors to present the data in the most comprehensible form. The editorial team has worked tirelessly to provide valuable and valid information to help people across the globe.

Every chapter published in this book has been scrutinized by our experts. Their significance has been extensively debated. The topics covered herein carry significant findings which will fuel the growth of the discipline. They may even be implemented as practical applications or may be referred to as a beginning point for another development. Chapters in this book were first published by Hindawi Publishing Corporation; hereby published with permission under the Creative Commons Attribution License or equivalent.

The editorial board has been involved in producing this book since its inception. They have spent rigorous hours researching and exploring the diverse topics which have resulted in the successful publishing of this book. They have passed on their knowledge of decades through this book. To expedite this challenging task, the publisher supported the team at every step. A small team of assistant editors was also appointed to further simplify the editing procedure and attain best results for the readers.

Our editorial team has been hand-picked from every corner of the world. Their multi-ethnicity adds dynamic inputs to the discussions which result in innovative outcomes. These outcomes are then further discussed with the researchers and contributors who give their valuable feedback and opinion regarding the same. The feedback is then collaborated with the researches and they are edited in a comprehensive manner to aid the understanding of the subject.

Apart from the editorial board, the designing team has also invested a significant amount of their time in understanding the subject and creating the most relevant covers. They scrutinized every image to scout for the most suitable representation of the subject and create an appropriate cover for the book.

The publishing team has been involved in this book since its early stages. They were actively engaged in every process, be it collecting the data, connecting with the contributors or procuring relevant information. The team has been an ardent support to the editorial, designing and production team. Their endless efforts to recruit the best for this project, has resulted in the accomplishment of this book. They are a veteran in the field of academics and their pool of knowledge is as vast as their experience in printing. Their expertise and guidance has proved useful at every step. Their uncompromising quality standards have made this book an exceptional effort. Their encouragement from time to time has been an inspiration for everyone.

The publisher and the editorial board hope that this book will prove to be a valuable piece of knowledge for researchers, students, practitioners and scholars across the globe.

List of Contributors

Aiman Yusoff, Andery Lim and Kushan U. Tennakoon
Faculty of Science & Institute for Biodiversity & Environmental Research, Universiti Brunei Darussalam, Tungku Link, Gadong, BE1410, Brunei Darussalam

N. T. R. N. Kumara and Piyasiri Ekanayake
Applied Physics Program, Faculty of Science, Universiti Brunei Darussalam, Jalan Tungku Link, Gadong BE1410, Brunei Darussalam

Neetu Srivastava
Department of Mathematics, Amrita Vishwa Vidyapeetham (University), Karnataka 560 035, India

Wim Bras and James Torbet
Netherlands Organisation for Scientific Research, Dutch-Belgian Beamlines, European Synchrotron Radiation Facility, BP 220, 38043 Grenoble, France

Gregory P. Diakun
Science and Technology Facility Council (STFC), Daresbury Laboratory, Cheshire WA4 4AD, UK

Geert L. J. A. Rikken
National Centre for Scientific Research (CNRS), National High Magnetic Field Laboratory, 143 Avenue de Rangueil, 31400 Toulouse, France

J. Fernando Diaz
CIB Centro de Investigaciones Biologicas, Ramiro de Maeztu 9, 28040 Madrid, Spain

Stephen Juma Mulware
Ion Beam Modification and Analysis Laboratory, Physics Department, University of North Texas, 1155 Union Circle, No. 311427, Denton, TX 76203, USA

Benjamin C. Yan
Yan Research, P.O. Box 4115, Federal Way, WA 98063, USA
University of Illinois College of Medicine, 190 Medical Sciences Building, MC-714, 506 Mathews Avenue, Urbana, IL 61801, USA

Johnson F. Yan
Yan Research, P.O. Box 4115, Federal Way, WA 98063, USA

Meixiang Xu and Liang Ma
Department of Neuroscience and Cell Biology, University of Texas Medical Branch, Galveston, TX 77555, USA

Paul J. Bujalowski
Department of Neuroscience and Cell Biology, University of Texas Medical Branch, Galveston, TX 77555, USA
Department of Biochemistry and Molecular Biology, University of Texas Medical Branch, Galveston, TX 77555, USA

Feng Qian
Department of Medicine, Division of Nephrology, University of Maryland School of Medicine, Baltimore, MD 21201, USA

R. Bryan Sutton
Department of Cell Physiology and Molecular Biophysics, Texas Tech University Health Sciences Center, Lubbock, TX 79430, USA

Andres F. Oberhauser
Department of Neuroscience and Cell Biology, University of Texas Medical Branch, Galveston, TX 77555, USA
Department of Biochemistry and Molecular Biology, University of Texas Medical Branch, Galveston, TX 77555, USA
Sealy Center for Structural Biology and Molecular Biophysics, University of Texas Medical Branch, Galveston, TX 77555, USA

Chi-Li Chiu
Department of Developmental and Cell Biology, University of California, Irvine, CA 92697, USA

Michelle A. Digman and Enrico Gratton
Department of Developmental and Cell Biology, University of California, Irvine, CA 92697, USA
Laboratory for Fluorescence Dynamics, Department of Biomedical Engineering, University of California, Irvine, CA 92697, USA

Julio A. Hernández
Seccion Biofísica, Facultad de Ciencias, Universidad de la Republica, Igua esq. Mataojo, 11400 Montevideo, Uruguay

V. Makarov
Department of Physics, UPR, San Juan, PR 00931, USA

L. Kucheryavykh Y. Kucheryavykh and M. J. Eaton
Department of Biochemistry, UCC, Bayamon, PR 00960, USA

A. Rivera and M. Inyushin
Department of Physiology, UCC, Bayamon, PR 00960, USA

S. N. Skatchkov
Department of Biochemistry, UCC, Bayamon, PR 00960, USA
Department of Physiology, UCC, Bayamon, PR 00960, USA

Walter Simmons
Department of Physics and Astronomy, University of Hawaii at Manoa, Honolulu, HI 96822, USA

Joel L. Weiner
Department of Mathematics, University of Hawaii at Manoa, Honolulu, HI 96822, USA

Daniel J. Belton
Department of Chemical and Biological Sciences, University of Huddersfield, Queensgate, Huddersfield HD1 3DH, UK

Aline F. Miller
School of Chemical Engineering and Analytical Science and Manchester Interdisciplinary Biocentre, University of Manchester, 131 Princess Street, Manchester M1 7DN, UK

Belisario Domínguez Mancera
Laboratorio de Neuroendocrinología, Instituto de Fisiología, Benemerita Universidad Autonoma de Puebla, CP 7200, PUE, Mexico
Laboratorio de Biología Celular, Facultad de Medicina Veterinaria y Zootecnia, Universidad Veracruzana, CP 91710, VER, Mexico

Eduardo Monjaraz Guzman and Jorge L. V. Flores-Hernández
Laboratorio de Neuroendocrinología, Instituto de Fisiología, Benemerita Universidad Autonoma de Puebla, CP 7200, PUE, Mexico

Manuel Barrientos Morales, José M. Martínez Hernandez, Antonio Hernández Beltran and Patricia Cervantes Acosta
Laboratorio de Biología Celular, Facultad de Medicina Veterinaria y Zootecnia, Universidad Veracruzana, CP 91710, VER, Mexico

Sofia Unnerstale and Lena Maler
Department of Biochemistry and Biophysics, Center for Biomembrane Research, The Arrhenius Laboratories for Natural Sciences, Stockholm University, 106 91 Stockholm, Sweden

Michael M. Varughese and Jay Newman
The Department of Physics and Astronomy, Union College, Schenectady, NY 12308, USA

María Elena Chánez-Cardenas
Laboratorio de Patologia Vascular Cerebral, Instituto Nacional de Neurologia y Neurocirugia, insurgentes sur 3877 col. la fama, 14269 Mexico, DF, Mexico

Edgar Vazquez-Contreras
Departamento de Ciencias Naturales, CNI, Universidad Autonoma Metropolitana Cuajimalpa, Pedro Antonio de los Santos 84 Col. Sn. Miguel Chapultepec Deleg, Miguel Hidalgo, 11851 Mexico, DF, Mexico

H. Ariel Alvarez, Andres N. McCarthy and J. Raul Grigera
Instituto de Fisica de Liquidos y Sistemas Biologicos (IFLYSIB), CONICET y Departamento de Ciencias Biologicas, Facultad de Ciencias Exactas, Universidad Nacional de La Plata, 49-789, cc 565, B1900BTE La Plata, Argentina

Sean S. Kohles
Regenerative Bioengineering Laboratory, Department of Biology, Science Research & Teaching Center (SRTC), Portland State University, P.O. Box 751, Portland, OR 97207, USA
Department of Surgery, Oregon Health & Science University, Portland, OR 97239, USA

Yu Liang and Asit K. Saha
Center for Allaying Health Disparities through Research and Education (CADRE) and Department of Mathematics & Computer Science, Central State University, Wilberforce, OH 45384, USA

Grigory G. Martinovich, Elena N. Golubeva, Irina V. Martinovich and Sergey N. Cherenkevich
Department of Biophysics, Belarusian State University, Nezavisimosti Avenue 4, 220030 Minsk, Belarus

Thomas R. Caulfield and Geoffrey C. Rollins
School of Chemistry & Biochemistry, Georgia Institute of Technology, 901 Atlantic Avenue, Atlanta, GA 30332-0230, USA

Batsal Devkota
School of Biology, Georgia Institute of Technology, 901 Atlantic Avenue, Atlanta, GA 30332-0230, USA

R. M. Dunki
Physics Institute, CAP, University of Zurich, CH 8057 Zurich, Switzerland

M. Dressel
Research Department, Cantonal Psychiatric Hospital, CH 8462 Rheinau, Switzerland
Department of Psychology, University of Konstanz, 78457 Konstanz, Germany
Verhaltenstherapie, Post-Straße 3, 79098 Freiburg, Germany